Simple Brownian Diffusion

Simple Brownian Diffusion

An Introduction to the Standard Theoretical Models

Daniel T. Gillespie & Effrosyni Seitaridou

OXFORD
UNIVERSITY PRESS

Great Clarendon Street, Oxford, OX2 6DP,
United Kingdom

Oxford University Press is a department of the University of Oxford.
It furthers the University's objective of excellence in research, scholarship,
and education by publishing worldwide. Oxford is a registered trade mark of
Oxford University Press in the UK and in certain other countries

British Library Cataloguing in Publication Data

Data available

Library of Congress Cataloging in Publication Data

Library of Congress Control Number: 2012942946

ISBN 978–0–19–966450–4

Printed and bound by
CPI Group (UK) Ltd, Croydon, CR0 4YY

Links to third party websites are provided by Oxford in good faith and
for information only. Oxford disclaims any responsibility for the materials
contained in any third party website referenced in this work.

About this book

The reader deserves to know up front what the authors think the three words in the book's title mean. *Diffusion* for us is the motion of one or more *solute* molecules in a sea of much more numerous *solvent* molecules. The solute and solvent molecules together make up a *solution*. Collisions between a solute molecule and a solvent molecule, and also collisions between two solvent molecules, are assumed to be non-reactive, in that they leave the chemical identity of both molecules unchanged. Collisions between two solute molecules might or might not result in such a change.

By *Brownian* diffusion, we mean we are assuming that the solvent molecules are *very much smaller and lighter* than the solute molecules. That was the situation when English botanist Robert Brown, looking through his one-lens microscope in 1827, observed the diffusional motion of some tiny particles that had been ejected from pollen grains in water. Brown did not know, nor would he ever learn, that the diffusing particles he was watching were tiny packets of starch and fat, nor that their erratic movements were being caused by collisions of those particles with many much smaller water molecules. This pronounced size difference we are assuming between solute and solvent molecules means that whenever a solvent molecule collides with a solute molecule, the velocity of the latter will change by only a very small amount. Brownian diffusion thus differs from self-diffusion, where the solvent molecules are physically identical to the solute molecules, and the change in the velocity of a solute molecule produced by a collision with a solvent molecule can be substantial.

Finally, the word *simple* has for us several meanings. Mainly, it means we are assuming that the *solute* molecules move about *approximately independently of each other*. Two conditions need to be satisfied in order for this assumption to be valid: First, although the system is presumed to be crowded with solvent molecules, it must be dilute enough in the solute molecules that most of the time the distance between nearest neighbor solute molecules is large compared to their diameters. Of course, that will not be true when two solute molecules are colliding with each other. But the diluteness condition ensures that a solute molecule will collide with other solute molecules much less frequently than it will collide with solvent molecules. Second, except for occasional short-duration collisional forces between two solute molecules, all other forces between solute molecules, including hydrodynamic forces that are transmitted indirectly by intervening solvent molecules, must be so weak as to be ignorable. As we will note with more specificity later, we will also lean on the word "simple" to avoid dealing with many not-so-simple aspects of diffusion.

We intend this book to be *a self-contained, tutorial introduction to simple Brownian diffusion for scientists*. Although there is no biology in the book, we hope it will be especially useful to scientists whose work deals with the chemistry of living cells. In cellular chemistry the role of the solvent molecules is usually played by water molecules, while the solute molecules are much larger, and much less numerous, organic structures. The smallness of the molecular counts of some solute species inside a cell often requires taking seriously the facts that molecules come in whole numbers and behave in ways that are to some degree random or stochastic. Those two aspects of cellular diffusion are given special attention in this book. Of course, not all molecular motion inside a cell can be characterized as simple Brownian diffusion—e.g., active molecular transport along cytoskeletal pathways. But to a reasonable approximation, much of it can. Among the several complicating features of cellular diffusion that we have *not* addressed in this book, under the cover of staying simple, are the crowding of solute molecules by other solute molecules, the non-spherical shapes of both solute and solvent molecules, and the electrically polar character of many of those molecules.

Our focus in this book is on the four most widely used theoretical models of simple Brownian diffusion: the classical *Fickian model*, the *Einstein model*, the *discrete-stochastic model*, and the *Langevin model*. Our primary goal is to achieve a thorough understanding of these four models so that we can clearly appreciate their respective domains of applicability and their relative strengths and weaknesses. Chapter 1 reviews the traditional Fickian model; that review has been kept brief since the Fickian model is widely known. Chapters 3 and 4 give a more detailed exposition and critique of the more sophisticated Einstein model. Chapters 5 and 6 do the same for the discrete-stochastic model, which we regard as an approximate formulation of the Einstein model that is often more convenient for numerical work. Chapters 8, 9, and 10 give a detailed presentation of the even more sophisticated but less widely known Langevin model. The concluding Chapter 11 returns to the Einstein model to develop some additional results, which fortuitously shed more light on the discrete-stochastic model. We take a brief excursion into the area of chemical reaction kinetics in Sections 3.7 and 4.8, where we derive an explicit formula for the stochastic reaction rate of a bimolecular chemical reaction in a dilute, well-stirred solution. But we have not tried to address the more challenging problem of how reaction–diffusion systems that are not dilute and well-stirred should be handled—although several results for the pure diffusion problem developed in Chapters 5 and 6 have a bearing on that important problem.

A sound understanding of all but the Fickian model of diffusion requires fluency in certain mathematical areas that are not covered in the typical university calculus course. We have therefore devoted two entire chapters, namely Chapter 2 on *random variable theory* and Chapter 7 on *continuous Markov process theory*, to introductory expositions of those purely mathematical topics. We have tried to make those two chapters self-contained and accessible to students of science who have only a reasonably good grounding in ordinary calculus. Indeed, throughout the entire book we have tried to adopt a tutorial tone that assumes as little background knowledge as possible on the part of the reader. For we intend this book to be accessible to readers who want to understand the physics and mathematics of diffusion, but who may regard themselves as neither physicists nor mathematicians. Readers who are proficient in random variable theory and continuous Markov process theory should find, in the physics of simple Brownian diffusion, an application which provides intuitive anchors for many of the abstract definitions and results of those two mathematical theories.

As regards derivations and proofs—and there are many in this book—we have aimed to make them rigorous by the standards of a physicist; that is to say, we have always opted for an argument that is convincing and transparent over one that lacks those two attributes but might be considered more rigorous by a mathematician. Some of the results derived in the book are, so far as we can tell, new. Many of the other results have been framed in a new context and given a fresh interpretation or significance. But in keeping with the tutorial tone of the book, we have not tried to systematically document with reference citations the historical development of all major ideas and results, as might be appropriate in a review-type article. Our focus has instead been on organizing and presenting the subject matter itself in a way we think is most satisfying from a purely logical perspective.

Most chapters have a collection of numbered "Notes" at their end. While some of those could have been presented as conventional, bottom-of-the-page, small-font footnotes, most are extended discussions of technical or historical points that are too long to present in that format. Often a Note will cite original or helpful literature, so they also serve as an annotated Bibliography for the book. But the reader should adopt a pragmatic attitude toward these Notes: If the first few lines of a Note don't seem worthwhile, then one should bail out and return to the text. Two chapters have several Appendices at their end; as those are basically just very long Notes, the reader should treat them in the same brutally pragmatic way. One can always come back and read a skipped Note or Appendix later if the need arises.

We believe that the widely recognized synergism that exists in the physical sciences between experiment and theory can often be enhanced by *numerical simulation*, especially when dealing with phenomena that are stochastic. Our experience suggests that if one does not know how to accurately simulate a stochastic process on a computer, then one's understanding of that process cannot be considered complete. Indeed, just knowing how to do a simulation that faithfully replicates a process, even if a simulation is never carried out, can make for a better understanding of that process. But there are also circumstances where examining the results of a numerical simulation can genuinely deepen our appreciation of the dynamics of a model. For all these reasons, we have devoted considerable attention in this book to the simulation of diffusion. Some noteworthy examples of how simulation can inform theory and experiment in this area are: the discrete-stochastic simulations done in Chapter 6 in connection with a microfluidics experiment on diffusing polystyrene beads; the series of exact simulations done in Chapter 9 giving side-by-side comparisons of solute molecule trajectories in the Einstein and Langevin models; and the simulations done in Chapter 11 investigating the implications of first-passage time theory for the discrete-stochastic model.

We are of course aware that one can envision diffusion in contexts that have nothing to do with large solute molecules moving about in a sea of many much smaller solvent molecules. For example, under the heading of diffusion, one could rightly study various forms of parabolic differential equations, or exotic discrete-time random walks, or heat conduction in solids, or the behavior of financial markets. But as those topics lie outside our area of concern, they will not be discussed in this book. Furthermore, under the cover of our qualifier "simple" we will not discuss *anomalous diffusion*, diffusive-like motion that is in a certain sense either faster (super-diffusion) or slower (sub-diffusion) than ordinary diffusion. Anomalous diffusion has been observed in such diverse physical settings as the diffusion of proteins across membranes, the diffusion of contaminants in groundwater, the trajectories of certain foraging bacteria and animals, and transient photocurrents in amorphous thin films. Among the several generalizations of the mathematical formalism of ordinary diffusion that have been used to describe anomalous diffusion are fractional Brownian motion, fractional calculus, and continuous-time random walks with non-exponential waiting times. While progress on these fronts has been substantial and continues to look promising, there remain many open questions as to how the microphysics that underlies anomalous diffusion in each specific setting should be mathematically modeled. But before progressing to those more advanced theories of diffusion, we ought to have a thorough understanding of the simpler theory that we are moving beyond, and developing *that* understanding is the limited aim of this book.

Finally, for readers who would like to do some additional reading that meaningfully extends our prefatory remarks here, we suggest the following: For historical information and technical details on Robert Brown's famous experiment in 1827, see the article "What Brown saw and you can too" by P. Pearle, B. Collett, K. Bart, D. Bilderback, D. Newman, and S. Samuels in the *American Journal of Physics* **78**:1278–1289 (2010). For an introduction to anomalous diffusion, see the textbook by J. Klafter and I. M. Sokolov, *First Steps in Random Walks: From Tools to Applications* (Oxford University Press, 2011), and also the review chapter "Subdiffusion limited reactions" by S. B. Yuste, K. Lindenberg, and J. J. Ruiz-Lorenzo in the book *Anomalous Transport: Foundations and Applications*, edited by R. Klages, G. Radons, and I. M. Sokolov (Wiley-VCH, 2008).

Acknowledgements

It is our great pleasure to acknowledge a number of individuals who have, in various ways, helped make this book a reality. Topping that list is Carol Gillespie, whose contributions extend beyond keeping one of the authors in a functioning mode: she created over a third of the book's figures, and she initiated and carried out an innovative series of computer simulations that substantially deepened our understanding of diffusion. We very much appreciate the efforts of Farhan Kamili, who, as a sophomore physics major, used his computer skills to help us create the simulation figures in Chapter 9. We are grateful to Attila Szabo of the National Institutes of Health for some very helpful discussions on diffusion-limited reactions. We thank artist Kent Gamble for taking time away from his more comedic illustration gigs to create half of the figures in Chapter 2, thereby making it the only presentation of random variable theory we know of that accords a prominent role to buckets. DG has benefitted greatly from many discussions over the past decade with Linda Petzold and members of her research group at the University of California at Santa Barbara. Of those Petzold Group members, Brian Drawert, Jin Fu, Andreas Hellander, and Mike Lawson were especially helpful in connection with several topics explicitly treated in this book. Both of us have benefited from many discussions with Rob Phillips and members of his research group at the California Institute of Technology. It was as a member of the Phillips Group that ES carried out the microfluidic diffusion experiments described in this book, and crucial to that effort were many helpful discussions with Mandar Inamdar, also a member of the Phillips Group at the time, and collaborators Ken Dill and Kingshuk Ghosh, then at the University of California at San Francisco. We are grateful to Katja Lindenberg of the University of California at San Diego for being so generous with her time and expertise in helping us understand the current frontiers in the field of anomalous diffusion. Finally, we want to thank John Doyle of the California Institute of Technology for his initiative in establishing an informal network of interdisciplinary researchers, and an environment in which they could productively interact; without that, this book would never have come into being.

Contents

1
The Fickian theory of diffusion

The classical macroscopic theory of diffusion focuses on two physical entities: the solute *molecular density* $\rho(\mathbf{r}, t)$, and the solute *molecular flux* $\mathbf{J}(\mathbf{r}, t)$. The former is defined as the average number of solute molecules per unit volume at position $\mathbf{r}$ at time t. The latter is defined as the vector whose component $\hat{\mathbf{n}} \cdot \mathbf{J}(\mathbf{r}, t)$ in the direction of any unit vector $\hat{\mathbf{n}}$ gives the average net number of solute molecules per unit time crossing a unit area normal to $\hat{\mathbf{n}}$, in the direction of $\hat{\mathbf{n}}$, at position $\mathbf{r}$ at time t. An assumed empirical relationship between ρ and $\mathbf{J}$ called Fick's Law leads to a partial differential equation for ρ called the diffusion equation, and that equation is the centerpiece of the classical theory of diffusion. In this chapter we will review the traditional derivation of the diffusion equation, and then solve that equation for several simple physical problems. In the chapters that follow, we will refine our physical interpretation of the diffusion equation, and then go beyond that equation by developing a more comprehensive mathematical description of simple Brownian diffusion.

1.1 Fick's Law and the diffusion equation

An exact mathematical relation between the solute molecule density $\rho(\mathbf{r}, t)$ and flux $\mathbf{J}(\mathbf{r}, t)$ arises from the fact that {the average net increase in the number of solute molecules inside the infinitesimal volume element $dxdydz$ at position $\mathbf{r}$ during the infinitesimal time interval $[t, t + dt)$} must be equal to {the average net influx of solute molecules into that volume element through its six sides during that time interval} (see Fig. 1.1):

$$\rho(\mathbf{r}, t + dt) \cdot dxdydz - \rho(\mathbf{r}, t) \cdot dxdydz$$
$$= (J_x(\mathbf{r}, t) \cdot dydz \cdot dt - J_x(\mathbf{r} + \hat{\mathbf{x}}dx, t) \cdot dydz \cdot dt)$$
$$+ (J_y(\mathbf{r}, t) \cdot dxdz \cdot dt - J_y(\mathbf{r} + \hat{\mathbf{y}}dy, t) \cdot dxdz \cdot dt)$$
$$+ (J_z(\mathbf{r}, t) \cdot dxdy \cdot dt - J_z(\mathbf{r} + \hat{\mathbf{z}}dz, t) \cdot dxdy \cdot dt).$$

Dividing this equation through by $dxdydzdt$ and then letting all infinitesimals approach zero, we obtain what is called the *continuity equation*:

$$\frac{\partial \rho(\mathbf{r}, t)}{\partial t} = -\left(\frac{\partial J_x(\mathbf{r}, t)}{\partial x} + \frac{\partial J_y(\mathbf{r}, t)}{\partial y} + \frac{\partial J_z(\mathbf{r}, t)}{\partial z} \right). \tag{1.1}$$

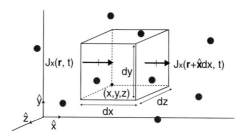

Fig. 1.1 *Pictorial representation of the continuity equation (1.1). The average net increase during the next infinitesimal time dt in the number of solute molecules (black spheres) inside the infinitesimal volume element dxdydz at point (x, y, z) (the small dot) is equal to the average net number of molecules entering the volume element through its left face in dt, $J_x(\mathbf{r}, t) \cdot dydz \cdot dt$, minus the average net number leaving the volume element through its right face in dt, $J_x(\mathbf{r} + \hat{\mathbf{x}}dx, t) \cdot dydz \cdot dt$, plus analogous contributions from the volume element's back and front faces, and its bottom and top faces.*

The classical diffusion equation is derived from the continuity equation (1.1) by assuming the validity of a *purely empirical* relation between ρ and the components (J_x, J_y, J_z) of $\mathbf{J}$ called *Fick's Law*.[1] Fick's Law asserts that

$$J_u(\mathbf{r}, t) = -D\frac{\partial \rho(\mathbf{r}, t)}{\partial u} \quad (u = x, y, z), \tag{1.2}$$

where D is some positive constant called the *diffusion coefficient*. When Eq. (1.2) is substituted into the right side of Eq. (1.1), the immediate result is the *diffusion equation*:

$$\frac{\partial \rho(\mathbf{r}, t)}{\partial t} = D\left(\frac{\partial^2 \rho(\mathbf{r}, t)}{dx^2} + \frac{\partial^2 \rho(\mathbf{r}, t)}{dy^2} + \frac{\partial^2 \rho(\mathbf{r}, t)}{dz^2}\right). \tag{1.3}$$

The solution to this partial differential equation depends on the imposed *initial condition* and *boundary conditions*. They are determined by the specific features of the physical problem at hand.

1.2 Some one-dimensional examples

The formula for the flux postulated by Fick's Law (1.2) implies that the diffusional motions of the solute molecules in the x-, y-, and z-directions are independent of each other. Consequently, considerable physical insight into diffusion can be gained with relative ease by considering "one-dimensional" problems, in which the average density of the solute molecules depends on only one component of $\mathbf{r}$, say the x-component. Physically, the assumption $\rho(\mathbf{r}, t) = \rho(x, t)$ is often applied to situations in which the containing volume is a right cylinder that is coaxial with the x-axis and has a cross–sectional area A. In that case, $A\rho(x, t)dx$ gives the average number of solute molecules at time t in the infinitesimally small "subcylinder" between x and $x + dx$.

The behavior of that average number is then controlled by the one-dimensional version of the diffusion equation (1.3), namely,

$$\frac{\partial \rho(x,t)}{\partial t} = D \frac{\partial^2 \rho(x,t)}{\partial x^2}. \tag{1.4}$$

Example 1 *Unrestricted diffusion*
Perhaps the most revealing solution of the one-dimensional diffusion equation (1.4) from a physical standpoint is the solution for the problem in which N solute molecules are initially very close together at $x = 0$, and then are allowed to diffuse unrestrictedly along the entire x-axis. The *initial condition* for that problem is $\rho(x,0) = N\delta(x)$, where δ is the Dirac delta function; this initial condition stipulates that the number of solute molecules at time 0 inside any x-axis interval (a,b) that contains the origin is $\int_a^b \rho(x,0)dx = N\int_a^b \delta(x)dx = N$. The two *boundary conditions* for this problem are $\rho(\pm\infty, t) = 0$, which are required in order that $\int_{-\infty}^{\infty} \rho(x,t)dx = N < \infty$. As can be verified by simply computing the indicated partial derivatives, the solution of Eq. (1.4) that satisfies these three conditions is

$$\rho(x,t) = \frac{N}{\sqrt{4\pi Dt}} \exp\left(-\frac{x^2}{4Dt}\right) \quad (t \geq 0; \ -\infty < x < \infty). \tag{1.5}$$

As will be discussed in more detail later, the function on the right side of Eq. (1.5) is a Gaussian or "normal" function in x with mean zero and variance $2Dt$. Figure 1.2a is a cartooned snapshot of this system at some instant $t > 0$; the figure shows the solute molecules, but not the many much smaller solvent molecules. Figure 1.2b plots the solution (1.5) as a function of x for four increasingly larger values of $t > 0$.

Upon substituting Eq. (1.5) into Fick's Law (1.2) and then evaluating the x-derivative, we find that the average net flux in the $+x$ direction at point x at time t is

$$J_x(x,t) = \frac{1}{2}\left(\frac{x}{t}\right)\frac{N}{\sqrt{4\pi Dt}} \exp\left(-\frac{x^2}{4Dt}\right). \tag{1.6}$$

This result has the intriguing form of half the molecular density at x times the average velocity x/t that a molecule must have had in order to get from the origin at time 0 to x at time t. Figure 1.2c shows plots of the flux (1.6) as a function of x corresponding to the four t-values in Fig. 1.2b.

It is important to understand that $J_x(x,t)$ measures the average *net* flux, i.e., {the average number of solute molecules at time t that pass per unit time through a unit area normal to the x-axis at x in the *positive* x-direction} *minus* {the average number of solute molecules at time t that pass per unit time through a unit area normal to the x-axis at x in the *negative* x-direction}. Separately, the positive and negative flux components might be relatively large. But their sum, the *net* flux $J_x(x,t)$, might be small, and even zero.

Example 2 *Diffusion in a semi-infinite volume with a reflecting face*
We now modify the problem in Example 1 by placing an impenetrable boundary at $x = 0^-$, just to the left of the initial N solute molecules at $x = 0$. This boundary

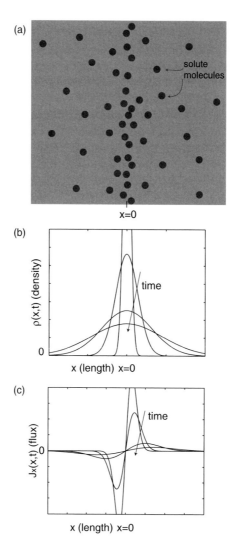

Fig. 1.2 *Unrestricted diffusion along the x-axis of N solute molecules, which at time $t = 0$ are all at point $x = 0$. (**a**) A cartooned snapshot of the system at some later instant $t > 0$. (**b**) A plot of the average number of solute molecules per unit volume at point x at time t, as given by Eq. (1.5), for four increasingly larger values of t. As time increases, more of the solute molecules tend to be found further away from the initial position; however, the area under each curve stays equal to N, the total number of solute molecules. (**c**) A plot of the net flux, as given by Eq. (1.6), for the same times as in (b). At any given point $x > 0$, the magnitude of the flux first rises, then falls. Since positive flux indicates that the solute molecules are on average moving to the right, the flux is always negative for $x < 0$. As time increases and the solute molecule density becomes increasingly uniform, the flux approaches zero everywhere.*

keeps the solute molecules off the negative x-axis, *reflecting* all molecules that strike it from the positive x-axis. The initial condition now is $\rho(x,0) = 2N\delta(x)$, since $\int_0^\infty \delta(x)\,dx = \frac{1}{2}$. Figure 1.3a shows a cartooned snapshot of this system at some later time $t > 0$. For this problem, we retain the boundary condition $\rho(\infty, t) = 0$; however, at $x = 0$ we need a boundary condition that describes the effect of the reflecting boundary there. Since that boundary prevents the passage of molecules through $x = 0$,

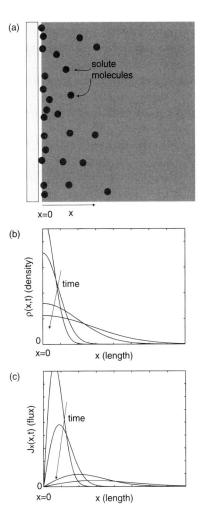

Fig. 1.3 *Diffusion along the positive x-axis of N solute molecules, all at $x = 0^+$ at time $t = 0$, with a perfectly reflecting boundary at $x = 0^-$. (a) A cartooned snapshot of the system at some later instant $t > 0$. (b) and (c) The average solute molecule density and net flux for the same four successively larger values of t as used in Fig. 1.2. These density and flux curves on the positive x-axis are exactly twice those for the unrestricted diffusion problem in Fig. 1.2.*

the net flux at $x = 0$ will always vanish. That implies by Fick's Law (1.2) the boundary condition,

$$\frac{\partial \rho(x,t)}{\partial x}\bigg|_{x=0} = 0, \ \forall t \geq 0 \quad \text{(for } x = 0 \text{ a perfectly reflecting boundary)}. \qquad (1.7)$$

It is straightforward to verify that the solution to the diffusion equation (1.4) for these initial and boundary conditions is the unrestricted solution (1.5) with its negative-x part "folded over" onto its symmetric positive-x part:

$$\rho(x,t) = \frac{2N}{\sqrt{4\pi Dt}} \exp\left(-\frac{x^2}{4Dt}\right) \quad (t \geq 0; \ 0 \leq x < \infty). \qquad (1.8)$$

Substituting this formula for $\rho(x,t)$ into Fick's Law (1.2) gives for the corresponding net flux on the positive x-axis exactly twice that in Eq. (1.6). Plots of the solution (1.8) for four successively larger values of t are shown in Fig. 1.3b, and the corresponding plots of the net flux $J_x(x,t)$ are shown in Fig. 1.3c.

This diffusion problem is often used to model the "doping" of a semiconductor. In that application, the role of the N solute molecules is played by the dopant molecules, which are initially deposited on the left face of a semiconductor slab at $x = 0$ and then allowed to diffuse into the slab on the positive x-axis. Of course, this semiconductor doping problem is not really "simple Brownian diffusion", because the scatterers of the dopant molecule are not small freely moving solvent molecules. The scatterers in this case are atoms or molecules that are bound in a fairly rigid geometric lattice. But the empirical Fick's Law has been found to be roughly as accurate for dopant molecules in a semiconductor as for solute molecules in a solution.

Example 3 *Diffusion in a semi-infinite volume with a buffered face*
Next we modify Example 2 to make the containing volume's left face at $x = 0$ permeable, and exposed to a *constant concentration* ρ_s of solute molecules via a buffered supply of those molecules on the negative x-axis. This system is cartooned in Fig. 1.4a, which shows the solute molecules at some time $t > 0$. The initial and boundary conditions for the diffusion equation (1.4) are now

$$\left.\begin{array}{l} \rho(x,0) = 0 \ (x > 0) \\ \rho(0,t) = \rho_s \text{ and } \rho(\infty,t) = 0 \ (t > 0) \end{array}\right\}. \qquad (1.9)$$

Note that the boundary condition at $x = \infty$ actually holds only for t finite.

The trick to solving the diffusion equation (1.4) for conditions (1.9) is to introduce the variable $\xi \equiv x/\sqrt{4Dt}$. Using the chain rule of differentiation, Eq. (1.4) becomes under this transformation of variable,

$$\frac{\partial \rho(\xi)}{\partial \xi} \frac{\partial \xi}{\partial t} = D \frac{\partial}{\partial \xi}\left(\frac{\partial \rho(\xi)}{\partial \xi} \frac{\partial \xi}{\partial x}\right) \frac{\partial \xi}{\partial x}$$

$$= D \frac{\partial^2 \rho(\xi)}{\partial \xi^2} \left(\frac{\partial \xi}{\partial x}\right)^2.$$

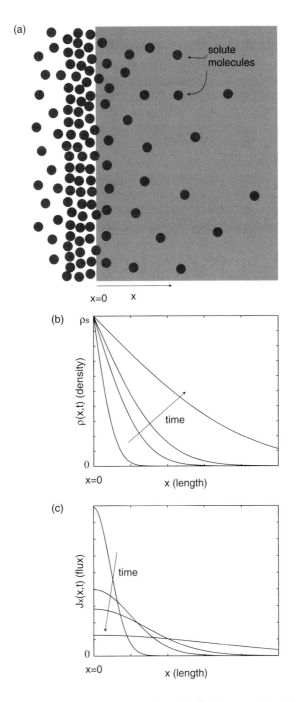

Fig. 1.4 *Diffusion along the positive x-axis with a "buffer" at $x = 0$ which keeps the solute molecule concentration there at the constant value ρ_s. It is assumed that at time $t = 0$ there are no solute molecules on the positive x-axis. (**a**) A cartooned snapshot of the system at some later instant $t > 0$. (**b**) The average solute molecule density, as given by Eq. (1.10), for four increasingly larger values of t. Since the supply of solute molecules from the buffer is inexhaustible, the total number of molecules on the positive x-axis, which is equal to the area under the concentration curve, grows without bound. (**c**) The net flux as given by Eq. (1.11) for the same times as in (b). At any point $x > 0$, the flux first rises from 0, and then eventually falls back to 0.*

When $\partial\xi/\partial t$ and $\partial\xi/\partial x$ are evaluated from the definition $\xi \equiv x/\sqrt{4Dt}$, this equation reduces to the following ordinary differential equation for $\rho(\xi)$:

$$\frac{d^2\rho(\xi)}{d\xi^2} = -2\xi\frac{d\rho(\xi)}{d\xi}.$$

Writing $d\rho/d\xi \equiv \rho'$, this equation becomes $d\rho'/d\xi = -2\xi\rho'$. Solving for ρ' gives

$$\rho' \equiv \partial\rho/\partial\xi = \alpha e^{-\xi^2},$$

where α is an integration constant. Integrating this equation with respect to ξ then gives

$$\rho(\xi) = \alpha \int_0^\xi e^{-\xi^2}\,d\xi + \beta,$$

where β is another integration constant. The values of the two integration constants are then fixed by two boundary conditions that follow from Eqs (1.9), namely

$$\rho(\xi = 0) = \rho_s \text{ and } \rho(\xi = \infty) = 0.$$

The result is

$$\rho(x,t) = \rho_s\left(1 - \text{erf}\,\frac{x}{\sqrt{4Dt}}\right) \quad (x \geq 0; t \geq 0), \tag{1.10}$$

where erf $\xi \equiv 2\pi^{-1/2}\int_0^\xi e^{-x^2}\,dx$ is the *error function.* The error function satisfies erf $0 = 0$ and erf $\infty = 1$, and it is easy to see that those two properties ensure satisfaction of the original initial and boundary conditions (1.9). Figure 1.4b shows plots of the solution (1.10) against x for four increasingly larger values of $t > 0$.

By substituting Eq. (1.10) into Fick's Law (1.2) and then evaluating the derivative with respect to x, we find for the average net flux in the $+x$-direction at point x at time t,

$$J_x(x,t) = \sqrt{\frac{D}{\pi t}}\,\rho_s e^{-x^2/(4Dt)} \quad (x \geq 0; t \geq 0). \tag{1.11}$$

Figure 1.4c shows plots of this flux function for the four t-values in Fig. 1.4b.

Like the problem in Example 2, this problem has applications in semiconductor doping. It has also been applied to several biological problems, such as for instance the transport of molecules across a cellular membrane of finite thickness, provided one is interested in the density profile of the diffusing molecules in the membrane for only relatively short times. With the diffusing molecules entering the membrane through its left side, their diffusive motion in the membrane is usually so slow that hardly any will reach the right side of the membrane for a relatively long time, even for thin membranes. That renders the semi-infinite slab approximation acceptable. Another biological application of this problem is tooth decay, with bacteria now playing the role of the solute molecules. Again, the process of tooth decay is so slow, and the thickness of the tooth is so large compared to the dimensions of the bacteria, that the tooth can be approximated by a semi-infinite volume for most time durations of practical interest. Of course, none of the aforementioned applications involve "solute molecules diffusing in a sea of smaller more numerous solvent molecules". But they serve to illustrate the wide range of practical problems that the simple Fickian model of diffusion has successfully been applied to.

Example 4 *Steady-state diffusion in a finite volume with two buffered faces*
Consider next the problem in which a finite cylindrical volume of unit cross-sectional area occupies the x-axis interval $[0, L]$, with its left boundary buffered to a constant solute molecule density ρ_0, and its right boundary buffered to a constant solute molecule density $\rho_L < \rho_0$. The situation is illustrated in Fig. 1.5a. For this system

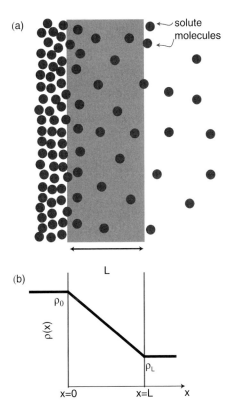

Fig. 1.5 *Steady-state diffusion in a finite segment of the x-axis between two buffered boundaries. The solute molecule concentration is kept at the constant values ρ_0 at $x = 0$ and ρ_L at $x = L$. (a) A cartooned snapshot of the system after a long enough time has elapsed that the average solute molecule density in $0 < x < L$ has effectively stopped changing with time. (b) A plot of the steady-state density in Eq. (1.14) shows that it varies linearly between the two buffered boundary values. The flux at any point between $x = 0$ and $x = L$, as given in Eq. (1.14), is equal to D times the negative of the slope of the concentration.*

we pose the question, what will be the density profile of the solute molecules along the x-axis *after a very long time has elapsed?* This question makes the implicit assumption that, regardless of the initial condition, the solution $\rho(x, t)$ to the diffusion equation (1.4) for the boundary conditions $\rho(0, t) = \rho_0$ and $\rho(L, t) = \rho_L$ will approach, as $t \to \infty$, a *time-independent steady-state* form $\rho_s(x)$. This limiting function $\rho_s(x) \equiv \rho(x, t \to \infty)$ must of course satisfy the diffusion equation (1.4). And since $\rho_s(x)$ is independent of t, the diffusion equation then simplifies in that limit to the ordinary differential equation

$$\frac{d^2 \rho_s(x)}{dx^2} = 0. \tag{1.12}$$

Notice that the diffusion coefficient D has dropped out of the picture. Quite generally, D influences the *rate of approach* to the steady state, but *not the form* of the steady state.

Two successive integrations of Eq. (1.12) yield

$$\rho_s(x) = \alpha + \beta x \quad (x \ge 0), \tag{1.13a}$$

where α and β are the constants of integration. This solution must satisfy the two boundary conditions

$$\rho_s(0) = \rho_0 \quad \text{and} \quad \rho_s(L) = \rho_L. \tag{1.13b}$$

Upon substituting Eq. (1.13a) into the two Eqs (1.13b) and then solving the resulting two equations simultaneously for α and β, we get

$$\rho_s(x) = \rho_0 - (\rho_0 - \rho_L)\frac{x}{L} \quad (0 \le x \le L). \tag{1.14}$$

A plot of this function is shown in Fig. 1.5b. The steady-state flux corresponding to this steady-state density is easily computed by substituting the latter into Fick's Law (1.2). That gives

$$J_x(x) = D\frac{\rho_0 - \rho_L}{L} \quad (0 \le x \le L). \tag{1.15}$$

This implies a flow of solute molecules from the higher buffered density region to the lower buffered density region that is constant in both space and time.

We mentioned earlier that the problem analyzed in Example 3 is sometimes used to model the transport of molecules across a cellular membrane over times that are small enough that a molecule is unlikely to traverse the full width of the membrane during that time, or equivalently, when the membrane is relatively *thick*. Our present example can be applied to the complementary long-time or *thin* membrane version of that problem, in which a molecule can diffuse across the membrane easily within the time of interest.

Example 5 *Diffusion in a finite volume with two reflecting faces*
For our final example of a one-dimensional diffusion problem, we consider a solution of solute and solvent molecules which is confined to a cylindrical volume Ω of length

L and unit cross-sectional area. The axis of this cylinder coincides with the x-axis, and the planar bounding surfaces of the cylinder are at $x = 0$ and $x = L$. We assume that the inner surfaces of Ω *reflect* any molecule striking it. Thus, following Eq. (1.7), we have the two *boundary conditions*

$$\left.\frac{\partial \rho(x,t)}{\partial x}\right|_{x=0} = 0 \quad (t \geq 0), \tag{1.16a}$$

$$\left.\frac{\partial \rho(x,t)}{\partial x}\right|_{x=L} = 0 \quad (t \geq 0). \tag{1.16b}$$

Our *initial condition* follows from the further stipulation that, up until time $t = 0$, the solute molecules are assumed to be confined by a thin wall at $x = \frac{1}{2}L$ to the *right half* of the cylinder, where they have a uniform average density $\rho_0 > 0$:

$$\rho(x,0) = \begin{cases} 0 & \text{if } 0 < x \leq \frac{1}{2}L, \\ \rho_0 & \text{if } \frac{1}{2}L < x < L. \end{cases} \tag{1.17}$$

We suppose that at time $t = 0$ the thin wall at $x = \frac{1}{2}L$ is removed, and for all $t > 0$ the solute molecules are allowed to diffuse freely throughout the cylinder. The average density $\rho(x,t)$ for $t > 0$ will therefore be the solution of the diffusion equation (1.4) that satisfies the boundary conditions (1.16) and the initial condition (1.17).

The mathematical procedure for finding that solution starts by provisionally assuming that $\rho(x,t)$ has the functional form

$$\rho(x,t) = X(x)T(t), \tag{1.18}$$

where X is some function of x only and T is some function of t only. Substituting Eq. (1.18) into the diffusion equation (1.4) gives

$$X(x)\frac{dT(t)}{dt} = DT(t)\frac{d^2 X(x)}{dx^2}.$$

When this is divided through by $DX(x)T(t)$, the result is

$$\frac{1}{DT(t)}\frac{dT(t)}{dt} = \frac{1}{X(x)}\frac{d^2 X(x)}{dx^2}. \tag{1.19}$$

Evidently, the left side of Eq. (1.19) is a function of t only, and the right side is a function of x only. The independence of those two variables implies that both sides must be equal to the same constant. A positive value for that constant would result in both $T(t)$ and $X(x)$ becoming unbounded as their arguments go to infinity. But that is unacceptable, because it would imply that $\rho(x,t)$ in Eq. (1.18) would become unbounded, and that is not allowed by the physical constraints on the problem. We therefore take that common constant to be $-k^2$, where k is a real number yet to be determined. So Eq. (1.19) implies that

$$\frac{dT(t)}{dt} = -k^2 DT(t) \quad \text{and} \quad \frac{d^2 X(x)}{dx^2} = -k^2 X(x). \tag{1.20}$$

These two ordinary differential equations are easily solved. As can be verified by simple substitution, the general solution of the first is

$$T(t) = \text{constant} \times e^{-k^2 Dt},$$

and the general solution of the second is

$$X(x) = \alpha_k \cos kx + \beta_k \sin kx,$$

where in the latter α_k and β_k can be any constants. The diffusion equation (1.4) will therefore be satisfied by the function

$$X(x)T(t) = (\alpha_k \cos kx + \beta_k \sin kx) e^{-k^2 Dt},$$

where k can be any real number and α_k and β_k can be any k-dependent constants. Notice that for $k = 0$ this function reduces to the single constant α_0, so the constant β_0 never actually appears. The linearity of the partial differential equation (1.4) ensures that any linear combination of solutions is also a solution; therefore,

$$\rho(x,t) = \frac{\alpha_0}{2} + \sum_{\text{all } k \neq 0} (\alpha_k \cos kx + \beta_k \sin kx) e^{-k^2 Dt} \tag{1.21}$$

also satisfies the diffusion equation (1.4). (To conform to conventional Fourier series notation, we have redefined α_0 to be $\alpha_0/2$.) We will show next that a particular set of values for all the constants in Eq. (1.21) can be found which will cause this solution of Eq. (1.4) to satisfy the two boundary conditions (1.16) and the initial condition (1.17).

The derivative of the function (1.21) with respect to x is evidently

$$\frac{\partial \rho(x,t)}{\partial x} = \sum_{\text{all } k \neq 0} k \left(-\alpha_k \sin kx + \beta_k \cos kx \right) e^{-k^2 Dt};$$

therefore, the two boundary conditions (1.16) become, respectively,

$$\left. \frac{\partial \rho(x,t)}{\partial x} \right|_{x=0} = \sum_{\text{all } k \neq 0} k \beta_k e^{-k^2 Dt} = 0, \tag{1.22a}$$

$$\left. \frac{\partial \rho(x,t)}{\partial x} \right|_{x=L} = \sum_{\text{all } k \neq 0} k \left(-\alpha_k \sin kL + \beta_k \cos kL \right) e^{-k^2 Dt} = 0. \tag{1.22b}$$

Condition (1.22a) can be satisfied for all $t \geq 0$ only by taking

$$\beta_k = 0 \text{ for all } k. \tag{1.23a}$$

That done, condition (1.22b) becomes

$$\sum_{\text{all } k \neq 0} (k\, \alpha_k \sin kL)\, e^{-k^2 Dt} = 0,$$

and it can be satisfied for all $t \geq 0$ by taking $kL = n\pi$ for any positive integer n:

$$k = \frac{n\pi}{L} \text{ for } n = 1, 2, \ldots. \qquad (1.23\text{b})$$

Under the conditions (1.23) on β_k and k, the solution (1.21) now takes the form

$$\rho(x, t) = \frac{\alpha_0}{2} + \sum_{n=1}^{\infty} \alpha_n \cos\left(\frac{n\pi x}{L}\right) e^{-\left(\frac{n\pi}{L}\right)^2 Dt}, \qquad (1.24)$$

where the α coefficients are now indexed by n instead of k. Notice that although negative values for n in (1.23b) would also satisfy the boundary condition (1.22b), they would add no generality to the solution (1.24) because the cosine is an even function.

It remains only to make the solution (1.24) satisfy the initial condition, i.e., to make

$$\rho(x, 0) = \frac{\alpha_0}{2} + \sum_{n=1}^{\infty} \alpha_n \cos\left(\frac{n\pi x}{L}\right) \qquad (1.25)$$

equal the function defined in Eq. (1.17). That the α constants in this formula can be chosen to make that happen follows from the mathematical theory of Fourier series. That theory tells us that any function $f(x)$ which is integrable and periodic on the interval $[0, 2L]$ can be written as the trigonometric series

$$f(x) = \frac{a_0}{2} + \sum_{n=1}^{\infty} \left(a_n \cos\frac{n\pi x}{L} + b_n \sin\frac{n\pi x}{L}\right), \qquad (1.26)$$

where

$$\left.\begin{array}{l} a_n = \dfrac{1}{L} \int_0^{2L} f(x) \cos\dfrac{n\pi x}{L} dx \\[3mm] b_n = \dfrac{1}{L} \int_0^{2L} f(x) \sin\dfrac{n\pi x}{L} dx \end{array}\right\} \quad (n = 0, 1, 2, \ldots). \qquad (1.27)$$

(That this is so can be verified by substituting Eq. (1.26) into each of the two formulas in Eqs (1.27) and then carrying out the integrations over x.) Now, the function $\rho(x, 0)$ specified in the initial condition (1.17) is defined only on the interval $[0, L]$. But we can make that function periodic on the larger interval $[0, 2L]$ simply by extending its definition from the form in Fig. 1.6a to the form in Fig. 1.6b:

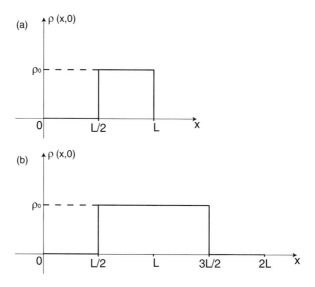

Fig. 1.6 *Periodically extending Eq. (1.17) to Eq. (1.28).* **(a)** *The assumed initial concentration profile of solute molecules on the x-axis between the two perfectly reflecting boundaries at $x = 0$ and $x = L$.* **(b)** *An extension of the function $\rho(x, 0)$ in (a) to values $x > L$ which makes that function periodic with period $2L$. This unphysical extension is necessary in order to construct a Fourier series representation of $\rho(x, 0)$ on the interval $[0, L]$, and eventually a Fourier series representation of $\rho(x, t > 0)$ on the interval $[0, L]$.*

$$\rho(x, 0) \rightarrow f(x) = \begin{cases} 0 & \text{if } 0 < x \le \tfrac{1}{2}L \text{ or } \tfrac{3}{2}L < x \le 2L, \\ \rho_0 & \text{if } \tfrac{1}{2}L < x \le \tfrac{3}{2}L. \end{cases} \tag{1.28}$$

If this formula for $f(x)$ is inserted into the formulas for a_n and b_n in Eqs (1.27) and the integrations over x are carried out, the result is

$$a_n = \begin{cases} \rho_0 & \text{if } n = 0 \\ (-1)^{\frac{n+1}{2}} \dfrac{2\rho_0}{n\pi} & \text{if } n = 1, 3, 5, \ldots \\ 0 & \text{if } n = 2, 4, 6, \ldots \end{cases} \tag{1.29a}$$

and

$$b_n = 0 \text{ for all } n \ge 1. \tag{1.29b}$$

Inserting these values for a_n and b_n into Eq. (1.26), we see that the resulting expression will agree with Eq. (1.25) if we simply replace α_n in Eq. (1.25) everywhere with the above values of a_n. Thus we conclude that the function in Eq. (1.24), which by construction satisfies the diffusion equation (1.4) and the boundary conditions (1.16), will also satisfy the initial condition (1.17) if we replace the α_n in Eq. (1.24) with the values for a_n in Eq. (1.29a):

$$\rho(x,t) = \frac{\rho_0}{2} + \sum_{n=1(\text{odd})}^{\infty} (-1)^{\frac{n+1}{2}} \frac{2\rho_0}{n\pi} \cos\left(\frac{n\pi x}{L}\right) e^{-\left(\frac{n\pi}{L}\right)^2 Dt} \qquad (x \in [0,L], t \ge 0). \quad (1.30)$$

Notice that although *each term* in the sum in Eq. (1.30) has the functional form $X(x)\,T(t)$ in Eq. (1.18), the *sum* $\rho(x,t)$ of those terms does *not* have that rather restrictive functional form. By leveraging the ability of the Fourier series to represent a wide class of functions, the separation-of-variables solution technique, which began with the assumption (1.18), has turned out to be much more broadly applicable than we might at first have guessed. The price we pay for this happy circumstance is that

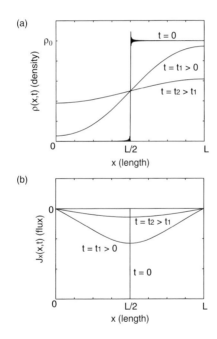

Fig. 1.7 *Diffusion on the x-axis between two perfectly reflecting boundaries at $x = 0$ and $x = L$. (a) Plots of the average solute molecule concentration $\rho(x,t)$ in Eq. (1.30) at times $t = 0, t = t_1 > 0$, and $t = t_2 > t_1$. Each curve was computed using only the first 200 terms in the summation over n (i.e., truncating the n-summation at $n = 399$). The peculiar behavior of $\rho(x,0)$ near $x = L/2$ is typical of the behavior of any Fourier series representation at a finite jump discontinuity. (b) Plots of the net flux $J_x(x,t)$ in Eq. (1.31) for the same times as in (a). The flux at $x = L/2$ dips infinitely sharply to $-\infty$ at time $t = 0$ owing to the immediate rush of solute molecules across that point in the negative x-direction at the instant the partition there is removed.*

the solution takes the form of an infinite series. But thanks to computers, that is not a high price for many numerical purposes.

Figure 1.7a shows plots of the solution (1.30) against x for three successively larger values of t. Each curve in that figure, including the one for $t = 0$, was computed by tallying the first 200 terms in the infinite sum (1.30). If we substitute Eq. (1.30) into Fick's Law (1.2) and evaluate the derivative with respect to x, we find for the average flux in the $+x$-direction at point x at time t,

$$J_x(x,t) = \frac{2D\rho_0}{L} \sum_{n=1(\text{odd})}^{\infty} (-1)^{\frac{n+1}{2}} \sin\left(\frac{n\pi x}{L}\right) e^{-\left(\frac{n\pi}{L}\right)^2 Dt} \quad (x \in [0, L], t \geq 0). \quad (1.31)$$

Figure 1.7b shows plots of this flux function against x for the values of t in Fig. 1.7a. Note that the plot for $J_x(x,0)$ is zero for all x except $x = L/2$, where it equals $-\infty$.

In Section 5.9 we will discuss in detail an actual laboratory experiment that tests formula (1.30) for $\rho(x,t)$.

1.3   The road ahead

One option for proceeding further into the subject of diffusion would be to continue along the path of Section 1.2, by finding solutions of the one-dimensional diffusion equation (1.4) for other initial and boundary conditions, and doing the same for the three-dimensional diffusion equation (1.3) for various geometries of practical interest. But that is not our plan here. Instead, we want to look deeper into the theory that underlies simple Brownian diffusion.

It might be thought that the best way to do that would be to somehow generalize Fick's Law (1.2). For example, we might allow the diffusion coefficient to depend on position and time. In the one-dimensional case, that would generalize Fick's Law to

$$J_x(x,t) = -D(x,t)\frac{\partial\rho(x,t)}{\partial x}, \quad (1.32a)$$

where $D(x,t)$ is the generalized diffusion coefficient. When this formula is substituted into the one-dimensional version of the continuity equation (1.1), we obtain what might seem to be a generalized form of the one-dimensional diffusion equation (1.4):

$$\frac{\partial\rho(x,t)}{\partial t} = \frac{\partial}{\partial x}\left(D(x,t)\frac{\partial\rho(x,t)}{\partial x}\right). \quad (1.32b)$$

But in fact, this is *not* a meaningful generalization of the conventional Fickian theory. All Eq. (1.32a) does is define a new function $D(x,t)$ to be $-J_x(x,t)/(\partial\rho(x,t)/\partial x)$. And Eq. (1.32b) merely rewrites the continuity equation in terms of that new function instead of the flux function $J_x(x,t)$. The real Fick's Law (1.2) does indeed define the diffusion coefficient D, but it also does more: It makes an *assertion* about how the underlying physics of molecular motion connects the solute density and the solute flux.

And that assertion can in principle be either verified or falsified by suitably designed laboratory experiments. In contrast, Eqs (1.32) are the results of simple mathematical manipulations which in the end convey no useful information about the physics that underlies diffusion. If by experimental measurement we found for some system that the function $D(x,t)$ defined by Eq. (1.32a) is not a constant, that would certainly falsify Fick's Law. But by itself, that finding would not provide any new insight into how solute and solvent molecules interact, or even what a better diffusion equation might be.

Surely though, we expect that Nature will deviate in some way from Fick's Law (1.2) to some degree, however small. We can imagine that such deviations might be described by some "corrected form" of Fick's Law, such as perhaps

$$J_x(x,t) = -D\frac{\partial\rho(x,t)}{\partial x} + \alpha\left(\frac{\partial\rho(x,t)}{\partial x}\right)^2, \tag{1.33}$$

which introduces a new parameter α, or maybe

$$J_x(x,t) = -D\frac{\partial\rho(x,t)}{\partial x} + \beta\frac{\partial^2\rho(x,t)}{\partial x^2}, \tag{1.34}$$

which introduces a different new parameter β, or maybe a combination of these. To the extent that accurate experimental measurements of J_x and ρ for specific systems can be performed, it would indeed be interesting to find out if the data so obtained can be accurately fitted by any such modified form of Fick's Law. Not only could that be useful for engineering applications, it would also provide a "prediction goal" for theories about the underlying microphysics of diffusion. But in the absence of solid experimental confirmation of any such modified form of Fick's Law, or some physics-based theory that implies that modification, it is hard to see what can be gained by pursuing this line of investigation.

A genuinely deeper theory of simple Brownian diffusion demands that we take account of *how solute and solvent molecules actually move about and collide in a solution.* The need for a deeper theory of diffusion than what is provided by the Fickian model is revealed by two simple observations: First, Fick's Law (1.2) succeeds in defining the diffusion coefficient D *only* when the gradient of ρ is non-zero. Does that mean that D is defined *only* in the presence of a density gradient in the solute molecules? Second, Fick's Law implies that the net flux **J** *vanishes* in the absence of a density gradient. Does that mean that diffusion is *not occurring* in systems where the average density of the solute molecules is constant?

The answer to both of these questions is no. The reasons why, and the larger story of simple Brownian diffusion, will unfold in Chapters 3 through 11. That journey will, however, require a comfortable fluency in the mathematics of "random variables". Readers who feel uneasy about that requirement should find the tutorial review in Chapter 2 to be a worthwhile diversion.

Note to Chapter 1

[1]Adolf Fick (1829–1901) was a German physiologist who made several major contributions to medical science, and who also had considerable facility with mathematics and physics. In 1855, motivated by his work in physiology, and perceiving an analogy between diffusion and Fourier's theory of heat conduction, Fick proposed the relation between the molecular flux and the gradient of the molecular density that is expressed in Eq. (1.2). Some authors refer to that equation as Fick's First Law, and to the consequent relation (1.3) as Fick's Second Law. But in this book, we will call Eq. (1.2) Fick's Law, and Eq. (1.3) the diffusion equation.

2
A review of random variable theory

In extending our analysis of diffusion beyond the traditional approach described in Chapter 1, we will find it necessary to work with random variables. Scientists typically fall into two groups in their knowledge of random variables: Either that knowledge is anecdotal, incomplete, and often in a few ways incorrect. Or it has been obtained from a fairly advanced math course, in which a random variable X has been defined as "a mapping of a sample space Ω onto the real numbers in the context of a probability space $(\Omega, \mathcal{F}, P)$ where $\mathcal{F}$ is a σ-algebra and P a probability measure". While the latter approach to random variables is precise and complete, it assumes a rather extensive background in pure mathematics. Here we will take a less formal, more intuitive approach to random variable theory, as we survey some definitions and results that we will be using repeatedly throughout the rest of the book. We will try to make this exposition self-contained, but to keep it reasonably brief we will not undertake any derivations that are tediously long.

2.1 Probability

Any quantitative discussion of things random will involve the notion of probability. For our purposes here, probability can be defined in terms of a series of "identical experimental tries": The *probability* $p(E)$ that a certain event E will occur on any one try is equal to the fraction of times E occurs, in the limit of infinitely many tries. It follows that $p(E) = 0$ if and only if E never occurs and hence is impossible, and $p(E) = 1$ if and only if E always occurs and hence is certain. In general, $0 \leq p(E) \leq 1$.

The calculus of probabilities is based entirely on two laws: The *multiplication law* states that, with $p(E_2|E_1)$ denoting the probability that event E_2 will occur *given* that event E_1 occurs, the probability $p(E_1 \text{ and } E_2)$ that both events will occur is equal to the product $p(E_1) \cdot p(E_2|E_1)$. The *addition law* states that if two events E_1 and E_2 are *mutually exclusive*, in that they never occur together, then the probability $p(E_1 \text{ or } E_2)$ that either event will occur is equal to the sum $p(E_1) + p(E_2)$. It is easy to derive these two laws directly from the above definition of probability.[1]

There are broader definitions of probability than the "frequency" definition that we are adopting here—definitions that would allow us to speak of such things as "the probability that I will get anything really useful out of this chapter". But the frequency definition is well suited to the physical sciences, where *repeated experiments* are considered to be the arbiter of what is true. And as a bonus, the frequency definition allows us to derive, instead of just postulate, the multiplication and addition laws.

2.2 Definition of a random variable

In general, a *variable* is a mathematical entity which we can, at our pleasure, "interrogate as to its value", or "sample". A real variable X is said to be a *random variable* if and only if the following is true: The probability that a sampled value of X will be found to lie in the infinitesimal interval $[x, x + dx)$ can be written

$$\text{Prob}\{X \in [x, x + dx)\} = P_X(x)\, dx, \tag{2.1}$$

where P_X is some non-negative function that satisfies the normalization condition $\int_{-\infty}^{\infty} P_X(x)dx = 1$. We will usually use an upper case letter to denote a random

Fig. 2.1 *Denizens of the World of Random Variables. We can think of a random variable X as an inexhaustible bucket of numbers, from which we may draw a sample value x whenever we please. We never know what value we will get on any one draw—it's rather like a box of chocolates. But we do know that if we draw many samples from the bucket and then make a normalized frequency histogram of those values, we will get, in the limit of infinitely many samples, exactly the curve $P_X(x)$-versus-x that is drawn on the outside of the bucket. This function P_X, which is non-negative and bounds unit area with the x-axis, is called the probability density function (PDF) of the random variable X. Through its formal definition in Eq. (2.1), it completely defines X. There are as many different random variables as there are different functions of x that are non-negative and bound unit area with the x-axis. Thus, the World of Random Variables consists of infinitely many buckets of numbers, each uniquely specified by its PDF. (Sketch by Kent Gamble. © 2012 Kent Gamble.)*

variable, and the corresponding lower case letter to denote the possible sample values of that random variable. Since by the addition law $\int_a^b P_X(x)dx$ gives the probability that a sample value of X will lie somewhere in the interval $[a, b)$, then the normalization condition simply asserts that any sample value of X will surely lie somewhere on the real axis. When (2.1) is true, the function P_X is called X's *probability density function* (PDF); it defines the random variable X as completely as is possible. Since probabilities are pure numbers, the right side of Eq. (2.1) shows that $P_X(x)$ has units of x^{-1}; thus, $P_X(x)$ by itself is not a probability.

Equation (2.1) tells us that, despite the unpredictability of the result of a *single* sampling of X, a normalized frequency histogram of the values found in many samplings will always approach the graph of $P_X(x)$-versus-x as the number of samplings goes to infinity. If that does *not* happen, i.e., if the histogram of X's sample values does *not* approach some fixed curve as the number of samplings gets larger and larger, then even though X might seem to be "random" in common parlance, it will *not* be a "random variable" from a mathematical point of view. So random variables are not totally chaotic: they possess a modicum of predictability.

It is sometimes helpful to think of a random variable X as a *bucket of numbers* which has an identifying plot of $P_X(x)$-versus-x drawn on its outside (see Fig. 2.1). Sampling X consists of drawing a number from the X-bucket, which magically never runs out of numbers. In general, we never know what value we will get when we do that. But we do know that a normalized frequency histogram of the values obtained in a sufficiently large number of drawings will follow the shape of the curve drawn on the side of the bucket. There are infinitely many different random variables, corresponding to all the differently shaped non-negative functions that bound unit area with the real axis—including all such functions that have no analytical representation.

2.3   Some commonly encountered random variables

The vast majority of random variables are unheralded, and will forever remain so. But a few are encountered so often in practical applications that they have been given names. Three that we will make repeated use of in this book are as follows: $X = \mathcal{U}(a, b)$, the *uniform* random variable on the interval $[a, b)$, is defined by the PDF

$$P_X(x) = \begin{cases} (b - a)^{-1}, & \text{if } a \leq x < b, \\ 0, & \text{otherwise,} \end{cases} \tag{2.2}$$

which is graphed in Fig. 2.2a; $X = \mathcal{E}(a)$, the *exponential* random variable with decay constant $a > 0$, is defined by the PDF

$$P_X(x) = \begin{cases} a \exp(-ax), & \text{if } x \geq 0, \\ 0, & \text{otherwise,} \end{cases} \tag{2.3}$$

which is graphed in Fig. 2.2b; and $X = \mathcal{N}(\mu, \sigma^2)$, the *normal* (or Gaussian) random variable with mean μ and variance σ^2, is defined by the PDF

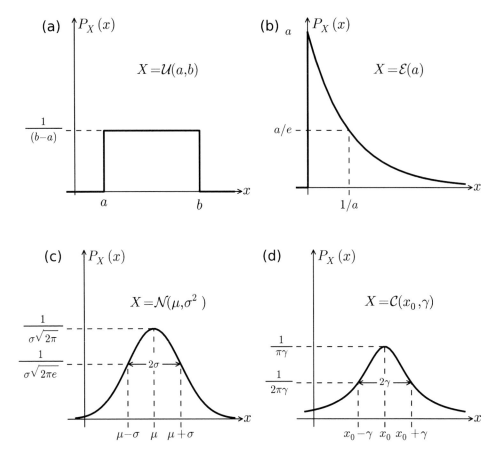

Fig. 2.2 *Probability density functions of several kinds of random variables that will be encountered often in this book.* **(a)** $\mathcal{U}(a,b)$ *is the* uniform *random variable on the interval* $[a,b)$. *A common misconception is that the uniform random variable is "truly random", since it is not biased in favor of any one sample value over another. But that is not true:* $\mathcal{U}(a,b)$ *is strongly biased against values of x that lie outside the interval* $[a,b)$, *and we cannot extend the boundaries of that interval to cover the entire real axis because of the normalization requirement. Every random variable is biased, and the nature of the bias is described by the PDF.* **(b)** $\mathcal{E}(a)$ *is the* exponential *random variable with decay constant a. Both its mean and its standard deviation are equal to* $1/a$. **(c)** $\mathcal{N}(\mu,\sigma^2)$ *is the* normal *random variable with mean* μ *and variance* σ^2. **(d)** $\mathcal{C}(x_0,\gamma)$ *is the* Cauchy *random variable centered at* x_0 *with half-width* γ. *Its tails are so heavy (i.e., they approach zero so slowly as* $|x| \to \infty$) *that the integral in Eq. (2.5) does not converge for any* $k \geq 1$; *thus, the Cauchy random variable has no mean or variance.*

$$P_X(x) = \frac{1}{\sqrt{2\pi\sigma^2}} \exp\left(-\frac{(x-\mu)^2}{2\sigma^2}\right) \quad -\infty < x < \infty, \tag{2.4}$$

which is graphed in Fig. 2.2c.

The normal random variable is not the only random variable that has a mean and a variance. Indeed, the uniform random variable $\mathcal{U}(a,b)$ has mean $(a+b)/2$ and variance $(b-a)^2/12$. And the exponential random variable $\mathcal{E}(a)$ has mean $1/a$ and variance $(1/a)^2$. To properly define the mean and variance of a random variable X, we must first define its kth moment:

$$\langle X^k \rangle \equiv \int_{-\infty}^{\infty} x^k P_X(x)\, dx \quad (k=0,1,2,\ldots). \tag{2.5}$$

The zeroth moment of X *always* exists and equals 1, owing to the normalization condition. But it is *not* required of a random variable that any of its higher order moments exist—i.e., that the integral (2.5) exist for any $k \geq 1$. The first moment $\langle X \rangle$ is called the *mean* of X. The difference between the second moment and the square of the first moment is called the *variance* of X:

$$\text{var}\{X\} \equiv \langle X^2 \rangle - \langle X \rangle^2 \equiv \left\langle (X - \langle X \rangle)^2 \right\rangle. \tag{2.6a}$$

The last expression shows that the variance is always non-negative The square root of the variance is called the *standard deviation* of X:

$$\text{sdev}\{X\} \equiv \sqrt{\langle X^2 \rangle - \langle X \rangle^2}. \tag{2.6b}$$

For some (though not all) random variables, $\langle X \rangle$ serves as a "best estimate" of what a sample value of X will be, and the smallness of $\text{sdev}\{X\}$ compared to $\langle X \rangle$ provides a measure of how good that estimate really is.

An example of a well-defined random variable for which none of the $k \geq 1$ moments exist, and which therefore has no mean, no variance, and no standard deviation, is the *Cauchy* random variable, $\mathcal{C}(x_0, \gamma)$. Its PDF is $\gamma\pi^{-1}/[(x-x_0)^2 + \gamma^2]$, where x_0 and $\gamma > 0$ are constants. A plot of this PDF is shown in Fig. 2.2d. Comparing it with the PDF of $\mathcal{N}(\mu, \sigma^2)$ in Fig. 2.2c might suggest that x_0 is the mean of $\mathcal{C}(x_0, \gamma)$ and γ^2 is its variance; however, applying the definition (2.5) will show that such is not the case.

A *sure variable* X is a special "non-random" variety of random variable for which $P_X(x)$ has the form $\delta(x-x_0)$, where δ is the Dirac delta function. That function is by definition zero for all $x \neq x_0$, but is infinite at $x = x_0$ in just such a way that

$$\int_{\alpha}^{\beta} \delta(x-x_0)\, dx = 1 \text{ for any } \alpha < x_0 < \beta. \tag{2.7}$$

Viewing the integrand here as a PDF, this relation implies that the probability that a sampling of the corresponding random variable will produce a value in any interval (α, β) is either 1 or 0, according to whether that interval does or does not contain x_0; that implies that the sample value will always be the value x_0. It is easy to show that this random variable has mean x_0 and variance 0. Such variables are the familiar variables of ordinary algebra and calculus. It can be proved that a necessary and sufficient condition for any random variable X to be "sure" is for its variance to vanish; thus, for example, $\mathcal{N}(\mu, 0)$ is the sure variable μ.

2.4   Multivariate random variables

A *multivariate random variable* $\mathbf{X}$ consists of a *set* of $n \geq 2$ random variables, $\mathbf{X} = (X_1, \ldots, X_n)$. A sample value of $\mathbf{X}$ thus consists of an n-tuple of "simultaneous" sample values of each of $\mathbf{X}$'s n components. The PDF of $\mathbf{X}$ is defined by the statement

$$P_{\mathbf{X}}(\mathbf{x})d^n\mathbf{x} \equiv P_{X_1,\ldots,X_n}(x_1, \ldots, x_n)dx_1 \cdots dx_n$$
$$= \text{Prob}\left\{X_i \in [x_i, x_i + dx_i) \text{ for all } i = 1, \ldots, n\right\}. \tag{2.8}$$

The graph of $P_{\mathbf{X}}$ on the outside of the "$\mathbf{X}$-bucket" that contains n-tuple samples $(x_1, \ldots, x_n)$ will be a surface over an n-dimensional hyperplane—not so easy to draw.

The function $P_{\mathbf{X}}$ is often referred to as the *joint PDF* of the n random variables $X_1, \ldots, X_n$. The *unconditioned* PDF P_{X_i} of X_i, which describes the behavior of X_i's sample values irrespective of the sample values of any of the other components of $\mathbf{X}$, is obtained by integrating (summing) $P_{\mathbf{X}}$ over all the other components, in accordance with the addition law of probability:

$$P_{X_i}(x_i) = \int_{-\infty}^{\infty} dx_1 \cdots \int_{-\infty}^{\infty} dx_{i-1} \int_{-\infty}^{\infty} dx_{i+1} \cdots \int_{-\infty}^{\infty} dx_n \, P_{\mathbf{X}}(x_1, \ldots, x_n). \tag{2.9}$$

If $P_{\mathbf{X}}(\mathbf{x})$ can be factored into the form $\prod_{i=1}^{n} P_{X_i}(x_i)$, then the random variables $X_1, \ldots, X_n$ are said to be *statistically independent*. In that case, a knowledge of the sample value of any one of those random variables will give us no information at all about the concomitant sample values of the other random variables. But if the functional dependence of $P_{\mathbf{X}}(\mathbf{x})$ on, say, x_1 and x_2 *cannot* be thusly factored, then X_1 and X_2 will be *statistically dependent*; in that case, a knowledge of the sample value of either X_1 or X_2 might allow us to sharpen our prediction of the concomitant sample value of the other.

2.5   Functional transformations of random variables: the RVT theorem

Most problems in applied random variable theory can be reduced to answering a question of the following general kind: Given a random variable $\mathbf{X} = (X_1, \ldots, X_n)$ with a known PDF $P_{\mathbf{X}}$, suppose we construct a new random variable $\mathbf{Y} = (Y_1, \ldots, Y_m)$, where n and m need not be equal, by doing the following: For each sample n-tuple $(x_1, \ldots, x_n)$ in the $\mathbf{X}$-bucket, we compute a sample m-tuple $(y_1, \ldots, y_m)$ for the $\mathbf{Y}$-bucket according to

$$y_i = f_i(x_1, \ldots, x_n) \quad (i = 1, \ldots, m), \tag{2.10}$$

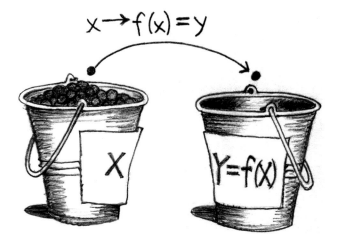

Fig. 2.3 *Functionally transforming a random variable. If we draw a sample value x from the* **X***-bucket, compute* $y = f(x)$ *where f is some given function, and then toss that y-value into a new bucket, we will call that new bucket "the random variable $Y = f(X)$". While this operational definition of what we mean by a function of a random variable is straightforward, it is not obvious what the PDF of the new random variable Y is. It turns out that the PDF P_Y of Y can be expressed in terms of the PDF P_X of X and the transforming function f by Eq. (2.12). More generally, for an n-variate random variable* $\mathbf{X} = (X_1, \ldots, X_n)$, *whose bucket is filled with sample n-tuples $(x_1, \ldots, x_n)$, and an m-component transformation function* $\mathbf{f} = (f_1, \ldots, f_m)$, *the transformation formulas $y_i = f_i(x_1, \ldots, x_n)$ $(i = 1, \ldots, m)$ will give a sample m-tuple of the m-variate random variable* $\mathbf{Y} \equiv \mathbf{f}(\mathbf{X}) = (Y_1, \ldots, Y_m)$ *whose PDF is given by Eq. (2.11). (Sketch by Kent Gamble. © 2012 Kent Gamble.)*

where each f_i is a given ordinary function of any or all of the n variables $x_1, \ldots, x_n$. We then say that Y_i is the random variable $f_i(X_1, \ldots, X_n)$. See Fig. 2.3. But what is the joint PDF $P_\mathbf{Y}$ of these new random variables $Y_1, \ldots, Y_m$?

The answer to this question is given by the *Random Variable Transformation (RVT) theorem*. It says that

$$P_\mathbf{Y}(y_1, \ldots, y_m) = \int_{-\infty}^{\infty} dx_1 \cdots \int_{-\infty}^{\infty} dx_n \, P_\mathbf{X}(x_1, \ldots, x_n) \prod_{i=1}^{m} \delta_i \left(y_i - f_i(x_1, \ldots, x_n)\right),$$

(2.11)

where each δ_i $(i = 1, \ldots, m)$ is a Dirac delta function.

We will not give here a general proof of the RVT theorem, but its proof for the simplest case $m = n = 1$ is both brief and instructive: For any two random variables X and Y, the probability $P_{X,Y}(x, y)dxdy$ that $X \in [x, x + dx)$ and also $Y \in [y, y + dy)$ can, by the multiplication law, be written as the product of $P_X(x)dx$ times $P_{Y|X}(y \,|\, x)dy$, where $P_{Y|X}(y \,|\, x)$ is the PDF of Y *given that* $X = x$; hence, the joint PDF of X and Y can always be written

$$P_{X,Y}(x,y) = P_X(x)P_{Y|X}(y\,|\,x).$$

Now suppose we *define* Y to be the random variable $f(X)$; i.e., a sample value y of Y can be obtained from a sample value x of X by evaluating $y = f(x)$. In that case, the condition $X = x$ implies that Y *surely* has the value $f(x)$; therefore, $P_{Y|X}(y\,|\,x) = \delta\,(y - f(x))$. Substituting this into the above identity, and then integrating $P_{X,Y}(x,y)$ over all values of x to get the unconditioned PDF of Y, we obtain

$$P_{Y=f(X)}(y) = \int_{-\infty}^{\infty} P_X(x)\,\delta\,(y - f(x))\,dx. \tag{2.12}$$

This is the $m = n = 1$ version of the RVT theorem. Proofs for $m, n > 1$ can be constructed using similar logic.[2]

Although carrying out the n integrations on the right side of Eq. (2.11) is often a very challenging task, many useful results can be deduced by doing so. We will now recount some of those results that will be needed in later chapters of this book.

2.6   Some useful consequences of the RVT theorem

If the function f in Eq. (2.12) is one-to-one or invertible on $(-\infty, \infty)$, we can make the integration variable change $x \to z \equiv f(x)$ on the right side to get

$$P_{Y=f(X)}(y) = \int_{-\infty}^{\infty} P_X(x)\,\delta\,(y - z)\left|\frac{dx}{dz}\right|dz.$$

In the integrand here, x is now the function of z obtained by solving $z = f(x)$ for x, i.e., $x = f^{-1}(z)$. The integration over z is now easily accomplished, with the delta function being eliminated and z being everywhere replaced by y. The result is

$$P_{Y=f(X)}(y) = P_X(x)\left|\frac{dx}{dy}\right|, \quad \text{where } x \equiv f^{-1}(y). \tag{2.13}$$

This result is most easily remembered via the mnemonic "$P_Y(y)dy = P_X(x)dx$".

Two simple but useful applications of Eq. (2.13) are as follows: First, if $y = f(x) = \alpha + x$ where α is a constant, then $x = y - \alpha$ and $dx/dy = 1$, so Eq. (2.13) becomes

$$P_{Y \equiv \alpha + X}(y) = P_X(y - \alpha). \tag{2.14a}$$

Second, if $y = f(x) = \alpha x$, then $x = y/\alpha$ and $dx/dy = 1/\alpha$, so Eq. (2.13) becomes

$$P_{Y \equiv \alpha X}(y) = |1/\alpha|\,P_X(y/\alpha). \tag{2.14b}$$

If we substitute into the right sides of Eqs (2.14a) and (2.14b) the PDF (2.4) of the *normal* random variable $X = \mathcal{N}(\mu, \sigma^2)$, we can easily deduce the respective identities

$$\alpha + \mathcal{N}(\mu, \sigma^2) = \mathcal{N}(\alpha + \mu, \sigma^2), \tag{2.15a}$$

$$\alpha \mathcal{N}(\mu, \sigma^2) = \mathcal{N}(\alpha\mu, \alpha^2\sigma^2). \tag{2.15b}$$

These relations in turn imply that $\alpha + \beta \mathcal{N}(0, 1) = \alpha + \mathcal{N}(0, \beta^2) = \mathcal{N}(\alpha, \beta^2)$, whence

$$\mathcal{N}(\mu, \sigma^2) = \mu + \sigma \mathcal{N}(0, 1). \tag{2.16}$$

If $n = m \geq 2$ and the $\mathbf{x} \to \mathbf{y}$ transformation (2.10) is invertible, it can be proved from Eq. (2.11) that the derivative in formula (2.13) gets replaced by the Jacobian:

$$P_{Y_1,\ldots,Y_n}(y_1,\ldots,y_n) = P_{X_1,\ldots,X_n}(x_1,\ldots,x_n) \left| \frac{\partial(x_1,\ldots,x_n)}{\partial(y_1,\ldots,y_n)} \right|. \tag{2.17}$$

Here it is understood that the x_i's on the right side are now functions of the y_i's which are obtained by solving the set of simultaneous equations (2.10) with $m = n$.

A great many useful results come from the RVT theorem with $m \neq n$. For example, suppose $n = 2$ and $m = 1$, and we define $Y = \alpha_1 X_1 + \alpha_2 X_2$, where the α_i's are constants. The RVT theorem then gives for the PDF of Y,

$$P_Y(y) = \int_{-\infty}^{\infty} dx_1 \int_{-\infty}^{\infty} dx_2 \, P_{X_1,X_2}(x_1, x_2) \, \delta\left(y - [\alpha_1 x_1 + \alpha_2 x_2]\right). \tag{2.18}$$

If we multiply this equation through by y and then integrate both sides over y, we get, because of the delta function,

$$\langle Y \rangle = \int_{-\infty}^{\infty} dx_1 \int_{-\infty}^{\infty} dx_2 \, (\alpha_1 x_1 + \alpha_2 x_2) \, P_{X_1,X_2}(x_1, x_2)$$

$$= \alpha_1 \int_{-\infty}^{\infty} dx_1 \, x_1 \int_{-\infty}^{\infty} dx_2 P_{X_1,X_2}(x_1, x_2) + \alpha_2 \int_{-\infty}^{\infty} dx_2 \, x_2 \int_{-\infty}^{\infty} dx_1 \, P_{X_1,X_2}(x_1, x_2)$$

$$= \alpha_1 \int_{-\infty}^{\infty} dx_1 \, x_1 \, P_{X_1}(x_1) + \alpha_2 \int_{-\infty}^{\infty} dx_2 \, x_2 \, P_{X_2}(x_2) \qquad \text{[by Eq. (2.9)]}$$

$$= \alpha_1 \langle X_1 \rangle + \alpha_2 \langle X_2 \rangle.$$

Thus, regardless of whether or not X_1 and X_2 are statistically independent,

$$\langle \alpha_1 X_1 + \alpha_2 X_2 \rangle = \alpha_1 \langle X_1 \rangle + \alpha_2 \langle X_2 \rangle. \tag{2.19a}$$

The generality of this result allows us to replace X_1 and X_2 by their squares:

$$\langle \alpha_1 X_1^2 + \alpha_2 X_2^2 \rangle = \alpha_1 \langle X_1^2 \rangle + \alpha_2 \langle X_2^2 \rangle.$$

By using these two identities, the formula for the variance of $\alpha_1 X_1 + \alpha_2 X_2$, namely

$$\text{var}\{\alpha_1 X_1 + \alpha_2 X_2\} \equiv \left\langle (\alpha_1 X_1 + \alpha_2 X_2)^2 \right\rangle - \langle \alpha_1 X_1 + \alpha_2 X_2 \rangle^2,$$

can be algebraically expanded to yield a companion to Eq. (2.19a):

$$\text{var}\{\alpha_1 X_1 + \alpha_2 X_2\} = \alpha_1^2 \, \text{var}\{X_1\} + \alpha_2^2 \, \text{var}\{X_2\} + 2\alpha_1 \alpha_2 \, \text{cov}\{X_1, X_2\}. \tag{2.19b}$$

Here we have defined the *covariance* of X_1 and X_2 by

$$\text{cov}\{X_1, X_2\} \equiv \langle X_1 X_2 \rangle - \langle X_1 \rangle \langle X_2 \rangle. \tag{2.20}$$

If X_1 and X_2 are statistically independent, in that their joint PDF factors as $P_{X_1,X_2}(x_1, x_2) = P_{X_1}(x_1)P_{X_2}(x_2)$, then it is easy to show that $\langle X_1 X_2 \rangle = \langle X_1 \rangle \langle X_2 \rangle$, so that $\text{cov}\{X_1, X_2\} = 0$. Thus we have the following often used corollary to Eqs (2.19): For X_1 and X_2 any two *statistically independent* random variables,

$$\langle X_1 \pm X_2 \rangle = \langle X_1 \rangle \pm \langle X_2 \rangle, \tag{2.21a}$$

$$\text{var}\{X_1 \pm X_2\} = \text{var}\{X_1\} + \text{var}\{X_2\}. \tag{2.21b}$$

It can be proved that the magnitude of the covariance of any two random variables is always bounded by the product of their standard deviations[3]

$$|\text{cov}\{X_1, X_2\}| \le \sqrt{\text{var}\{X_1\} \cdot \text{var}\{X_2\}}. \tag{2.22}$$

This prompts us to define the *correlation* of X_1 and X_2 by

$$\text{corr}\{X_1, X_2\} \equiv \frac{\text{cov}\{X_1, X_2\}}{\sqrt{\text{var}\{X_1\} \cdot \text{var}\{X_2\}}}. \tag{2.23}$$

Owing to Eq. (2.22), $\text{corr}\{X_1, X_2\}$ is always a pure number between -1 and $+1$:

$$-1 \le \text{corr}\{X_1, X_2\} \le +1. \tag{2.24}$$

As such, it often provides a revealing measure of the statistical dependency between X_1 and X_2. In particular, if X_1 and X_2 are statistically independent, then $\text{corr}\{X_1, X_2\} = 0$; if $X_2 = \alpha X_1$ where $\alpha > 0$, then $\text{corr}\{X_1, X_2\} = +1$; and if $X_2 = \alpha X_1$ where $\alpha < 0$, then $\text{corr}\{X_1, X_2\} = -1$. More generally, for two random variables with a *positive* correlation, larger (smaller) sample values of one tend to occur together with larger (smaller) sample values of the other. And for two *negatively* correlated random variables, larger (smaller) sample values of one tend to occur together with smaller (larger) sample values of the other.

Another useful consequence of the RVT theorem for $n = 2$ and $m = 1$ is the linear combination rule for statistically independent normal random variables: If $\mathcal{N}_1(\mu_1, \sigma_1^2)$ and $\mathcal{N}_2(\mu_2, \sigma_2^2)$ are *statistically independent* and α_1 and α_2 are any constants, then

$$\alpha_1 \mathcal{N}_1(\mu_1, \sigma_1^2) + \alpha_2 \mathcal{N}_2(\mu_2, \sigma_2^2) = \mathcal{N}\left(\alpha_1 \mu_1 + \alpha_2 \mu_2, \ \alpha_1^2 \sigma_1^2 + \alpha_2^2 \sigma_2^2\right). \tag{2.25}$$

The fact that the mean and variance of the linear combination on the left side of Eq. (2.25) is what is claimed on the right side could, of course, have been inferred from Eqs (2.19). The novelty of Eq. (2.25) is its assertion that the linear combination of two statistically independent normal random variables is itself normal. To prove Eq. (2.25), one takes $X_i = \mathcal{N}_i\left(\mu_i, \sigma_i^2\right)$ for $i = 1$ and 2, so that $P_{X_i}(x_i)$ is given by Eq. (2.4) with μ and σ^2 appropriately subscripted, and one then invokes the RVT theorem in the form of Eq. (2.18). Because of the statistical independence of X_1 and X_2, their joint PDF $P_{X_1,X_2}(x_1, x_2)$ will be the product $P_{X_1}(x_1)P_{X_2}(x_2)$. The two integrations in Eq. (2.18) are then performed with the help of a special Fourier-integral representation of the Dirac delta function, which we will not discuss here.[4] The result turns out to be the PDF of the normal random variable on the right side of Eq. (2.25).

The Fourier-integral delta function representation also allows the RVT theorem to be used for $n \gg 1$ and $m = 1$ to prove the famous *Central Limit Theorem* (CLT).[5] The CLT says, in one of its several forms, that if $X_1, \ldots, X_n$ are n *statistically independent* random variables with a common PDF P_X that has a well-defined mean μ and variance σ^2, then

$$X_1 + \cdots + X_n \xrightarrow[n \to \infty]{} \mathcal{N}\left(n\mu, n\sigma^2\right). \tag{2.26}$$

We already know from Eq. (2.25) that (2.26) would be exact for all n if the X_i's were themselves normal. The surprising message of the CLT is that, even if the X_i's are *not* normal, their *sum* will be approximately normal for n sufficiently large. The results (2.25) and (2.26) are a large part of the reason why the normal random variable occupies such a prominent place in practical applications of random variable theory.

Applying the identity (2.15b) to the result (2.26) gives a second version of the CLT:

$$\frac{X_1 + \cdots + X_n}{n} \xrightarrow[n \to \infty]{} \mathcal{N}\left(\mu, \sigma^2/n\right). \tag{2.27}$$

If we view each X_i here as a statistically independent copy of some random variable X, the result (2.27) provides a basis for many statistical estimation procedures: It tells us that the "average" of a sufficiently large number n of independent sample values of any random variable X which has a finite mean μ and a finite standard deviation σ will behave like a sample value of the *normal* random variable with the same mean μ but a *smaller* standard deviation $\sigma/\sqrt{n}$. Thus, the average of increasingly larger numbers of sample values of X will "estimate the mean of X" with increasingly greater accuracy, and in a way that can be quantified by the known properties of the normal random variable. But it is important to understand that this result does *not* hold if either the mean or the variance of P_X does not exist. For example, suppose the n X_i's on the left side of Eq. (2.27) are statistically independent versions of the Cauchy random variable $\mathcal{C}(x_0, \gamma)$. As mentioned earlier, $\mathcal{C}(x_0, \gamma)$ does not have a mean or variance. Nevertheless, the similarity in the shapes of the PDF of $\mathcal{C}(x_0, \gamma)$ in Fig. 2.2d and the PDF of $\mathcal{N}(\mu, \sigma^2)$ in Fig. 2.2c tempts us to suppose that the arithmetic average of n statistically independent versions of $\mathcal{C}(x_0, \gamma)$ should be some random variable whose PDF has a peak at x_0 which gets narrower as n is increased. But that's not so. Application of the RVT theorem, again using the Fourier-integral representation of the Dirac delta function, reveals that the left side of Eq. (2.27) is *exactly* $\mathcal{C}(x_0, \gamma)$ for *all* n; in other words, taking the arithmetic average of an arbitrarily large number of sample values of $\mathcal{C}(x_0, \gamma)$ estimates x_0 no better than a *single sampling* of $\mathcal{C}(x_0, \gamma)$!

2.7   The bivariate normal random variable

At several places in the following chapters, we will encounter two normal random variables $X_1 = \mathcal{N}(\mu_1, \sigma_1^2)$ and $X_2 = \mathcal{N}(\mu_2, \sigma_2^2)$ that are *correlated*, in that their covariance, cov $\{X_1, X_2\} \equiv c_{12}$, is not necessarily zero. The two-component random variable $\mathbf{X} = (X_1, X_2)$ is then said to be a *bivariate normal* random variable. It can be proved (but we won't bother to do so here since we will never require the formula in this book) that the joint PDF of X_1 and X_2 has the explicit form

$$P_{X_1, X_2}(x_1, x_2) = \frac{1}{2\pi\sqrt{\sigma_1^2 \sigma_2^2 - c_{12}^2}}$$

$$\times \exp\left(-\frac{\sigma_2^2(x_1 - \mu_1)^2 + \sigma_1^2(x_2 - \mu_2)^2 - 2c_{12}(x_1 - \mu_1)(x_2 - \mu_2)}{2\left(\sigma_1^2 \sigma_2^2 - c_{12}^2\right)}\right). \tag{2.28}$$

A way of representing two correlated normal random variables that will be more useful to us than this joint PDF is the following: If $N_1 = \mathcal{N}(0, 1)$ and $N_2 = \mathcal{N}(0, 1)$ are *statistically independent*, then the two random variables X_1 and X_2 defined by

$$X_1 = \mu_1 + \sigma_1 N_1, \tag{2.29a}$$

$$X_2 = \mu_2 + \left(\frac{c_{12}}{\sigma_1}\right) N_1 + \left(\sigma_2^2 - \frac{c_{12}^2}{\sigma_1^2}\right)^{1/2} N_2, \tag{2.29b}$$

will be *normal* with respective means μ_1 and μ_2, variances σ_1^2 and σ_2^2, and covariance c_{12}. Thus, formulas (2.29) provide a compact representation of any two correlated normal random variables whose joint PDF is Eq. (2.28). The proof of Eqs (2.29) is surprisingly easy: To show that X_1 and X_2 as thus defined are normal with the claimed means and variances, just apply Eqs (2.16) and (2.25) to each of the above two formulas. To show that X_1 and X_2 have covariance c_{12}, first write from Eq. (2.20)

$$\text{cov}\{X_1, X_2\} \equiv \langle X_1 X_2 \rangle - \langle X_1 \rangle \langle X_2 \rangle = \langle (X_1 - \mu_1)(X_2 - \mu_2) \rangle,$$

where the last equality can be verified by expanding the product inside the brackets on the right and then averaging the resulting terms separately. Next, substitute for the two factors $(X_i - \mu_i)$ on the right from Eqs (2.29). Finally, collect terms and use the relations $\langle N_1^2 \rangle = 1$ and $\langle N_1 N_2 \rangle = 0$.

An example of the usefulness of the representation (2.29) is the easy way it allows us to prove the following important generalization of Eq. (2.25): A linear combination of two normal random variables, *even if they are not statistically independent*, is normal. With much effort, this result can be proved from the RVT formula (2.18), using the expression (2.28) for the joint PDF of the two correlated normal random variables X_1 and X_2. But the result follows much more easily from formulas (2.29): they show clearly that any linear combination of X_1 and X_2 will necessarily be a linear combination of the two *statistically independent* normal random variables N_1 and N_2, and therefore will be normal by Eq. (2.25). This result, together with the general result (2.19b), shows in particular that the sum/difference of any two correlated normal random variables is the normal random variable

$$\mathcal{N}(\mu_1, \sigma_1^2) \pm \mathcal{N}(\mu_2, \sigma_2^2) = \mathcal{N}\left(\mu_1 \pm \mu_2, \sigma_1^2 + \sigma_2^2 \pm 2c_{12}\right). \tag{2.30}$$

Formulas (2.29) can also be used to prove a result that describes a bit more quantitatively the connection between the fluctuations in two correlated normal random variables. Here we use the word "fluctuation" to mean, loosely, the scattering of the sample values of a random variable about its mean. The connection between the fluctuations in two correlated *normal* random variables X_1 and X_2 is given by

$$[\text{corr}\{X_1, X_2\}]^2 = \left\{ \begin{array}{c} \text{the } \textit{fraction} \text{ of the variance of either } X_1 \text{ or } X_2 \text{ that} \\ \text{is associated with the fluctuations in the other} \end{array} \right\}. \tag{2.31}$$

We have used the cautious phrasing "is associated with" instead of "is caused by" because it is not always possible to say whether X_1 is causing the fluctuations in X_2 or X_2 is causing the fluctuations in X_1. In any case, to prove Eq. (2.31), we first solve Eq. (2.29a) for N_1 and then substitute the result into Eq. (2.29b) to get

$$X_2 = \frac{c_{12}}{\sigma_1^2} X_1 + \sqrt{\sigma_2^2 - \frac{c_{12}^2}{\sigma_1^2}} \, N_2 + \left(\mu_2 - \frac{\mu_1 c_{12}}{\sigma_1^2}\right). \tag{2.32}$$

This formula shows that the fluctuations in X_2 can be viewed as arising solely from the fluctuations in the two *statistically independent* random variables X_1 and N_2. Now, the prime measure of the fluctuations in any random variable is its variance. And by applying to Eq. (2.32) the variance rule (2.19b), we can write the variance of the linear combination (2.32) as follows (remember that the variance of any sure variable is zero):

$$\text{var}\{X_2\} = \left(\frac{c_{12}}{\sigma_1^2}\right)^2 \text{var}\{X_1\} + \left(\sigma_2^2 - \frac{c_{12}^2}{\sigma_1^2}\right) \text{var}\{N_2\}.$$

The two terms on the right evidently quantify the contributions to $\text{var}\{X_2\}$ coming from the fluctuations in the two statistically independent random variables X_1 and N_2. In particular, the contribution to $\text{var}\{X_2\}$ coming from the fluctuations in X_1 is seen to be

$$\left(\frac{c_{12}}{\sigma_1^2}\right)^2 \text{var}\{X_1\} = \frac{c_{12}^2}{\sigma_1^4} \sigma_1^2 = \left(\frac{c_{12}}{\sigma_1 \sigma_2}\right)^2 \text{var}\{X_2\}. \tag{2.33}$$

Since the factor in parentheses on the right side of Eq. (2.33) is $\text{corr}\{X_1, X_2\}$, we conclude that $[\text{corr}\{X_1, X_2\}]^2$ gives the *fraction* of $\text{var}\{X_2\}$ that is associated with the fluctuations in X_1. And of course, this entire argument can be repeated with X_1 and X_2 switching roles.

2.8 Generating numerical samples of random variables

Another important area of random variable theory is the development of methods for generating on a computer *sample values* of specific random variables, i.e., "random numbers". The ability to do that is key to all types of "Monte Carlo calculations", which indeed can be defined as any calculation that makes use of random numbers. Numerical analysts have developed several clever and efficient algorithms for generating sample values u of the *unit-interval uniform random variable* $\mathcal{U}(0,1) \equiv U$. Those sample values can in turn be used, in a variety of ways, to construct sample values x of *any* PDF P_X.

The classic *inversion method* for doing that is as follows: First compute X's *cumulative distribution function* (CDF):

$$F_X(x) \equiv \text{Prob}\,\{X \leq x\} = \int_{-\infty}^{x} P_X(x')dx'. \tag{2.34}$$

Next compute the *inverse* F_X^{-1} of F_X. Although that inverse may not be easy to compute, it always exists, because the function defined in Eq. (2.34) increases monotonically from 0 to 1 as x goes from $-\infty$ to ∞. Then, where u is a sample value of $U \equiv \mathcal{U}(0,1)$, compute a sample value x of X by setting $F_X(x) = u$ and then solving for x; i.e., take $x = F_X^{-1}(u)$. The correctness of this procedure follows from the RVT theorem: Let $\tilde{X}$ be the random variable whose sample value $\tilde{x}$ is computed by solving the equation $u = F_X(\tilde{x})$ for $\tilde{x}$. Then it follows from the RVT result (2.13) that

$$P_{\tilde{X}}(\tilde{x}) = P_U(u) \left| \frac{du}{d\tilde{x}} \right| = \frac{1}{(1-0)} \cdot \left| \frac{dF_X(\tilde{x})}{d\tilde{x}} \right| = P_X(\tilde{x}).$$

The last step here follows from the definition (2.34) of F_X. Since $P_{\tilde{X}}$ is the same function as P_X, then $\tilde{X}$ must be the same random variable as X.

The simplest application of the inversion method is generating a sample value of the *uniform* random variable $X = \mathcal{U}(a,b)$; applying the rule $x = F_X^{-1}(u)$ to the CDF corresponding to the PDF in Eq. (2.2) gives the generating formula $x = a + (b-a)u$, a totally unsurprising result. Less obvious is the inversion generating formula for the *exponential* random variable, $X = \mathcal{E}(a)$. From Eqs (2.3) and (2.34), we obtain for this random variable, $F_X(x) = 1 - e^{-ax}$. Setting that equal to u and then solving for x gives[6]

$$x = \frac{1}{a} \ln\left(\frac{1}{1-u}\right) \quad \text{for } X = \mathcal{E}(a). \tag{2.35}$$

Generating samples of the *normal* random variable $\mathcal{N}(\mu, \sigma^2)$ using the inversion method is a bit more complicated, because the inverse of the CDF of the normal random variable is not analytically simple. But suppose we set ourselves the seemingly more difficult task of generating *two statistically independent* samples of $\mathcal{N}(\mu, \sigma^2)$, i.e., a sample (x_1, x_2) of the random variable $\mathbf{X} = (X_1, X_2)$ whose PDF is

$$P_{X_1,X_2}(x_1, x_2) = \frac{1}{\sqrt{2\pi\sigma^2}} \exp\left(-\frac{(x_1-\mu)^2}{2\sigma^2}\right) \cdot \frac{1}{\sqrt{2\pi\sigma^2}} \exp\left(-\frac{(x_2-\mu)^2}{2\sigma^2}\right)$$

$$= \frac{1}{2\pi\sigma^2} \exp\left(-\frac{(x_1-\mu)^2 + (x_2-\mu)^2}{2\sigma^2}\right).$$

To accomplish this task, we first transform to a new pair of random variables S and Θ defined by

$$X_1 = \mu + S \cos \Theta, \quad X_2 = \mu + S \sin \Theta.$$

The full-axis domains of X_1 and X_2 are secured by requiring $S \geq 0$ and $0 \leq \Theta < 2\pi$. By the RVT theorem result (2.17), the joint PDF of S and Θ can be computed from the joint PDF of X_1 and X_2 as

$$P_{S,\Theta}(s,\theta) = P_{X_1,X_2}(x_1,x_2) \left| \frac{\partial(x_1,x_2)}{\partial(s,\theta)} \right|,$$

where on the right $x_1 = \mu + s \cos \theta$ and $x_2 = \mu + s \sin \theta$. The Jacobian here is easily shown to be equal to s, so by the above formula for P_{X_1,X_2} we have

$$P_{S,\Theta}(s,\theta) = (2\pi\sigma^2)^{-1} e^{-s^2/2\sigma^2} s$$

$$\equiv \left(\sigma^{-2} s\, e^{-s^2/2\sigma^2} \right) \cdot \left(\frac{1}{2\pi} \right),$$

where $0 \leq s$ and $0 \leq \theta < 2\pi$. The factored form of this result tells us that S and Θ are statistically independent of each other. And since the second factor is a properly normalized PDF for Θ, which is evidently just the random variable $\mathcal{U}(0, 2\pi)$, then the first factor must be the properly normalized PDF of S. The corresponding CDFs of S and Θ are then easily computed from these PDFs using the definition (2.34):

$$F_S(s) = 1 - e^{-s^2/2\sigma^2}, \quad F_\Theta(\theta) = \frac{\theta}{2\pi}.$$

Now we apply the inversion rule separately to these two statistically independent CDFs: Where u_1 and u_2 are two statistically independent samples of U, we solve $F_S(s) = u_1$ for s, and $F_\Theta(\theta) = u_2$ for θ. That gives

$$s = \sigma \sqrt{2 \ln (1/(1 - u_1))} \quad \text{and} \quad \theta = 2\pi u_2. \tag{2.36a}$$

From these joint samples of S and Θ, we may now compute the joint samples of X_1 and X_2 from the transformation formulas relating those two sets of random variables:

$$x_1 = \mu + s \cos \theta \quad \text{and} \quad x_2 = \mu + s \sin \theta. \tag{2.36b}$$

Formulas (2.36) are known as the Box–Muller generating formulas. Their derivation assures that x_1 and x_2 will be statistically independent samples of $\mathcal{N}(\mu, \sigma^2)$. If only one sample is needed, either x_1 or x_2 can be chosen and the other can be discarded. But if, as is usually the case, more samples of $\mathcal{N}(\mu, \sigma^2)$ will be needed later, the sample not chosen can be stored for later use. Figure 2.4 shows a normalized frequency histogram of 2000 sample values generated by formulas (2.36) with $\mu = 0$ and $\sigma = 1$, together with a plot of the PDF of the *unit normal random variable* $\mathcal{N}(0, 1) \equiv N$ as prescribed by Eq. (2.4). Sample values of that random variable will be used frequently in later chapters of this book.

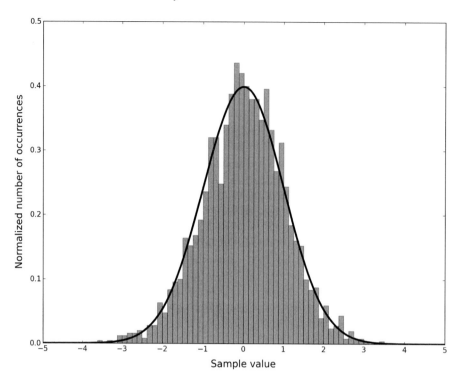

Fig. 2.4 *A plot of the PDF of $\mathcal{N}(0,1)$, the normal random variable with mean 0 and variance 1, superimposed on a normalized frequency histogram of 2000 sample values of that random variable. The sample values were generated from sample values of $\mathcal{U}(0,1)$ using formulas (2.36). The frequency histogram was "normalized" by dividing the number of sample values falling into a given histogram bin by the product of the bin width (here 0.125) and the total number of sample values (here 2000). Sample values of $\mathcal{N}(0,1)$ will be used many times later in this book, and this figure gives a feeling for what those sample values will typically be.*

So we can generate a sample of $\mathcal{E}(a)$ by drawing a number u from the $\mathcal{U}(0,1)$-bucket and transforming it into a number x for the $\mathcal{E}(a)$-bucket through formula (2.35). And we can generate samples of $\mathcal{N}(\mu,\sigma^2)$ by drawing two numbers u_1 and u_2 from the $\mathcal{U}(0,1)$-bucket and transforming them into two numbers x_1 and x_2 for the $\mathcal{N}(\mu,\sigma^2)$-bucket through formulas (2.36). In practice, numerical analysts have developed other generating algorithms for the exponential, normal, and many other random variables that are computationally a bit faster than these classic inversion generating procedures, and those more efficient algorithms have found their way into many popular software packages.

2.9   Integer-valued random variables

The foregoing discussion has assumed that X is a *real*-valued random variable. If X is confined to *integer* values, the definition of the PDF $P_X(x)$ changes to

$P_X(x) \equiv \text{Prob}\{X = x\}$, where x is now restricted to integer values. The PDF P_X is then sometimes called a probability mass function instead of a probability density function. From a practical point of view, there are only two other changes from the real-variable formalism: First, integrals over real x get everywhere replaced by sums over integer x. For instance, the definition (2.5) of the kth moment of X becomes

$$\langle X^k \rangle \equiv \sum_{x=-\infty}^{\infty} x^k P_X(x) \quad (k = 0, 1, \ldots). \tag{2.37}$$

For $k = 1$, this formula gives the often used rule-of-thumb that *the mean of an integer random variable is the sum $\sum_x x \cdot P_X(x)$ of all its possible values "weighted" by their probabilities*. The second change from the real-variable theory is that the Dirac delta function $\delta(x - x_0)$ gets everywhere replaced by the Kronecker delta function δ_{x,x_0}, which is defined to be 1 if $x = x_0$ and 0 otherwise; e.g., $P_X(x) = \delta_{x,x_0}$ implies that X is the sure (integer) variable x_0.

A frequently encountered integer-valued random variable is the *binomial* random variable, $X = \mathcal{B}(p, N)$, where p can be any real number in $[0, 1]$ and N can be any positive integer. The PDF of $X = \mathcal{B}(p, N)$ is

$$P_X(x) = \begin{cases} \dfrac{N!}{x!\,(N-x)!} p^x (1-p)^{N-x}, & \text{if } x = 0, 1, \ldots, N, \\ 0, & \text{otherwise .} \end{cases} \tag{2.38}$$

From this definition it can be shown, using Eq. (2.37) and some subtle algebraic manipulations, that the mean and variance of $\mathcal{B}(p, N)$ are

$$\langle X \rangle = Np \text{ and } \text{var}\{X\} = Np(1-p). \tag{2.39}$$

If p is the probability that the toss of a weighted coin will produce heads, then $P_X(x)$ will give the probability that, in a sequence of N tosses, heads will be obtained exactly x times; indeed, this result can be read off directly from the right side of Eq. (2.38): The factor $p^x(1-p)^{N-x}$ is the probability that, in N tosses, a particular set of x tosses will give heads and the other $(N - x)$ won't, and the binomial factor is the number of distinct ways of choosing x identical objects out of N.

If X assumes only integer values $i \geq x_0$, then with $P_X(i) = p_i$, X's CDF is $F_X(x) = \sum_{i=x_0}^{x} p_i$. The inversion generating procedure for generating a sample value x of this random variable from a sample value u of $\mathcal{U}(0, 1)$, which for real-valued random variables would be to solve $F_X(x) = u$ for x, now becomes: find the *smallest integer x for which $F_X(x)$ exceeds u*:[7]

$$x = \text{the smallest integer for which } \sum_{i=x_0}^{x} p_i > u. \tag{2.40}$$

For example, if X is equally likely to be any of the integers between n_1 and n_2 inclusively—or in other words if $P_X(x)$ is equal to $(n_2 + 1 - n_1)^{-1}$ for all the integers in the interval $[n_1, n_2]$ and zero everywhere else—then application of the generating rule (2.40) leads to the formula

$$x = [n_1 + (n_2 + 1 - n_1)u], \tag{2.41}$$

where $[z]$ means "the greatest integer in z". The correctness of this formula can be seen by first noting that, since u is randomly uniform in the interval $[0, 1)$, then the real number x in brackets in Eq. (2.41) will be randomly uniform in the interval $[n_1, n_2 + 1)$. Therefore, the integer $n \equiv [x]$ identifies the interval $[n, n + 1)$ in which x lies, and those intervals can be put into a one-to-one correspondence with the integers $n_1, \ldots, n_2$.

Formula (2.41) evidently provides a fast way of "randomly choosing an integer" in the interval $[n_1, n_2]$. However, if the p_i values are not simply related to the index i, it will usually be necessary to compute the individual partial sums in Eq. (2.40) in succession in order to determine x. That can be a time-consuming task if the number of possible values of X is large.

The generating procedure (2.40) can be used to construct random samples of $X = \mathcal{B}(p, N)$ by taking $x_0 = 0$. But that generating procedure turns out to be efficient only if N is small. Numerical analysts have developed more efficient binomial generating procedures for large N, and these have been implemented in many popular software packages.

Notes to Chapter 2

[1]To derive the multiplication and addition laws from the frequency definition of probability, let $m_n(E)$ denote the number of times event E occurs in n tries. By definition, the conditional probability $p(E_2 \mid E_1)$ is the limit as $n \to \infty$ of the ratio $m_n(E_1 \text{ and } E_2)/m_n(E_1)$, the fraction of tries yielding E_1 in which E_2 also occurs. If we divide the numerator and the denominator of this fraction by n before we take the limit $n \to \infty$, the limit yields the multiplication law. The addition law, unlike the multiplication law, requires E_1 and E_2 to be mutually exclusive; therefore, the number of tries in which either event occurs will be $m_n(E_1 \text{ or } E_2) = m_n(E_1) + m_n(E_2)$. Dividing that through by n and then taking the limit $n \to \infty$ gives the addition law.

[2]For details on how the RVT theorem is proved for $m, n > 1$, see Section 1.6 of the book *Markov Processes: An Introduction for Physical Scientists* (Academic Press, 1992) by D. Gillespie.

[3]To prove Eq. (2.22), expand the right side of the inequality $0 \leq \text{var}\{\alpha_1 X_1 - \alpha_2 X_2\}$ using Eq. (2.19b) to get

$$0 \leq \alpha_1^2 \, \text{var}\{X_1\} + \alpha_2^2 \, \text{var}\{X_2\} - 2\alpha_1\alpha_2 \, \text{cov}\{X_1, X_2\},$$

and then take $\alpha_1 = \text{cov}\{X_1, X_2\}$ and $\alpha_2 = \text{var}\{X_1\}$.

[4]The Fourier-integral representation of the Dirac delta function is

$$\delta(x_1 - x_2) = \frac{1}{2\pi} \int_{-\infty}^{\infty} du \, \exp[i u(x_1 - x_2)],$$

where $i \equiv \sqrt{-1}$. Although the integral over u here is actually defined in conjunction with a special double-limit, in all applications of the RVT theorem, that integral can be manipulated as an ordinary integral. For further information on this formula, see Appendix B of the book cited in Note 2. (For readers who happen to be familiar with the "characteristic function" of a random variable: using the above integral representation of the Dirac delta function in applications of the RVT theorem is mathematically equivalent to using characteristic functions.)

[5]A detailed proof of the Central Limit Theorem from the RVT theorem can be found in Section 1.6 of the book cited in Note 2. Derivations of several other useful results from the RVT theorem can be found in the article "A theorem for physicists in the theory of random variables", *American Journal of Physics* **51**:520–533 (1983), by D. Gillespie.

[6]The exponential generating formula (2.35) is often written with $(1 - u)$ replaced by u, on the grounds that if u is uniformly distributed over the unit interval then $(1 - u)$ should be too. Although that replacement is usually okay, it can sometimes lead to numerical problems. That's because the formal definition of the random variable $\mathcal{U}(0, 1)$ puts its values in the interval $0 \leq u < 1$, and modern random number generators occasionally produce the allowed sample value 0 but never the disallowed value 1. Formula (2.35) thus avoids the rare but very annoying divide-by-zero.

[7]For detailed derivations of Eqs (2.40) and (2.41), see for example pp. 52–53 of the book cited in Note 2.

3
Einstein's theory of diffusion

Before it was universally accepted that a fluid consists of many moving molecules, Fick's Law (1.2) and the diffusion equation (1.3) were widely regarded as statements in continuum mechanics. The kinetic molecular hypothesis gained widespread acceptance only after Albert Einstein's famous analysis of Brownian motion in 1905 and Jean Perrin's subsequent experimental confirmation of that analysis in 1908.[1] Einstein essentially derived the diffusion equation (1.3) from a model of random molecular motion instead of from the continuity equation (1.1) and Fick's Law (1.2). In this chapter we will present Einstein's derivation, examine its strengths and weaknesses, and deduce some of its implications. We will then take the first step toward deriving a formula for the diffusion-controlled stochastic rate of a bimolecular chemical reaction, a derivation that we will complete in Chapter 4.

3.1   Einstein's derivation of the diffusion equation

Our presentation in this section is not meant to be a faithful reproduction of the arguments in Einstein's published papers, but merely a distillation of their key points. Focusing on a one-dimensional system, Einstein began by assuming the existence of a time interval δt which, on the one hand, could be considered infinitesimally small on a macroscopic scale, yet on the other hand was large enough that a solute molecule typically experiences many collisions with solvent molecules in that time. He then introduced a probability density function $\phi(\xi; \delta t)$, defined so that $\phi(\xi; \delta t)\, d\xi$ gives the probability that the x-coordinate of a solute molecule will change during the next δt by an amount between ξ and $\xi + d\xi$. This stochastic change in the position of a solute molecule is assumed to represent the combined effects of the many random collisions that molecule will have with the many smaller solvent molecules in time δt. Einstein also assumed that the change in position of a solute molecule during each successive δt interval is independent of the changes in previous δt intervals; this assumption is another reason why δt cannot be taken arbitrarily small.

Armed with the displacement PDF ϕ, Einstein observed that the average number of solute molecules with x-coordinate between x and $x + dx$ at time $t + \delta t$ can be written in terms of that number at the slightly earlier time t as follows:

$$\rho(x, t + \delta t)\, dx = \int_{\xi=-\infty}^{\infty} [\rho(x - \xi, t)\, dx] \times [\phi(\xi; \delta t)\, d\xi]. \tag{3.1}$$

The rationale for Eq. (3.1) is this: The first factor in brackets on the right is the average number of solute molecules in the dx-interval at $x - \xi$ at time t. The second factor is the average fraction of those that will, as a result of collisions with solvent molecules in the next δt, suffer a change in x-coordinate between ξ and $\xi + d\xi$, and will therefore end up in the dx-interval at x at time $t + \delta t$. The product is then summed (integrated) over all possible values of ξ to get the average number of molecules with x-coordinate between x and $x + dx$ at time $t + \delta t$. Canceling the dx's in Eq. (3.1), and then expanding $\rho(x - \xi)$ under the integral sign in a Taylor series about x, Einstein obtained

$$\rho(x, t + \delta t) = \int\limits_{-\infty}^{\infty} \phi(\xi; \delta t) \rho(x - \xi, t)\, d\xi$$

$$= \int\limits_{-\infty}^{\infty} \phi(\xi; \delta t) \left[\rho(x, t) + \sum_{k=1}^{\infty} \frac{(-\xi)^k}{k!} \frac{\partial^k \rho(x, t)}{\partial x^k} \right] d\xi$$

$$\rho(x, t + \delta t) = \rho(x, t) \int\limits_{-\infty}^{\infty} \phi(\xi; \delta t)\, d\xi + \sum_{k=1}^{\infty} \frac{\partial^k \rho(x, t)}{\partial x^k} \left[\frac{1}{k!} \int\limits_{-\infty}^{\infty} (-\xi)^k \phi(\xi; \delta t)\, d\xi \right]. \quad (3.2)$$

Since ϕ is a proper PDF in ξ, the integral in the first term on the right is unity. And since diffusive moves in the positive and negative x-directions should be equally likely, then $\phi(\xi; \delta t)$ should be an *even* function of ξ, so the integral under the summation vanishes for all odd integers k. Transposing the first term on the right and dividing through by δt, Einstein obtained the following expression for the time derivative of $\rho(x, t)$—which we note is really a "macroscopic" time derivative since δt cannot be taken arbitrarily small:

$$\frac{\partial \rho(x, t)}{\partial t} = \sum_{k=1}^{\infty} \left[\frac{1}{\delta t} \frac{1}{(2k)!} \int\limits_{-\infty}^{\infty} \xi^{2k} \phi(\xi; \delta t)\, d\xi \right] \frac{\partial^{2k} \rho(x, t)}{\partial x^{2k}}. \quad (3.3)$$

Einstein next made the assumption that the infinite series (3.3) converges sufficiently rapidly that all terms beyond the first ($k = 1$) can be ignored. That approximation gives

$$\frac{\partial \rho(x, t)}{\partial t} = D \frac{\partial^2 \rho(x, t)}{\partial x^2}, \quad (3.4)$$

where D is defined by

$$D \equiv \frac{1}{2\delta t} \int\limits_{-\infty}^{\infty} \xi^2 \phi(\xi; \delta t)\, d\xi. \quad (3.5)$$

Since Eq. (3.4) is the classical one-dimensional diffusion equation (1.3), Einstein had thus derived that equation by reasoning from the expected effects on the solute molecules of collisional impacts from the surrounding solvent molecules.

3.2 A critique of Einstein's derivation

Einstein's derivation of the diffusion equation is disarmingly simple, but in its day its probabilistic reasoning was innovative and groundbreaking. However, a closer inspection reveals that it has a couple of weaknesses. One is the casual truncation of the infinite series in Eq. (3.3) at the second term. Although the factor $1/(2k)!$ encourages the hope that some approximating truncation might be reasonable, the assumption that the sum of all terms in that series beyond the first is negligible has no obvious justification.

The second weakness of Einstein's analysis is its conclusion that the quantity on the right side of Eq. (3.5), which involves δt, should be equal to the experimentally observed diffusion coefficient D, which we would not expect to depend upon the somewhat arbitrary value of δt. That is, while δt will surely have some upper bound and some lower bound, it seems unlikely that those two bounds will be so tight that there will be only one acceptable value for δt. A closer inspection of the right side of Eq. (3.5) reveals that the only way that the right side could be *independent* of δt is if the integral there, which is evidently the second moment of the PDF $\phi(\xi; \delta t)$, were *directly proportional to* δt. Thus it might be argued that the definition (3.5) is nothing more than the *assumption* that that is so. In that case, all Einstein's analysis really proves is that, if we *assume* that the variance in the displacement is $\propto t$ for $t = \delta t$, then (as we will deduce from Eq. (3.3) shortly) that will also be true for all larger t.

This circularity in Einstein's analysis becomes even more jarring when we recognize that Einstein's PDF ϕ is really just a special case of the solute density function ρ. To see this, suppose there are N solute molecules in the system at the origin at time $t = 0$, i.e., $\rho(x, 0) = N\delta(x)$. Then $N^{-1}\rho(x, \delta t)\, dx$ will give the *average fraction* of those N molecules whose x-coordinates will be in the interval $[x, x + dx)$ at time $t = \delta t$. But since the solute molecules move about *independently* of each other, that average fraction must *also* be the *probability* that a *single* solute molecule, initially at the origin, will at time δt have its x-coordinate in the interval $[x, x + dx)$. And that probability is, by definition, $\phi(x; \delta t)\, dx$. Thus we see that ϕ is related to ρ by

$$\phi(x; \delta t) = N^{-1}\rho(x, \delta t), \quad \text{when } \rho(x, 0) = N\delta(x). \tag{3.6}$$

Therefore, in assuming various properties for the PDF ϕ, Einstein was in some sense assuming that those properties were possessed by the solute density function ρ, which was the target of his analysis.

3.3 Einstein's new perspective

Although Eq. (3.6) highlights a disquieting logical circularity in Einstein's derivation of Eqs (3.4) and (3.5), it also opens up a new perspective on diffusion. In effect, Einstein represented the x-coordinate of a *single* solute molecule at time t by a *random variable* (see Chapter 2), which we will denote by $X(t)$, and he focused attention on the probability density function (PDF) of that random variable, which we will denote by $P_X(x, t \,|\, x_0, t_0)$. This function is defined by the statement that

$$P_X(x, t \mid x_0, t_0)\, dx \equiv \text{ the probability that } X(t) \text{ will lie between}$$

$$x \text{ and } x + dx, \text{ given that } X(t_0) = x_0. \tag{3.7}$$

In fact, Einstein's definition of his function ϕ shows that ϕ is none other than this function P_X for $t - t_0 = \delta t$:

$$P_X(x + \xi, t + \delta t \mid x, t) = \phi(\xi; \delta t). \tag{3.8a}$$

The validity of Eq. (3.8a) can be seen by multiplying each side by $d\xi$ to get the probability that a solute molecule, which is at x at time t, will be found in the interval $[x + \xi, x + \xi + d\xi)$ at time $t + \delta t$. Furthermore, by a simple extension of the reasoning that led to Eq. (3.6), it follows that the functions P_X and ρ are also intimately related; specifically, P_X is just ρ normalized to 1 instead of to N:

$$P_X(x, t \mid x_0, t_0) = N^{-1} \rho(x, t), \text{ when } \rho(x, t_0) = N\delta(x - x_0). \tag{3.8b}$$

With Eqs (3.8) connecting Einstein's two functions ϕ and ρ to the PDF P_X by simple proportionalities, all of Einstein's equations (3.1) through (3.5) become equations that are satisfied by P_X. And as we will see later in Chapter 7, those equations are in fact simple versions of some foundational equations of modern *Markov process theory*: Eq. (3.1) is a simple version of what is now called a "Chapman–Kolmogorov equation"; Eq. (3.3) is a simple version of what is now called a "forward Kramers–Moyal equation"; and the diffusion equation (3.4) is a simple version of what is now called a "forward Fokker–Planck equation". Of particular interest to us here is the last. Owing to the simple proportionality between ρ and P_X in Eq. (3.8b), Eq. (3.4) implies that

$$\frac{\partial P_X(x, t \mid x_0, t_0)}{\partial t} = D \frac{\partial^2 P_X(x, t \mid x_0, t_0)}{\partial x^2}. \tag{3.9}$$

This can be regarded as the "Einstein version" of the classical diffusion equation (1.4). It is important to appreciate the logical shift that this equation makes in our perspective on diffusion: Whereas the diffusion equation (1.4) allowed us to determine $\rho(x, \delta t)\, dx$, the *average number* of solute molecules in the interval $[x, x + dx)$ at time t, Eq. (3.9) allows us to determine $P_X(x, t \mid x_0, t_0)dx$, the *probability* that a *particular* solute molecule will be in the interval $[x, x + dx)$ at time t.

As we noted earlier in connection with Eq. (1.5), the solution to Eq. (3.9) for the initial condition $P_X(x, t_0 \mid x_0, t_0) = \delta(x - x_0)$ and boundary conditions $P_X(\pm\infty, t \mid x_0, t_0) = 0$ is

$$P_X(x, t \mid x_0, t_0) = \frac{1}{\sqrt{4\pi D(t - t_0)}} \exp\left(-\frac{(x - x_0)^2}{4D(t - t_0)}\right) \quad (t \geq t_0), \tag{3.10}$$

as can easily be verified by direct substitution. Comparing Eq. (3.10) to the canonical form (2.4) of the PDF of the normal random variable $\mathcal{N}(\mu, \sigma^2)$ with mean μ and variance σ^2, we conclude that if $X(t_0) = x_0$ and there are no restrictions on the movement of the diffusing molecule along the x-axis, then $X(t)$ for all $t > t_0$ will be the *normal* random variable with mean x_0 and variance $2D(t - t_0)$:

$$X(t) = \mathcal{N}_x\left(x_0, 2D(t - t_0)\right). \tag{3.11a}$$

Similarly, the y- and z- components of the solute molecule's position are

$$Y(t) = \mathcal{N}_y\left(y_0, 2D(t - t_0)\right), \quad Z(t) = \mathcal{N}_z\left(z_0, 2D(t - t_0)\right). \tag{3.11b}$$

Furthermore, these three normal random variables are *statistically independent* of each other. The probabilistic characterization in Eqs (3.11) of the position of a single solute molecule that is diffusing unrestrictedly with diffusion coefficient D can be regarded as the central result of Einstein's theory of diffusion.

It is easy to see from Eqs (3.11a) and (2.15a) that

$$X(t) - x_0 = \mathcal{N}\left(0, 2D(t - t_0)\right).$$

Now, for any random variable Y, it follows from the definition (2.6a) of the variance that if $\langle Y \rangle = 0$, then $\langle Y^2 \rangle \equiv \mathrm{var}\{Y\}$. Therefore, the above relation implies that the *mean of the square* of the x-displacement $X(t) - x_0$ is equal to its variance $2D(t - t_0)$. Analogous reasoning applied to the other components gives

$$\left\langle (X(t) - x_0)^2 \right\rangle = \left\langle (Y(t) - y_0)^2 \right\rangle = \left\langle (Z(t) - z_0)^2 \right\rangle = 2D(t - t_0). \tag{3.12}$$

So, since $(\mathbf{R}(t) - \mathbf{r}_0)^2 = (X(t) - x_0)^2 + (Y(t) - y_0)^2 + (Z(t) - z_0)^2$, we conclude that the *mean-square displacement*—i.e., the mean of the squared displacement—of the diffusing molecule over the time interval $[t_0, t]$ is equal to

$$\left\langle (\mathbf{R}(t) - \mathbf{r}_0)^2 \right\rangle = 6D(t - t_0) \quad \text{(in 3 dim)}, \tag{3.13a}$$

or, if the motion is confined to just the xy-plane,

$$\left\langle (\mathbf{R}(t) - \mathbf{r}_0)^2 \right\rangle = 4D(t - t_0) \quad \text{(in 2 dim)}. \tag{3.13b}$$

It was these predictions for the mean-square displacement of a solute molecule in a macroscopic time t that were confirmed by Perrin's experiments on Brownian motion in 1908. These mean-square displacement formulas are often taken to be the operational definition of the diffusion coefficient D for experimental purposes—which is very much in accord with the way Einstein originally defined D in Eq. (3.5).

3.4 The covariance and correlation

For any $t_2 \geq t_1 \geq t_0$, the probability that $X(t_1)$ will be in $[x_1, x_1 + dx_1)$ *and* $X(t_2)$ will be in $[x_2, x_2 + dx_2)$ is, by the multiplication law of probability theory, equal to the product of {the probability $P_X(x_1, t_1 \mid x_0, t_0)\, dx_1$ for the first eventuality} times {the probability $P_X(x_2, t_2 \mid x_1, t_1)\, dx_2$ for the subsequent eventuality}. That implies that the *joint PDF* of the random variables $X(t_1)$ and $X(t_2)$ is $P_X(x_2, t_2 \mid x_1, t_1) \cdot P_X(x_1, t_1 \mid x_0, t_0)$. Now, we know that the Einstein PDF $P_X(x, t \mid x_0, t_0)$ in Eq. (3.10) satisfies the following normalization, mean, and variance conditions:

$$
\left.\begin{aligned}
\int_{-\infty}^{\infty} P_X(x,t \,|\, x_0, t_0)\, dx &= 1 \\
\int_{-\infty}^{\infty} x\, P_X(x,t \,|\, x_0, t_0)\, dx &= x_0 \\
\int_{-\infty}^{\infty} (x - x_0)^2\, P_X(x,t \,|\, x_0, t_0)\, dx &= 2D(t - t_0)
\end{aligned}\right\}. \tag{3.14}
$$

Using these relations, it is easy (and instructive) to show that[2]

$$
\langle X(t_i) \rangle \equiv \int_{-\infty}^{\infty} dx_1 \int_{-\infty}^{\infty} dx_2\, x_i\, P_X(x_2, t_2 \,|\, x_1, t_1) P_X(x_1, t_1 \,|\, x_0, t_0)
$$

$$
= x_0 \quad (i = 1 \text{ and } 2), \tag{3.15a}
$$

and

$$
\text{var}\{X(t_i)\} \equiv \int_{-\infty}^{\infty} dx_1 \int_{-\infty}^{\infty} dx_2\, (x_i - \langle X_i \rangle)^2\, P_X(x_2, t_2 \,|\, x_1, t_1) P_X(x_1, t_1 \,|\, x_0, t_0)
$$

$$
= 2D(t_i - t_0) \quad (i = 1 \text{ and } 2), \tag{3.15b}
$$

just as we expect. However, $X(t_1)$ and $X(t_2)$ are *not* statistically independent of each other. To quantify their statistical dependence, we first compute, with the help of Eqs (3.14) and (3.15),

$$
\langle X(t_1) X(t_2) \rangle \equiv \int_{-\infty}^{\infty} dx_1 \int_{-\infty}^{\infty} dx_2\, x_1 x_2\, P_X(x_2, t_2 \,|\, x_1, t_1) P_X(x_1, t_1 \,|\, x_0, t_0)
$$

$$
= \int_{-\infty}^{\infty} dx_1 \left(\int_{-\infty}^{\infty} dx_2\, x_2\, P_X(x_2, t_2 \,|\, x_1, t_1) \right) x_1\, P_X(x_1, t_1 \,|\, x_0, t_0)
$$

$$
= \int_{-\infty}^{\infty} dx_1\, (x_1)\, x_1\, P_X(x_1, t_1 \,|\, x_0, t_0)
$$

$$
= \langle X^2(t_1) \rangle \equiv \text{var}\{X(t_1)\} + \langle X(t_1) \rangle^2
$$

$$
\langle X(t_1) X(t_2) \rangle = 2D(t_1 - t_0) + x_0^2.
$$

Subtracting from this the quantity $\langle X(t_1) \rangle \langle X(t_2) \rangle = x_0^2$, we conclude from the definition (2.20) of the covariance that

$$
\text{cov}\{X(t_1), X(t_2)\} = 2D(t_1 - t_0) \quad (t_0 \le t_1 \le t_2). \tag{3.16}
$$

The fact that cov $\{X(t_1), X(t_2)\}$ is not identically zero shows that $X(t_1)$ and $X(t_2)$ are statistically dependent. But the results that cov $\{X(t_1), X(t_2)\}$ *increases* with $t_1 - t_0$ and is *independent* of t_2 seem strange. As discussed in connection with Eq. (2.23), a better descriptor of the statistical dependency of any two random variables than their covariance is their *correlation*, which is obtained by dividing their covariance by the square root of the product of their variances. That gives, in light of formula (3.15b),

$$\mathrm{corr}\left\{X(t_1), X(t_2)\right\} = \sqrt{\frac{t_1 - t_0}{t_2 - t_0}} \quad (t_0 \le t_1 \le t_2). \tag{3.17}$$

Equation (3.17) shows that $X(t_1)$ and $X(t_2)$ are *positively* correlated, with the degree of that correlation approaching zero as $(t_2 - t_1) \to \infty$. (To see the last point, write the denominator in the above equation as $t_2 - t_1$ plus $t_1 - t_0$.)

Somewhat more intriguing is this: Since $X(t_1)$ and $X(t_2)$ are both *normal* random variables, then we can infer from Eq. (2.31) that $[\mathrm{corr}\left\{X(t_1), X(t_2)\right\}]^2$ is the fraction of the variance of either that "is associated with" the fluctuations in the other. In fact, since $X(t_1)$ can influence the later value $X(t_2)$ but not vice versa, it seems reasonable to make the somewhat stronger statement,

$$\left\{ \begin{array}{c} \text{the fraction of var} \left\{X(t_2)\right\} \text{ that is} \\ \text{caused by the fluctuations in } X(t_1) \end{array} \right\} = \frac{t_1 - t_0}{t_2 - t_0} \quad (t_0 \le t_1 \le t_2). \tag{3.18}$$

The result (3.18) quantifies the dependency of the solute molecule's positional variance at time t_2 on its positional fluctuations at the earlier time t_1. The implication of Eq. (3.18) that the influence of the fluctuations in $X(t_1)$ on the fluctuations in $X(t_2 \ge t_1)$ is *total* at $t_2 = t_1$ and *diminishes* as t_2 is taken larger compared to t_1 is of course quite reasonable. But the more quantitative assertion of Eq. (3.18) that, when $(t_2 - t_0) = 2(t_1 - t_0)$, 50% of the variance of $X(t_2)$ is attributable to the earlier fluctuations in $X(t_1)$, is not something we would have confidently predicted.

3.5 The relative diffusion coefficient

Suppose a solution contains two solute molecules 1 and 2 with respective diffusion coefficients D_1 and D_2 (which might be equal). In the rest frame of the solution, the x-coordinates of these two molecules evolve in time independently of each other according to Eq. (3.11a), so

$$X_1(t) = \mathcal{N}_{1x}\left(x_{10}, 2D_1(t - t_0)\right), \quad X_2(t) = \mathcal{N}_{2x}\left(x_{20}, 2D_2(t - t_0)\right),$$

and similarly for the y- and z-components. Since these two normal random variables are statistically independent, it follows from the linear combination property (2.25) for statistically independent normal random variables (taking $\alpha_1 = 1$ and $\alpha_2 = -1$) that

$$X_1(t) - X_2(t) = \mathcal{N}\left(x_{10} - x_{20}, 2D_1(t - t_0) + 2D_2(t - t_0)\right).$$

Thus, the position of molecule 1 *relative to molecule 2* evolves according to

$$X_{12}(t) \equiv X_1(t) - X_2(t) = \mathcal{N}\left(x_{10} - x_{20}, 2(D_1 + D_2)(t - t_0)\right). \tag{3.19}$$

The import of Eq. (3.19) is that, *as seen by an observer in the rest frame of molecule 2,* molecule 1 moves about by ordinary diffusion with diffusion coefficient $D_1 + D_2$.

3.6   The probability flux: boundary conditions

Suppose we define

$$\hat{J}_x(x, t \,|\, x_0, 0) \equiv -D \frac{\partial P_X(x, t \,|\, x_0, 0)}{\partial x}. \tag{3.20}$$

Although obviously inspired by Fick's Law (1.2), Eq. (3.20) is *not* an empirical hypothesis. It merely defines a new function $\hat{J}_x(x, t \,|\, x_0, 0)$, which for reasons that will become clear shortly we will call the *probability flux in the x-direction*. This function is actually the x-component of a vector function $\hat{\mathbf{J}}(\mathbf{r}, t \,|\, \mathbf{r}_0, 0)$, whose y- and z-components, $\hat{J}_y(y, t \,|\, y_0, 0)$ and $\hat{J}_z(z, t \,|\, z_0, 0)$, are analogously defined in terms of the PDFs of the random variables $Y(t)$ and $Z(t)$.

With the definition (3.20), Einstein's diffusion equation (3.9) can be written

$$\frac{\partial P_X(x, t \,|\, x_0, 0)}{\partial t} = -\frac{\partial \hat{J}_x(x, t \,|\, x_0, 0)}{\partial x}. \tag{3.21}$$

Equation (3.21) has the form of a "continuity equation", as can be seen by comparing it to the classical continuity equation (1.1). To understand the implications of this, let δx and δt be positive infinitesimals, and write Eq. (3.21) as

$$\frac{P_X(x, t + \delta t \,|\, x_0, 0) - P_X(x, t \,|\, x_0, 0)}{\delta t} = -\frac{\hat{J}_x(x + \delta x, t \,|\, x_0, 0) - \hat{J}_x(x, t \,|\, x_0, 0)}{\delta x}.$$

Focusing on the case in which the system volume Ω is a right cylinder of cross-sectional area A whose axis is parallel to the x-axis, we multiply this equation through by $A \cdot \delta x \cdot \delta t$ to get

$$P_X(x, t + \delta t \,|\, x_0, 0) \cdot A\delta x - P_X(x, t \,|\, x_0, 0) \cdot A\delta x$$
$$= \hat{J}_x(x, t \,|\, x_0, 0) \cdot A\delta t - \hat{J}_x(x + \delta x, t \,|\, x_0, 0) \cdot A\delta t. \tag{3.22}$$

As we will now show, we can make physical sense of this last relation if we simply adopt the following *interpretation* of $\hat{J}_x$:

$$\hat{J}_x(x, t \,|\, x_0, 0) \cdot A\delta t = \text{the net amount of the solute molecule's}$$
$$\text{\textit{position probability} that crosses point } x,$$
$$\text{from left to right, in time } [t, t + \delta t). \tag{3.23a}$$

Note for later reference that an immediate corollary of this interpretation is[3]

$$\hat{J}_x(x, t \,|\, x_0, 0) \cdot A\delta t = \text{the \textit{increase} in time } [t, t + \delta t) \text{ in the}$$
$$\text{probability that the solute molecule}$$
$$\text{\textit{is to the right of} point } x. \tag{3.23b}$$

Under the interpretation (3.23a), Eq. (3.22) evidently makes the following statement: {The *net increase* during time $[t, t + \delta t)$ in the *probability of the solute molecule being inside the interval* $[x, x + \delta x)$} is equal to {the net amount of that probability that *flows into* the interval $[x, x + \delta x)$ across its left edge x during that time} *minus* {the net amount of that probability that *flows out of* the interval $[x, x + \delta x)$ across its right edge $x + \delta x$ during that time}. Thus, in a reversal of the logic we used in Section 1.1 to derive the classical diffusion equation (1.3) from the continuity equation and Fick's Law, we conclude that by adopting the interpretation of $\hat{J}_x$ in Eq. (3.23a), we can view the time-evolution of the probability distribution of the x-coordinate of a single solute molecule as arising solely from "an incompressible flow of position probability over the x-axis", a flow that is quantified by the probability flux $\hat{J}_x(x, t \mid x_0, 0)$ defined in Eq. (3.20). That probability flux differs from the conventional molecular flux $J_x(x, t \mid x_0, 0)$ in Chapter 1 in that what is "flowing" here is the *position probability of a single solute molecule* instead of the average number of solute molecules.

Since we cannot experimentally observe a "flow of position probability", the probability flux might seem to be little more than an academic curiosity. However, at the boundary of the containing volume Ω, the probability flux can be usefully informative. In general, the boundary of Ω might be *perfectly reflecting*, in that it redirects any solute molecule that strikes it back into Ω; or *perfectly absorbing*, in that it permanently removes from Ω any solute molecule that strikes it; or *mixed*, sometimes reflecting an impinging solute molecule and sometimes absorbing it. To appreciate the utility of the probability flux at a boundary surface, suppose the cylindrical volume Ω considered above has its *right boundary*, a planar surface of area A, at $x = b_R$. Then if a solute molecule inside Ω strikes that boundary from the left, it will either be reflected back to the left, or it will be absorbed. But "absorption" in this circumstance can be viewed as the solute molecule *crossing* $x = b_R$ *from left to right and never returning to* Ω; it therefore follows from Eq. (3.23b) that

$$\hat{J}_x(b_R, t \mid x_0, 0) \cdot A\,\delta t = \text{the } increase \text{ in time } [t, t + \delta t) \text{ in the}$$
$$\text{probability that the solute molecule}$$
$$has\ been\ absorbed \text{ at } x = b_R. \qquad (3.23)$$

Now let $Q(t)$ denote {the probability that the solute molecule gets absorbed at $x = b_R$ *before* time t}. Then by the addition law of probability for mutually exclusive events, we have

$$Q(t + \delta t) = Q(t) + \{\text{the probability that the solute molecule}$$
$$\text{gets absorbed at } x = b_R \text{ } during \text{ } [t, t + \delta t)\}.$$

Rearranging this to read

$$Q(t + \delta t) - Q(t) = \{\text{the probability that the solute molecule}$$
$$\text{gets absorbed at } x = b_R \text{ during } [t, t + \delta t)\},$$

we observe that the *difference* on the *left side* of this last equation is precisely the *probability increase* that appears on the *right side* of Eq. (3.24). We may therefore join those two equations to obtain the result

$$\hat{J}_x(b_{\mathrm{R}}, t \,|\, x_0, 0) \cdot A\delta t = \{\text{the probability that the solute molecule}$$
$$\text{gets absorbed at the right boundary}$$
$$x = b_{\mathrm{R}} \text{ in the time interval } [t, t + \delta t)\}. \qquad (3.25a)$$

In an analogous way, we can prove that for a *left* boundary $x = b_{\mathrm{L}}$,

$$-\hat{J}_x(b_L, t \,|\, x_0, 0) \cdot A\delta t = \{\text{the probability that the solute molecule}$$
$$\text{gets absorbed at the left boundary}$$
$$x = b_{\mathrm{L}} \text{ in the time interval } [t, t + \delta t)\}. \qquad (3.25b)$$

The results (3.25) tell us that {the *inflow* of position probability flux at any boundary surface area element} multiplied by {the product of the area of that element times δt} is equal to {the probability that the solute molecule will be absorbed by that area element in the next δt}. Note that this applies to a *single* solute molecule, namely, the one whose PDF is the function P_X on the right side of the flux's definition (3.20).

One useful consequence of the boundary absorption probability formulas (3.25) is the condition that must be satisfied by the solution of the Einstein diffusion equation (3.9) at any perfectly reflecting boundary $x = x_{\mathrm{refl}}$. That *perfectly reflecting boundary condition* is

$$\left. \frac{\partial P_X(x, t \,|\, x_0, 0)}{\partial x} \right|_{x = x_{\mathrm{refl}}} = 0. \qquad (3.26)$$

This follows from the fact that at a perfectly reflecting boundary $x = x_{\mathrm{refl}}$ the absorption probability must by definition vanish, so Eqs (3.25) mandate that $\hat{J}_x(x_{\mathrm{refl}}, t \,|\, x_0, 0) = 0$, and that in turn requires, through the probability flux definition (3.20), the result (3.26). The boundary condition for a perfectly absorbing boundary $x = x_{\mathrm{abs}}$ is even more transparent: Since such a boundary immediately removes any solute molecule that strikes it, there will be *zero* probability of ever finding a solute molecule *on* that boundary. Therefore, the *perfectly absorbing boundary condition* for the Einstein diffusion equation (3.9) is

$$P_X(x_{\mathrm{abs}}, t \,|\, x_0, 0) = 0. \qquad (3.27)$$

In the theory of partial differential equations, the perfectly reflecting boundary condition (3.26) is known as the Dirichlet boundary condition, while the perfectly absorbing boundary condition (3.27) is known as the Neumann boundary condition. The boundary surface absorption probability formula (3.25b) will play a crucial role in the analysis of the following section.

3.7 The stochastic bimolecular chemical reaction rate: Part I

One important application of the theory of diffusion is the estimation of the *stochastic reaction rate* of the generic bimolecular chemical reaction

$$S_1 + S_2 \to \text{products} \qquad (3.28)$$

in the circumstance that the molecules of both reactant species are *solute* molecules in a common bath of very many, much smaller, chemically inert solvent molecules. The specific quantity that we ultimately want to estimate here is

$P_{\delta t} \equiv$ the *probability* that reaction (3.28) will occur somewhere inside the

system in the next macroscopically infinitesimal time interval δt. (3.29)

More specifically, we want to compute this quantity under the following five *assumptions*:

(i) We have at the present time x_1 S_1 molecules and x_2 S_2 molecules, all contained, along with the solvent molecules, inside some volume Ω at some absolute temperature T.

(ii) The *solute* molecules are *dilute* and *macroscopically well-stirred* inside Ω; exactly what these terms mean will be explained shortly.

(iii) The S_1 and S_2 molecules have respective diffusion coefficient D_1 and D_2.

(iv) If the distance between the center of an S_1 molecule and the center of an S_2 molecule decreases to σ_{12}, then those two molecules can be considered to be *colliding*. So, for example, if the S_1 and S_2 molecules were hard spheres with respective radii r_1 and r_2, we would have $\sigma_{12} = r_1 + r_2$.

(v) There exists a well-defined (but usually hard to compute) probability q that an S_1–S_2 *collision* will result in a *reaction* (3.28). Thus, q is the *collision-conditioned reaction probability* for reaction (3.28); it also gives the average fraction of S_1–S_2 molecular collisions that will result in reaction (3.28). Although we will not need to make any assumptions about the specific form of q, we should note that q will always have some value between 0 and 1, and it will usually depend on the system temperature. As an example, if the condition for a collision to result in a reaction is that the kinetic energy associated with the relative motion of the colliding molecules along their line of centers at contact exceeds some threshold value E_{th}, then under some plausible physical assumptions it can be proved that $q = \exp\left(-E_{\text{th}}/k_B T\right)$;[4] this is called the *Arrhenius factor*.

We begin our quest for the probability (3.29) by randomly choosing *one* S_1 molecule and *one* S_2 molecule, as shown in Fig. 3.1. By "randomly", we mean that the selection of these two molecules pays no attention to their *positions* inside Ω. Next we ascribe about the S_2 molecule an imaginary *action sphere* of radius σ_{12}, so called because if the center of the S_1 molecule happens to be touching the surface of that sphere, then by definition the two molecules will be colliding.

Our assumption that the solute molecules are *dilute* means this: It is possible to ascribe about our S_2 molecule a *second* imaginary sphere (see again Fig. 3.1) whose radius Σ_{12} satisfies

$$\sigma_{12} \ll \Sigma_{12} \ll \Omega^{1/3}, \tag{3.30}$$

and which *only infrequently* contains the center of *any* S_1 molecule or *any other* S_2 molecule. Since this Σ_{12}-sphere is large compared to the σ_{12}-sphere, this requirement ensures that the solute molecules will be relatively isolated from each other (but not of course from the solvent molecules), and hence that the interaction between any particular S_1–S_2 pair of molecules will be approximately independent of what is happening to the other solute molecules.

Our assumption that the S_1 and S_2 molecules be *macroscopically well-stirred* means this: On the scale of the system volume Ω, the solute molecules are distributed over Ω in a *randomly uniform* way. More precisely, the *probability* that a randomly chosen solute molecule will be found to lie inside any subvolume ω *that is not extremely small compared to* Ω can be estimated as ω/Ω, regardless of where ω is situated inside Ω. (Notice that this definition of well-stirred does *not* require there to be a large number of solute molecules present.) But in *very small* subvolumes of Ω, and more particularly in the region between the σ_{12}-sphere and the Σ_{12}-sphere around our S_2 molecule, our chosen S_1 molecule need *not* be uniformly distributed. One might

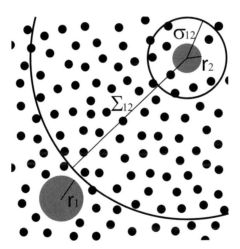

Fig. 3.1 *A randomly chosen pair of S_1 and S_2 solute molecules, with respective radii r_1 and r_2, in a sea of very many much smaller solvent molecules (the small darker spheres). Ascribed about the S_2 molecule is the "action sphere" of radius $\sigma_{12} = r_1 + r_2$; if the center of the S_1 molecule lies on that sphere, the two solute molecules are considered to be* colliding. *Concentric with the action sphere is a second sphere of radius Σ_{12}, which is assumed to be large compared to σ_{12} yet small compared to the diameter of the containing volume Ω. The system is assumed to be dilute in the solute molecules in the sense that the region between the σ_{12}-sphere and the Σ_{12}-sphere is only rarely occupied by the center of any* solute *molecule.*

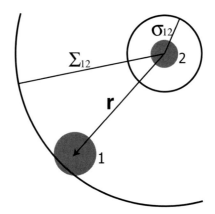

Fig. 3.2 *Dynamical variables for the diffusion equation (3.33). The vector **r** locates the center of the chosen S_1 molecule relative to the center of the chosen S_2 molecule. The magnitude r of **r** is obviously bounded below by σ_{12}. Although r could be as large as the diameter $\Omega^{1/3}$ of the system, for the purpose of solving the steady-state isotropic diffusion equation (3.36) to get $P_1(r)$ near the surface of the σ_{12}-sphere, we can bound r from above by the much smaller distance $\Sigma_{12} \gg \sigma_{12}$ if we impose the macroscopically well-stirred boundary condition (3.39a) at the surface of the Σ_{12}-sphere.*

wonder if it is physically possible for the solute molecules to be distributed in this way—uniformly on a large scale (here of size $\Omega^{1/3}$) but not uniformly on a small scale (here of size Σ_{12}). An example which suggests that this is not only possible but rather commonplace is provided by our universe: Cosmologists tell us that matter is distributed throughout the visible universe approximately uniformly on the *universe's* scale. But as we move to the progressively smaller scales of our galaxy, and then our solar system, and then our planet, etc., the distribution of matter can become increasingly, and often dramatically, non-uniform.

Our strategy for computing $P_{\delta t}$ in Eq. (3.29) will be to first compute the *single-pair* reaction probability

$$p_{\delta t} \equiv \text{the probability that our randomly chosen } S_1 - S_2 \text{ molecular}$$
$$\text{pair will react according to (3.28) in the next } \delta t. \tag{3.31}$$

By summing this probability over all $x_1 x_2$ pairs of S_1 and S_2 molecules inside Ω, we will get, by the addition law of probability, the probability $P_{\delta t}$ in Eq. (3.29).[5] Since that summation simply multiplies $p_{\delta t}$ by $x_1 x_2$, we thus have

$$P_{\delta t} = p_{\delta t} \, x_1 x_2. \tag{3.32}$$

Turning now to the task of estimating the single-pair reaction probability $p_{\delta t}$, we set up a coordinate frame whose origin is at the center of our randomly chosen S_2 molecule, and we let $\mathbf{R}_1(t)$ be the position vector of the *center* of our randomly chosen S_1 molecule in that reference frame. This position vector is a random variable, and we let $P_1(\mathbf{r}, t)$ denote its PDF; i.e., $P_1(\mathbf{r}, t)\, d\omega$ gives the probability that $\mathbf{R}_1(t)$ will lie in the infinitesimal volume element $d\omega$ around point $\mathbf{r} = (x, y, z)$ at time t. See Fig. 3.2. Notice that $P_1(\mathbf{r}, t) = 0$ everywhere inside the σ_{12}-sphere, since that region is not accessible to the center of the S_1 molecule. Since the S_1 molecule moves relative to the S_2 molecule with diffusion coefficient $D_{12} \equiv D_1 + D_2$ (see Section 3.5), the PDF $P_1(\mathbf{r}, t)$ of the S_1 molecule's center will satisfy the three-dimensional version of the Einstein diffusion equation (3.9):

$$\frac{\partial P_1(\mathbf{r}, t)}{\partial t} = D_{12} \nabla_{\mathbf{r}}^2 P_1(\mathbf{r}, t) \quad (|\mathbf{r}| \geq \sigma_{12}). \tag{3.33}$$

Here, $\nabla_{\mathbf{r}}^2$ is the Laplacian operator whose Cartesian form is displayed in Eq. (1.3).

Our plan for computing $p_{\delta t}$ is this: First we will compute $P_1(\mathbf{r}, t)$ by solving Eq. (3.33); then we will compute the probability flux defined in Eq. (3.20),

$$\hat{\mathbf{J}}_1(\mathbf{r}, t) = -D_{12} \nabla_{\mathbf{r}} P_1(\mathbf{r}, t); \tag{3.34}$$

finally, we will estimate the probability (3.31) as

$$p_{\delta t} = \oiint_{\text{act sph}} \left(-\hat{\mathbf{J}}_1(\mathbf{r}, t) \bullet d\boldsymbol{\sigma}_{12} \right) \delta t, \tag{3.35}$$

where the integration is taken over the surface of the σ_{12}-sphere. The rationale for Eq. (3.35) is as follows: $-\hat{\mathbf{J}}_1(\mathbf{r}, t)$ for $\mathbf{r}$ on the surface of the σ_{12}-sphere is the incoming probability flux at that surface; therefore, by Eq. (3.25b), $-\hat{\mathbf{J}}_1(\mathbf{r}, t) \bullet d\boldsymbol{\sigma}_{12} \delta t$ gives the probability that the S_1 molecule will be *absorbed* by the infinitesimal surface element $d\boldsymbol{\sigma}_{12}$ of the action sphere at $\mathbf{r}$ in the next δt. Since absorption by the σ_{12}-sphere means that the S_1 molecule has chemically reacted with the S_2 molecule, then integrating (summing) the probability $-\hat{\mathbf{J}}_1(\mathbf{r}, t) \bullet d\boldsymbol{\sigma}_{12} \delta t$ over the σ_{12}-sphere gives, by the addition law of probability, the probability that our S_1 molecule will *react* with our S_2 molecule in the next δt.

Evaluating the integral in Eq. (3.35) evidently requires knowing $\hat{\mathbf{J}}_1(\mathbf{r}, t)$ only on the surface of the σ_{12}-sphere; thus, it will suffice to solve the partial differential equation (3.33) only in the relatively small region of Ω between the σ_{12}-sphere and the Σ_{12}-sphere. Note that $P_1(\mathbf{r}, t) = 0$ everywhere inside the σ_{12}-sphere, but it need not vanish on the surface of the σ_{12}-sphere. In the region between the two spheres, it is physically reasonable to assume that $P_1(\mathbf{r}, t)$ will depend only on the magnitude r of $\mathbf{r}$, and not the direction of $\mathbf{r}$. We will thus write $P_1(\mathbf{r}, t)$ as simply $P_1(r, t)$.[6] The form of the Laplacian operator $\nabla_{\mathbf{r}}^2$ in spherical coordinates in that circumstance then gives for Eq. (3.33),

$$\frac{\partial P_1(r, t)}{\partial t} = D_{12} \frac{1}{r^2} \frac{\partial}{\partial r} \left(r^2 \frac{\partial P_1(r, t)}{\partial r} \right) \equiv D_{12} \frac{1}{r} \frac{\partial^2}{\partial r^2} (r P_1(r, t)). \tag{3.36}$$

To solve Eq. (3.36) for $\sigma_{12} \leq r \leq \Sigma_{12}$ and $t \geq 0$, we must specify *an initial condition* and *two boundary conditions*. Physical reasoning must be our guide in choosing those conditions; we are not free to simply "assume" them. To formulate the initial condition, we must ask ourselves this question: What is *physically special* about time $t = 0$ for our problem? The honest answer: Practically nothing! Since the purpose of the probability $P_{\delta t}$ in Eq. (3.29) is to take us from one chemical reaction event in the system to the next, then our "time zero" will be the instant immediately after the last chemical reaction event inside Ω. Inasmuch as our chosen S_2 molecule was not a reactant in that event (otherwise it would no longer be present), then within the restricted confines of its Σ_{12}-sphere, practically nothing will have changed: the instant just after the system's last reaction event will seem no different from the instant just after the previous reaction event, and the reaction event before that, etc. It should therefore be a reasonable approximation to choose for $P_1(r, t)$ at $t = 0$ the *steady-state solution* of Eq. (3.36)—i.e., the solution to the equation obtained by setting the time-derivative on the left side of Eq. (3.36) to zero.[7] But doing that will mean that $P_1(r, t)$ for *all* $t > 0$ will also be that steady-state solution. Thus we conclude that $P_1(r, t \geq 0)$ is the time-independent solution $P_1(r)$ of the ordinary differential equation

$$0 = \frac{d^2}{dr^2}(rP_1(r)) \quad (\sigma_{12} \leq r \leq \Sigma_{12}). \tag{3.37}$$

Two integrations of this equation give $rP_1(r) = \alpha r + \beta$, where α and β are integration constants; therefore,

$$P_1(r) = \alpha + \frac{\beta}{r} \quad (\sigma_{12} \leq r \leq \Sigma_{12}). \tag{3.38}$$

The two integration constants here must now be determined by the two boundary conditions.

One boundary condition is provided by the "macroscopically well-stirred" condition, using the following reasoning: The requirement that the S_1 molecule be distributed uniformly over Ω on a *macroscopic* scale means that at *most* points inside Ω, the PDF of the center of the S_1 molecule will be constant. And since the integral of the PDF over Ω must equal 1, the value of that constant must be Ω^{-1}. But we will *not* require $P_1(r)$ to be equal to Ω^{-1} *everywhere*, in particular not between the σ_{12}-sphere and the Σ_{12}-sphere around each S_2 molecule.[8] However, because of the macroscopically well-stirred condition, we will require the solution (3.38) to be equal to Ω^{-1} *on the surface of the Σ_{12}-sphere*, which is "far" from the surface of the σ_{12}-sphere because $\Sigma_{12} \gg \sigma_{12}$. Thus we have for our first boundary condition on Eq. (3.38),

$$P_1(r = \Sigma_{12}) = \Omega^{-1}. \tag{3.39a}$$

For our second boundary condition, we will simply stipulate that the value of $P_1(r)$ on the surface of the σ_{12}-sphere must be equal to some *yet-to-be-determined constant*, which we shall simply denote by $P_1(\sigma_{12})$:

$$P_1(r = \sigma_{12}) = P_1(\sigma_{12}). \tag{3.39b}$$

We might wish that we could be more definite than this, but the theory of diffusion as we have developed it thus far gives us no *physical* license for asserting any specific value for $P_1(\sigma_{12})$.

When the two boundary conditions (3.39) are used to determine the integration constants α and β in Eq. (3.38), and account is then taken of the diluteness condition $\Sigma_{12} \gg \sigma_{12}$, the solution (3.38) takes the form

$$P_1(r) = \left(\Omega^{-1} - \frac{\sigma_{12}P_1(\sigma_{12})}{\Sigma_{12}} \right) - \frac{\sigma_{12}\left(\Omega^{-1} - P_1(\sigma_{12})\right)}{r} \quad (\sigma_{12} \le r \le \Sigma_{12} \gg \sigma_{12}).$$

(3.40)

Substituting this into Eq. (3.34), we find for the radially inward component of the probability flux at the surface of the σ_{12}-sphere,

$$-\hat{J}_{1r}(\sigma_{12}) = -\left(-D_{12} \left.\frac{\partial P_1(r)}{\partial r}\right|_{r=\sigma_{12}} \right) = \frac{D_{12}\left(\Omega^{-1} - P_1(\sigma_{12})\right)}{\sigma_{12}}.$$

(3.41)

The integral of this function over the surface of the σ_{12}-sphere is easily computed:

$$\oiint_{\text{act sph}} \left(-\hat{\mathbf{J}}_1(\mathbf{r}, t) \bullet \mathbf{d\sigma}_{12} \right) = -J_{1r}(\sigma_{12}) \cdot 4\pi\sigma_{12}^2 = 4\pi D_{12}\sigma_{12}\left(\Omega^{-1} - P_1(\sigma_{12})\right).$$

(3.42)

Substituting this result into Eq. (3.35), we conclude that the probability that our chosen S_1–S_2 molecular pair will undergo reaction (3.28) in the next macroscopically infinitesimal time interval δt is

$$p_{\delta t} = 4\pi D_{12}\sigma_{12}\left(\Omega^{-1} - P_1(\sigma_{12})\right)\delta t.$$

(3.43)

Unfortunately, the result (3.43) does not by itself enable us to reach our goal of a fully explicit formula for $P_{\delta t}$ by way of Eq. (3.32). The problem is, we do not know the value of $P_1(\sigma_{12})$. The question of the correct value for $P_1(\sigma_{12})$ is a long-standing one in the deterministic version of this problem, where the focus is on the average S_1 molecular *density* $\rho_1(r)$ instead of the single-molecule PDF $P_1(r)$, and the aim is to compute a traditional deterministic reaction *rate* instead of a reaction probability in time δt.[9] That approach was initiated in 1917 by Marian Smoluchowski (about whom we will learn more in Chapter 10). He simply took $\rho_1(\sigma_{12}) = 0$, which made the σ_{12}-sphere perfectly absorbing. In our case that would be equivalent to setting $P_1(\sigma_{12}) = 0$, so that Eq. (3.43) would reduce to

$$p_{\delta t}^{\text{Smol}} \equiv 4\pi D_{12}\sigma_{12}\Omega^{-1}\delta t.$$

(3.44)

An inspection of Eq. (3.43) shows that, since $p_{\delta t}$ and $P_1(\sigma_{12})$ must both be positive, then as $P_1(\sigma_{12})$ increases from its smallest possible value 0 to its largest possible value Ω^{-1}, $p_{\delta t}$ decreases from its largest possible value $p_{\delta t}^{\text{Smol}}$ to its smallest possible value 0. But although there might be circumstances in which $P_1(\sigma_{12})$ has, at least approximately, its minimum value zero, we have no license for assuming that to be true generally. The simple truth is this: there is nothing in the theory of diffusion, at least as we have developed it thus far, that justifies our assigning *any* specific value to $P_1(\sigma_{12})$ in formula (3.43). In the next chapter, we will discover the reason for this impasse, and a way around it.[10]

Notes to Chapter 3

[1]Albert Einstein (1879–1955) actually wrote two papers on Brownian motion in 1905: "Über die von der molekularkinetischen Theorie der Wärme geforderte Bewegung von in ruhenden Flüssigkeiten suspendierten Teilchen", *Annalen der Physik* **17**:549–560 (1905); and "Zur Theorie der Brownshen Bewegung", *Annalen der Physik* **19**:371–381 (1906). English translations of those, and three more papers he wrote on the same subject over the next three years, can be found in *Albert Einstein, Investigations on the Theory of the Brownian Movement*, edited by R. Fürth and translated by A. Cowper (Dover, 1956). Einstein set a record for productivity in 1905 that remains the envy of physicists everywhere. That year, in addition to his work on Brownian motion, he introduced the concept of the photon to explain the photoelectric effect, which had been discovered 17 years earlier; also, he formulated his special theory of relativity, which asserted totally unexpected connections between space and time, and mass and energy. All three of these bold theoretical forays in 1905 were confirmed experimentally. But it was only for his work on the photoelectric effect that Einstein was awarded a Nobel Prize, in 1921. Yes, they still owe him two more – actually three more, in view of his 1916 general theory of relativity, which remains today our standard theory of gravity.

Jean Perrin (1870–1942) was a French physicist. He received his education in Paris at the prestigious ENS and Sorbonne. He was a lecturer at the Sorbonne in 1908 when he performed the experiments on Brownian motion that confirmed Einstein's theory of diffusion. He received a letter in 1909 from Einstein saying, "I had believed it to be impossible to investigate Brownian motion so precisely". Einstein's and Perrin's works together convinced many eminent doubters of the era that matter indeed consists of molecules in rapid motion. Those works also provided an independent way to measure Avogadro's number. In 1910 Perrin became a Professor at the Sorbonne, and he enjoyed a distinguished and productive career there over the next three decades, winning the Nobel Prize in 1926 for work that elaborated and extended his earlier Brownian motion experiments. But Perrin's career came to an end in 1940, when he fled to the United States to escape the Nazi invasion of France; he died in New York City two years later. A readable overview of the Einstein–Perrin work can be found in Abraham Pais's book *Subtle Is the Lord: The Science and the Life of Albert Einstein* (Oxford University Press, 1982).

[2]The only tricky part in deriving Eqs (3.15) is in getting the formula for $\text{var}\{X_2\}$. The trick for that is to write $(x_2 - \langle X_2 \rangle)^2$ as (remembering that $\langle X_2 \rangle = x_0$)

$$(x_2-x_0)^2 = ((x_2-x_1) + (x_1-x_0))^2 = (x_2-x_1)^2 - 2(x_2-x_1)(x_1-x_0) + (x_1-x_0)^2,$$

and then integrate over x_1 and x_2 term by term with the help of Eqs (3.14).

[3]The fact that Eq. (3.23a) implies Eq. (3.23b) can be most easily appreciated by considering a simple example: Suppose that at time t the probability that a solute molecule lies to the right of point x is 0.2, and in time $[t, t + \delta t)$ a *net* amount 0.05 of the molecule's position probability crosses point x from left to right. Then

at time $t + \delta t$, the probability of finding the molecule to the right of point x will be $0.2 + 0.05 = 0.25$.

[4]The "plausible physical assumptions" needed here are that the reacting molecules are spherical, and are uniformly distributed inside a fixed volume with velocities that are distributed according to the Maxwell–Boltzmann distribution corresponding to absolute temperature T. For details of the derivation, see D. Gillespie, "A rigorous derivation of the chemical master equation", *Physica A* **188**:403–425 (1992).

[5]The addition law of probability tells us that the probability for any one of several *mutually exclusive* events to occur can be computed as the sum of their individual occurrence probabilities. Mutual exclusivity here demands that the probability of more than one S_1–S_2 pair reacting in time δt be negligibly small. Satisfying that condition is the reason why we require δt to be small, and also why we require the system to be dilute in the S_1 and S_2 molecules.

[6]But note that $P_1(r, t)$ is still the PDF of $\mathbf{R}_1(t)$, and has not become the PDF of $|\mathbf{R}_1(t)|$; thus, the probability that $|\mathbf{R}_1(t)|$ will lie between r and $r + dr$ is given not by $P_1(r, t) \cdot dr$, but by $P_1(r, t) \cdot 4\pi r^2 dr$.

[7]In choosing the initial condition, it might seem tempting to take $P_1(r, 0)$ to be the *uniform* distribution, with the constant value Ω^{-1} everywhere between the σ_{12}-sphere and the Σ_{12}-sphere. However, because of the partially absorbing nature of the σ_{12}-sphere, that uniform distribution would rapidly collapse to the steady-state distribution $P_1(r)$, which, as will be seen later, takes smaller values at points that are closer to the surface of the partially absorbing σ_{12}-sphere. And, being time-independent, that steady-state distribution would *persist* until the chosen S_2 molecule is removed by a reaction event. But the system has no mechanism for reestablishing a uniform distribution in the region between the σ_{12}-sphere and the Σ_{12}-sphere after each subsequent reaction event in the system. These considerations show that the uniform distribution is *not* a physically reasonable candidate for the initial condition in this problem.

[8]Earlier in our analysis we assumed that the probability distribution of the center of the S_1 molecule in the region between the σ_{12}-sphere and the Σ_{12}-sphere is *isotropic*, in that $P_1(\mathbf{r}, t)$ does not depend on the *direction* of $\mathbf{r}$. That is *not* the same as assuming that the S_1 molecule distribution is *uniform* in that region, i.e., that $P_1(\mathbf{r}, t)$ has the same value at every point in that region. A uniform distribution is necessarily an isotropic distribution, but not vice versa. As will be seen later, $P_1(\mathbf{r}, t)$ has smaller values at points that are nearer the surface of the partially absorbing σ_{12}-sphere, so the S_1 molecule's distribution is in fact *not* uniform between the σ_{12}-sphere and the Σ_{12}-sphere. The symmetry assumptions we have made here can be more clearly appreciated by imagining that our observable universe is just the region between those two spheres: we cannot tell the difference between one radial direction and another, but we can tell how far we are from the surfaces of the two spheres.

[9]We have deliberately avoided the conventional reaction *rate* in our analysis here, because ascribing a physically sensible meaning to "the rate at which a *single* S_2

molecule is reacting" is problematic: For most purposes, an S_2 molecule will either react completely at some instant in time or else it will not react at all; it will not react by gradual degrees. A stochastic derivation that focuses on the *average* reaction rate of a single S_2 molecule is given in D. Gillespie, "A diffusional bimolecular propensity function" *Journal of Chemical Physics* **131**:16409 (2009); that derivation is logically equivalent to the derivation presented here, but its use of the average reaction rate necessitates a more convoluted line of reasoning. But both derivations refine, correct, and stochastically extend the deterministic analysis given in the pioneering paper on this subject by F. Collins and G. Kimball, "Diffusion-controlled reaction rates" *Journal of Colloid Science* **4**:425–437 (1949). Details on that connection are given in Section VI of the 2009 paper.

[10]Smoluchowski's pioneering work on this problem was actually done in the context of diffusion-limited coagulation in a colloidal solution. It was reported in his paper with the (translated) title "Attempt at a mathematical theory of aggregation kinetics in colloidal solutions" in *Zeitschrift für Physikalische Chemie* **92**:129–168 (1917). The adaptation of Smoluchowski's work to chemical kinetics was made much later, in 1935, by B. Sveshnikoff. However, Sveshnikoff's adaptation contained a modification which was in fact incorrect, as was pointed out in 1949 by F. Collins and G. Kimball in their paper cited in Note 9. Sveshnikoff's modification can be summarized in our stochastic context as follows: Since $p_{\delta t}$ must lie somewhere between 0 and $p_{\delta t}^{\mathrm{Smol}}$, we can always write $p_{\delta t}$ as $p_{\delta t}^{\mathrm{Smol}}$ multiplied by some factor that lies between 0 and 1. Sveshnikoff supposed that factor to be the conditional probability q of a reaction given a collision, which we introduced at the beginning of this section but have not yet used. Writing $p_{\delta t}$ as $p_{\delta t}^{\mathrm{Smol}} \cdot q$ would have the added benefit of ensuring that $p_{\delta t} \to 0$ as $q \to 0$, which we expect to be true on purely physical grounds. But from the vantage point of our stochastic analysis, it is easy to see the flaw in Sveshnikoff's reasoning: If $p_{\delta t}^{\mathrm{Smol}}$ were the probability that the S_1 and S_2 molecules will *collide* in the next δt, then indeed its product with the probability q of a reaction given a collision would, by the multiplication law of probability, give $p_{\delta t}$. But $p_{\delta t}^{\mathrm{Smol}}$ is a *reaction* probability, not a collision probability; therefore, its product with the collision-conditioned reaction probability q is meaningless. In the next chapter (Section 4.8), we will see how Sveshnikoff's intuitive idea can be implemented correctly.

4

Implications and limitations
of the Einstein theory of diffusion

In the preceding chapter, we saw how Einstein's analysis of Brownian motion shifted the focus of the classical diffusion equation from the average behavior of many solute molecules to the probabilistic behavior of an individual solute molecule. In the present chapter we will examine further implications of the Einstein theory of diffusion, using not only analytical reasoning but also numerical simulation. We will first show how simulation can be used to construct plots of a diffusing molecule's probability density function $P_X(x, t \mid x_0, t_0)$ in the presence of a single reflecting or absorbing boundary. Then we will see how simulation can be used to construct "snapshots" of an unrestricted solute molecule's position along its trajectory. But we will discover that such trajectory snapshots expose a serious physical limitation of the Einstein theory. We will rationalize a quick fix that allows us to complete the derivation we began in Section 3.7 of a formula for the stochastic rate of a diffusion-controlled bimolecular chemical reaction. But from a broader view, it will become apparent that, while the Einstein theory of diffusion has a wide range of practical utility, a physically more accurate theory is needed. Developing such a theory will be our aim in Chapters 7 through 10.

4.1   Numerical simulation strategies

In Chapter 3, we saw that the solution $P_X(x, t \mid x_0, t_0)$ of the Einstein diffusion equation (3.9) is the probability density function (PDF) of a single solute molecule's x-position $X(t)$, subject to the initial condition $P_X(x, t_0 \mid x_0, t_0) = \delta(x - x_0)$ and whatever boundary conditions apply. We also saw that when there are no impediments to the molecule's motion along the x-axis, $P_X(x, t \mid x_0, t_0)$ will be the function in Eq. (3.10). That function implies that $X(t)$ is the normal random variable $\mathcal{N}(x_0, 2D(t - t_0))$ with mean x_0 and variance $2D(t - t_0)$. Now, the normal random variable property (2.16) tells us that

$$\mathcal{N}(x_0, 2D(t - t_0)) = x_0 + \sqrt{2D(t - t_0)} \mathcal{N}(0, 1). \tag{4.1}$$

Since sample values n of the normal random variable $\mathcal{N}(0, 1)$ are easily generated on modern computers—e.g., via Eqs (2.36)—then for given values of D, x_0, and $t - t_0$ we should be able to construct a *sample value* x_t of the random variable $X(t)$ by simply evaluating

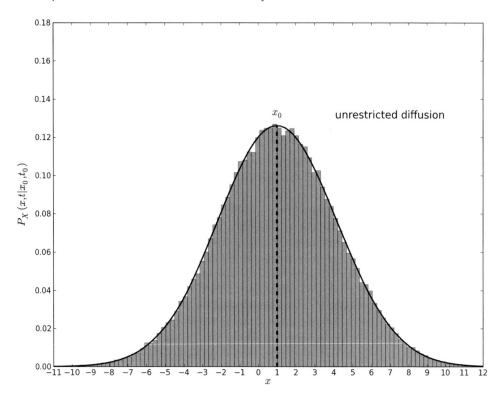

Fig. 4.1 *Unrestricted diffusion along the x-axis of a single solute molecule in the Einstein theory of diffusion. The molecule has diffusion coefficient $D = 1$ and is initially at $x_0 = 1$. The solid curve plots the PDF in Eq. (3.10) of the molecule's position at time $t = t_0 + 5$. The superimposed normalized histogram is for 10^5 samplings of that position as computed from formula (4.2). (The histogram was normalized by dividing the number of x_t-values in each bin by the product of the bin width and the total number of x_t-values.)*

$$x_t = x_0 + \sqrt{2D(t - t_0)}\, n. \tag{4.2}$$

In Fig. 4.1, the solid curve is a plot of the PDF in Eq. (3.10) for $D = 1$, $x_0 = 1$, and $t - t_0 = 5$. The superimposed histogram contains 10^5 numerical values of x_t generated according to formula (4.2) for these same parameter values. Evidently, the generating formula (4.2) provides an easy alternative for estimating, with as much accuracy as we might wish, the PDF of $X(t)$ for specified values of all the relevant parameters—at least in this unrestricted case.

But practical problems in diffusion nearly always involve *boundaries*. In Section 3.6, we discussed how perfectly reflecting boundaries and perfectly absorbing boundaries can be specified mathematically—see Eqs (3.26) and (3.27). How would the foregoing analysis be changed if there were a perfectly *reflecting* boundary at $x = x_r > x_0$? The delta-function initial condition for the Einstein diffusion equation (3.9) will be the same as before, but now the solution would be confined to the region $x \leq x_{\mathrm{r}}$, and we must impose the perfectly reflecting boundary condition (3.26):

$$\left. \frac{\partial P_X(x,t \mid x_0,t_0)}{\partial x} \right|_{x=x_{\mathrm{r}}} = 0. \tag{4.3}$$

The solution to Eq. (3.9) for these initial and boundary conditions can be obtained by using some standard analytical techniques in the theory of partial differential equations.[1] As can be verified by straightforward substitution, that solution is in fact

$$P_X(x,t \mid x_0,t_0) = \frac{1}{\sqrt{4\pi D(t-t_0)}} \left[\exp\left(-\frac{(x-x_0)^2}{4D(t-t_0)} \right) + \exp\left(-\frac{(x-(2x_{\mathrm{r}}-x_0))^2}{4D(t-t_0)} \right) \right]$$

$$(t \geq t_0; x \leq x_{\mathrm{r}}). \tag{4.4}$$

But there is also an exact algorithm for sampling $X(t)$ for these initial and boundary conditions. As we prove in Appendix 4A, that sampling algorithm is this: First generate a *tentative* time-t position x_t^{tent} according to the unrestricted sampling formula (4.2):

$$x_t^{\mathrm{tent}} = x_0 + \sqrt{2D(t-t_0)}\, n. \tag{4.5}$$

Then take

$$x_t = \begin{cases} x_t^{\mathrm{tent}}, & \text{if } x_t^{\mathrm{tent}} < x_{\mathrm{r}}, \\ 2x_{\mathrm{r}} - x_t^{\mathrm{tent}}, & \text{if } x_t^{\mathrm{tent}} \geq x_{\mathrm{r}}. \end{cases} \tag{4.6}$$

Notice that in the case $x_t^{\mathrm{tent}} \geq x_{\mathrm{r}}$, x_t is taken to be the reflection of x_t^{tent} about x_{r}, namely $x_{\mathrm{r}} - (x_t^{\mathrm{tent}} - x_{\mathrm{r}})$.

In Fig. 4.2, the solid curve is a plot of the function (4.4) for $D=1$, $x_0=1$, $x_{\mathrm{r}}=4$, and $t-t_0=5$. The normalized histogram contains 10^5 numerical values of x_t generated according to Eqs (4.5) and (4.6) for these same parameter values. The agreement between the histogram and the curve shows that the algorithm of Eqs (4.5) and (4.6) indeed generates sample values of the PDF in Eq. (4.4), thus corroborating the independent derivations of both. And as in Fig. 4.1, we again see that the simulation approach for estimating the PDF of $X(t)$ offers an easy alternative to solving the diffusion equation in this case.

Finally, let us consider what would happen if we had instead a perfectly *absorbing* boundary at $x = x_{\mathrm{a}} > x_0$. In that case we would have to impose, instead of the boundary condition (4.3), the perfectly absorbing boundary condition (3.27):

$$P_X(x = x_{\mathrm{a}}, t \mid x_0,t_0) = 0. \tag{4.7}$$

Again the solution to Eq. (3.9) for this boundary condition and the delta-function initial condition can be obtained by using some standard analytical techniques in the theory of partial differential equations.[1] As can be verified by direct substitution, that solution is

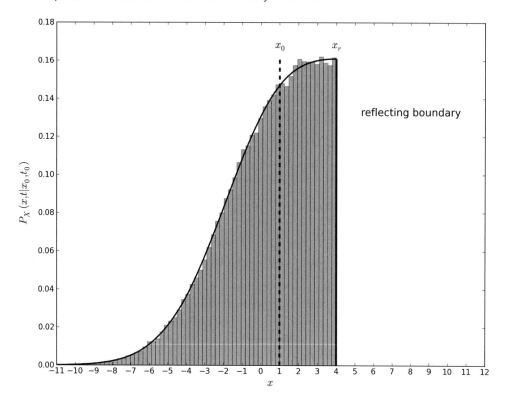

Fig. 4.2 *Einstein diffusion along the x-axis of a single solute molecule with a perfectly reflecting boundary at $x = x_r = 4$. The solid curve plots the solution (4.4) of the Einstein diffusion equation (3.9) subject to the boundary condition (4.3) for $D = 1$, $x_0 = 1$, and $t = t_0 + 5$. The superimposed normalized histogram is for 10^5 samplings of that position as computed from the algorithm in Eqs (4.5) and (4.6).*

$$P_X(x,t \mid x_0, t_0) = \frac{1}{\sqrt{4\pi D(t - t_0)}} \left[\exp\left(-\frac{(x - x_0)^2}{4D(t - t_0)} \right) - \exp\left(-\frac{(x - (2x_a - x_0))^2}{4D(t - t_0)} \right) \right]$$

$$(t \geq t_0; x \leq x_a). \qquad (4.8)$$

In Appendix 4B, we prove that an exact algorithm for generating sample values of $X(t)$ exists also in this case. That algorithm is as follows: First generate a tentative time-t position x_t^{tent} according to the unrestricted sampling formula (4.5). If $x_t^{\text{tent}} \geq x_a$, consider the molecule to have been absorbed at x_a by time t, so that no x_t value exists. If on the other hand $x_t^{\text{tent}} < x_a$, generate a sample value u of the uniform random variable on the unit interval, $\mathcal{U}(0, 1)$, and compare it to

$$\bar{p} = \exp\left(-\frac{(x_a - x_t^{\text{tent}})(x_a - x_0)}{D(t - t_0)} \right). \qquad (4.9)$$

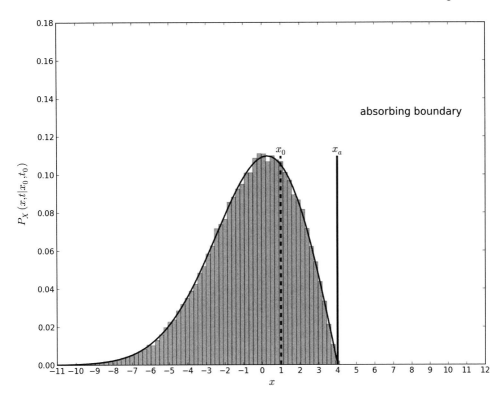

Fig. 4.3 *Einstein diffusion along the x-axis of a single solute molecule with a perfectly absorbing boundary at $x = x_a = 4$. The solid curve plots the solution (4.8) of the Einstein diffusion equation (3.9) subject to the boundary condition (4.7) for $D = 1$, $x_0 = 1$, and $t = t_0 + 5$. The superimposed normalized histogram is for 10^5 samplings of that position as computed from the algorithm described at Eq. (4.9). Notice that whereas the area under the PDFs in Figs 4.1 and 4.2 is 1, signifying that at time t the molecule is for sure somewhere on the x-axis, the area under the PDF here is only 0.6566; that is the probability that the molecule at time t will not yet have been absorbed at $x = x_a$.*

If $\bar{p} \geq u$, consider the molecule to have been absorbed at x_a by time t; otherwise, take $x_t = x_t^{\text{tent}}$.

In Fig. 4.3, the solid curve is a plot of the function (4.8) for $D = 1$, $x_0 = 1$, $x_a = 4$, and $t - t_0 = 5$. The normalized histogram contains 10^5 numerical values of x_t generated according to the foregoing procedure for these same parameter values. The agreement between the curve and the histogram confirms that the above simulation procedure for generating sample values of $X(t)$ offers an alternative to solving the diffusion equation to get the PDF of $X(t)$ in this case. Notice that the area under the curve and histogram in Fig. 4.3 is less than it is in Figs 4.2 and 4.1. In fact, integrating the function (4.8) over $-\infty < x \leq x_a$ yields $\text{erf}\left((x_a - x_0)/\sqrt{4D(t - t_0)}\right)$, and for the parameter values being used here that evaluates to 0.6566. This tells us that the probability that the solute molecule has been absorbed at the absorbing boundary $x_a = 4$ by time $t_0 + 5$ is $1 - 0.6566 = 0.3434$.

Extending the foregoing simulation procedures to accommodate a second boundary point on the x-axis is problematic, because the proofs in Appendices 4A and 4B rely on symmetry arguments that get spoiled by the presence of a second boundary point. Extending the one-dimensional simulation procedures to higher dimensions can be done; however, there are limitations that preclude doing so, at least exactly, when finite or non-planar boundaries are involved, as of course they are in most practical situations.[2]

We turn now to consider a second way to use the unrestricted diffusion result (4.1) for simulation. If in Eq. (4.1) we make the replacements $x_0 \to x_t$ and $t - t_0 \to \Delta t$, we get

$$X(t + \Delta t) = \mathcal{N}(x_t, 2D\Delta t) = x_t + \sqrt{2D\Delta t}\, \mathcal{N}(0, 1). \tag{4.10}$$

Since analogous relations hold for the y- and z-components of the solute molecule's position, it follows that if the molecule is at any point (x_t, y_t, z_t) at time t, then its coordinates at any later time $t + \Delta t$ can be computed as

$$x_{t+\Delta t} = x_t + \sqrt{2D\Delta t}\, n_x, \tag{4.11a}$$

$$y_{t+\Delta t} = y_t + \sqrt{2D\Delta t}\, n_y, \tag{4.11b}$$

$$z_{t+\Delta t} = z_t + \sqrt{2D\Delta t}\, n_z, \tag{4.11c}$$

where n_x, n_y, and n_z are statistically independent samples of the normal random variable $\mathcal{N}(0, 1)$ with mean 0 and variance 1.

Formulas (4.11) allow us to construct what we will call an "n-point representation" of the diffusing molecule's trajectory over any finite time interval $[0, t]$: With the solute molecule at some known initial point $P_0 = (x_0, y_0, z_0)$ at time 0, we take $\Delta t = t/n$ and then compute from formulas (4.11) the molecule's positions $P_i = (x_i, y_i, z_i)$ at the successive instants $t_i = i \cdot \Delta t$ for $i = 1, \ldots, n$; i.e., for each i we take $x_i = x_{i-1} + \sqrt{2D\Delta t}\, n_x$, and analogously for the y- and z-components. In Fig. 4.4 we show a 10-point representation of the trajectory of a molecule diffusing in two dimensions generated according to this procedure with $x_0 = y_0 = 0$, $\Delta t = 1$, and $D = 1$. We can think of the points $P_1, P_2, \ldots, P_{10}$ in this figure as "snapshots" of the molecule's location in the xy-plane at the successive times $1, 2, \ldots, 10$. We have connected successive points by straight lines to show their sequencing order; however, this is *not* meant to suggest that the molecule actually follows a straight-line path between successive points. We will exhibit more examples of n-point representations of solute molecule trajectories constructed in this way in Chapter 9.

It is important to recognize that formulas (4.11) are *exact* consequences of the Einstein model of diffusion—i.e., exact consequences of the solution (3.10) of the Einstein diffusion equation (3.9) for unrestricted motion—*for all* $\Delta t > 0$. Therefore, the complete trajectory of the diffusing molecule over the time interval $[0, t]$ should in principle be obtainable by taking the limit $\Delta t \to 0$, or equivalently the limit $n \to \infty$.

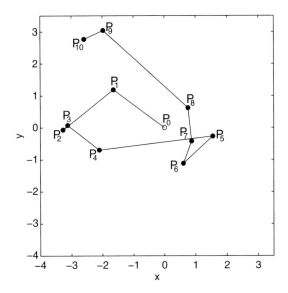

Fig. 4.4 *A "10-point representation" of a solute molecule's trajectory in the* xy-*plane over the time interval* $0 \le t \le 10$ *as predicted by the Einstein theory of diffusion, assuming that* $D = 1$ *and the molecule started at point* $P_0 = (x_0, y_0) = (0, 0)$. *The points* $P_1, P_2, \ldots, P_{10}$ *are the positions* (x_t, y_t) *of the solute molecule at the successive instants* $t = 1, 2, \ldots, 10$, *as computed from successive applications of formulas (4.11a) and (4.11b) with* $\Delta t = 10/10 = 1$. *The connecting lines are to make clear the sequencing order of the points; they do* not *represent the actual paths of the solute molecule between successive sampling times.*

In other words, we should be able to obtain as accurate a picture of the diffusing molecule's trajectory as we desire simply by taking Δt "small enough", or equivalently n "large enough". But that turns out not to be true.

4.2   A serious problem

Einstein's derivation in Section 3.1 is predicated on the assumption that all considered time intervals—in particular his δt in Eq. (3.1)—are *large enough that each solute molecule experiences many collisions with solvent molecules during that time.* For many purposes this restriction does not pose a problem. Even time intervals Δt that are very small from a macroscopic point of view will usually contain very many collisions of the solute molecule with solvent molecules. But the mere existence of a positive lower bound on Δt in Eqs (4.11) makes it impossible to construct a trajectory for the solute molecule that is resolved arbitrarily finely; because, if we use Eqs (4.11) with Δt smaller than whatever the lower bound on Δt is, then even though the resulting trajectory will be mathematically exact so far as the Einstein diffusion equation (3.9) is concerned, we will have no guarantee that the trajectory will be *physically* accurate.

It might be supposed that any such physical inaccuracies will be minor and inconsequential. But that turns out not to be so. To see why, suppose a solute molecule is at the origin $O = P_0$ at time 0, and we use Eqs (4.11) with $\Delta t = t/n$ to compute the successive positions P_i of that molecule at the n equally spaced times $t_i = i \cdot \Delta t$ $(i = 1, \ldots, n)$ between time 0 and some fixed final time t. Then with $\langle \cdots \rangle$ denoting an average over infinitely many such n-point constructions, we can prove two results: First, the *mean net displacement* of the solute molecule in time t is

$$\langle \overline{P_0 P_n} \rangle \equiv \langle R_t \rangle = \begin{cases} \sqrt{\pi D t} & \text{(in 2D)}, \\ 4\sqrt{Dt/\pi} & \text{(in 3D)}. \end{cases} \tag{4.12}$$

Second, the mean length of the broken line $\overline{P_0 P_1 \cdots P_n}$ connecting the successive points with straight-line segments, which will obviously be a *lower bound* on the mean distance actually traveled by the solute molecule in time t, is

$$\langle \overline{P_0 P_1 \cdots P_n} \rangle = \sqrt{n} \langle R_t \rangle. \tag{4.13}$$

Since the *true* mean distance traveled by the solute molecule in time t is the limit of $\langle \overline{P_0 P_1 \cdots P_n} \rangle$ as $n \to \infty$, it follows from these two results that *the solute molecule travels an infinite distance in the finite time t.* Before we examine the implications of this prediction of Eqs (4.11), let us prove Eqs (4.12) and (4.13) in the two-dimensional case; the three-dimensional proof is only slightly more complicated.

4.3   Proof of Eqs (4.12) and (4.13) in two dimensions

Let $\mathbf{R}_t = (X_t, Y_t)$ denote the diffusing molecule's position in the xy-plane at any time $t > 0$, given that $\mathbf{R}_0 = (0, 0)$. Equations (3.11), from which formulas (4.11) were derived, tell us that X_t and Y_t are statistically independent normal random variables with means 0 and variances $2Dt$. Their joint PDF is therefore

$$P_{XY}(x_t, y_t; t) = (4\pi Dt)^{-1} \exp\left(-\frac{x_t^2}{4Dt}\right) \exp\left(-\frac{y_t^2}{4Dt}\right) \quad (-\infty < x_t, y_t < \infty). \tag{4.14}$$

Define the polar coordinates R_t and Θ_t of $\mathbf{R}_t$ in the usual way:

$$X_t = R_t \cos \Theta_t, \quad Y_t = R_t \sin \Theta_t.$$

According to Eq. (2.17), the joint PDF of the two random variables R_t and Θ_t will be related to the joint PDF of X_t and Y_t by

$$P_{R\Theta}(r_t, \theta_t; t) = P_{XY}(x_t, y_t; t) \left| \frac{\partial(x_t, y_t)}{\partial(r_t, \theta_t)} \right|, \tag{4.15}$$

where the Jacobian is for the transformation $x_t = r_t \cos \theta_t$, $y_t = r_t \sin \theta_t$. That Jacobian is easily found to be equal to r_t; thus, with Eq. (4.14), this last equation becomes

$$P_{R\Theta}(r_t, \theta_t; t) = (4\pi Dt)^{-1} r_t \exp\left(-\frac{r_t^2}{4Dt}\right) \quad (0 \le r_t < \infty; 0 \le \theta_t < 2\pi). \tag{4.16}$$

Integrating $P_{R\Theta}$ over θ_t then gives the PDF of the net displacement R_t:[3]

$$P_R(r_t; t) = (2Dt)^{-1} r_t \exp\left(-\frac{r_t^2}{4Dt}\right) \quad (0 \leq r_t < \infty). \tag{4.17}$$

From this formula, it is straightforward to show that the *mean* net displacement of the molecule in time t is

$$\langle R_t \rangle \equiv \int_0^\infty r_t P_R(r_t; t) dr_t = \sqrt{\pi Dt}. \tag{4.18}$$

Equation (4.18) establishes the two-dimensional version of Eq. (4.12). It also tells us that the mean distance between any two successive points in an n-point representation of the trajectory is

$$\langle |\overline{P_{i-1} P_i}| \rangle = \sqrt{\pi D \Delta t} = \sqrt{\pi D(t/n)}.$$

Therefore, the mean length of the broken line segment $\overline{P_0 P_1 \cdots P_n}$ is

$$\left\langle \sum_{i=1}^n |\overline{P_{i-1} P_i}| \right\rangle = \sum_{i=1}^n \langle |\overline{P_{i-1} P_i}| \rangle = n\sqrt{\pi D(t/n)} = \sqrt{n\pi Dt}.$$

This result, together with Eq. (4.18), establishes Eq. (4.13) for the two-dimensional case.

4.4 Implications of Eqs (4.12) and (4.13)

Taken together, Eqs (4.12) and (4.13) imply that the average length of the n-point representation of a diffusing molecule's trajectory over any finite time interval increases without bound as $n \to \infty$. Since the true length of the molecule's trajectory is the length of the infinitely resolved path obtained by letting $n \to \infty$, it follows that formulas (4.11) predict that the average length of the trajectory of a diffusing molecule over any finite time interval is *infinite*. But the only way a molecule could travel an infinite distance in a finite time would be for it to move with *infinite speed*, and hence also with infinite kinetic energy.

Is this disquieting prediction of Einstein's diffusion equation merely a harmless reflection of the fact that the Maxwell–Boltzmann velocity distribution (see Appendix 4C) has tails that extend to $\pm\infty$? No. Although it is true that the Maxwell–Boltzmann distribution imposes no upper bound on the speed of a molecule, the *average* speed of the Maxwell–Boltzmann distribution is *finite*—in fact, on the order of $\sqrt{k_B T/m}$, where k_B is Boltzmann's constant, T is the absolute temperature of the system, and m is the molecule's mass. In contrast, the results (4.12) and (4.13) imply an *average* speed that is *infinite*. Confirmation of this prediction of formulas (4.11) can be obtained by using those formulas to directly compute the velocity of the solute molecule at any instant t: Thus for example, the x-component of the molecule's velocity at time t is, according to Eq. (4.11a),

$$\frac{dx}{dt} \equiv \lim_{\Delta t \to 0} \frac{x_{t+\Delta t} - x_t}{\Delta t} = \lim_{\Delta t \to 0} \frac{\sqrt{2D\Delta t}\, n_x}{\Delta t} = n_x \lim_{\Delta t \to 0} \sqrt{\frac{2D}{\Delta t}} = \pm\infty. \tag{4.19}$$

Here the last equality follows because the value of the random number n_x will equally likely be positive or negative (but practically never zero).

The physical picture conveyed by these results—of a solute molecule moving with infinite kinetic energy, its velocity components switching back and forth between $-\infty$ and $+\infty$ infinitely rapidly—is of course absurd. It is not physically possible for the small solvent molecules to cause a solute molecule to behave in that way. We must conclude that the Einstein theory of diffusion fails for very small time intervals in a quite serious way. This failure implies more particularly that the Einstein theory is incapable of giving us a physically sensible description of both the *velocity* and the *true trajectory* of a diffusing molecule, because the definitions of both of those physical variables require taking the limit of infinitely small time intervals.

This situation seems puzzling from the perspective of traditional mathematical physics. There we are accustomed to numerically solving ordinary differential equations like $dx(t)/dt = A(x(t))$ by using the Euler formula,

$$x(t + \Delta t) = x(t) + A(x(t))\Delta t,$$

confident that it will give us as much accuracy as we desire if only we take Δt *small* enough. But the situation is fundamentally different with formulas (4.11). Even though those formulas are mathematically exact consequences of the diffusion equation for *all* Δt, they provide a *physically* accurate description of diffusion only when Δt is *large* enough that a solute molecule experiences in Δt many collisions with solvent molecules. The inconvenience of this restriction is exacerbated by the fact that the theory of diffusion, at least as we have developed it thus far, gives us *no quantitative guide* for deciding when Δt *is* large enough. It thus appears that the ability of the traditional diffusion equation to describe the *physical phenomenon* of molecular diffusion is more circumscribed than one might reasonably infer from that equation's name.

4.5   A hint of a quantitative lower bound on Δt in Eqs (4.11)

We can get at least a hint of a quantitative lower bound on Δt in Eqs (4.11) by reasoning as follows: Thermodynamics tells us that any molecule of mass m which is "at temperature T" will be moving with an *average speed* on the order of $\sqrt{k_B T/m}$ (see Appendix 4C). Therefore, in any finite time t, such a molecule will on average move a total distance that is on the order of $\sqrt{k_B T/m} \cdot t$. But the broken line $\overline{P_0 P_1 \cdots P_n}$ in an "n-point trajectory" of a diffusing molecule will always *underestimate* the total distance the diffusing molecule travels in visiting the n successive points $P_1, \ldots, P_n$ in time t, since the straight line between P_i and P_{i+1} is the shortest possible distance between those two points. Therefore, the average length of the broken line segment $\overline{P_0 P_1 \cdots P_n}$, which according to Eqs (4.12) and (4.13) is on the order of $\sqrt{n}\sqrt{Dt}$, must always be *less* than the average length of the true trajectory:

$$\sqrt{n}\sqrt{Dt} < \sqrt{k_B T/m} \cdot t.$$

Solving this inequality for $t/n = \Delta t$ yields the following lower bound on Δt:

$$\Delta t > \frac{mD}{k_{\mathrm{B}}T}. \tag{4.20}$$

In Chapter 9, we will see by way of a more rigorous analysis that the inequality (4.20) must in fact be *strongly* satisfied in order for Eqs (4.11) to be *physically* accurate. But for now, we will not invoke that yet-to-be-established result.

4.6   The small-scale motion of a solute molecule

Einstein's derivation in Section 3.1 of the updating formulas (4.11) requires the solute molecule to experience a large enough number of collisions with solvent molecules during Δt that the position of the solute molecule at the *end* of that time interval will appear to be *randomly different* from its position at the *beginning*. It therefore seems reasonable to attribute the physical breakdown of formulas (4.11) when Δt is taken too small to this requirement *not* being satisfied. The question then arises: When a solute molecule is *not* having enough collisions with solvent molecules to effectively randomize its position, just how will it be moving?

Once asked, this question is disarmingly easy to answer: On spatial-temporal scales so small that the solute molecule's displacement is *not* being randomized by collisions with the solvent molecules, the solute molecule will be moving in a *nearly straight line* at a *nearly constant speed*. In other words, the solute molecule will be moving approximately *ballistically*, as in a *dilute gas*. The then well-defined velocity of the solute molecule can be expected to be distributed according to the Maxwell–Boltzmann distribution (see Appendix 4C); i.e., each rectilinear component of the solute molecule's velocity can be regarded as a statistically independent sample of the normal random variable with mean 0 and variance $k_{\mathrm{B}}T/m$, where k_{B} is Boltzmann's constant, T is the absolute temperature of the system, and m is the solute molecule's mass.

In short, on sufficiently small spatial and temporal scales, a solute molecule should move approximately like a dilute-gas molecule. Happily, the theory of dilute gases has been well developed.

4.7   Collision probability of a solute molecule with a surface

An established result in the kinetic theory of dilute gases in equilibrium is that

{the average number of gas molecules striking the walls

of the containing volume per unit area and per unit time}$= \frac{1}{4}\rho\bar{v}$,

where ρ is the average number of molecules per unit volume, and $\bar{v}$ is the average speed of those molecules. In this section, we will modify the proof of that result to obtain an analogous result for a *single solute molecule in solution*.

More specifically, we consider a small, stationary, one-sided surface element δA, whose area we will also denote by δA, which is immersed in a solution. This surface element is "imaginary" in that every point on it is reachable by the *center* of a solute molecule. For example, δA might be an area element on the imaginary surface resulting from translating inwardly a boundary wall by the radius of a solute molecule; or δA might be an area element on the imaginary "action sphere" surrounding some other solute molecule. Let p_0 denote the value of the PDF of the position of a randomly chosen solute molecule at δA; more precisely, if $d\omega$ is an infinitesimal volume element abutting δA, then $p_0 \, d\omega$ is the probability that the center of the solute molecule will be inside $d\omega$. Finally, let $\bar{v}$ denote the average speed of the solute molecule; as discussed in the preceding section, $\bar{v}$ is just the average speed the solute molecule would have if it were in a dilute gas at the same temperature as the solution. Under these assumptions we will prove the following: the *probability* $p_{\delta t}^{\mathrm{coll}}$ that the center of the solute molecule will collide with the surface element δA *in the next very small time* δt is

$$p_{\delta t}^{\mathrm{coll}} = \tfrac{1}{4} p_0 \, \bar{v} \delta t \delta A. \tag{4.21}$$

To prove Eq. (4.21), we start by defining λ to be the *smaller* of the following two lengths: (i) the radius of the largest neighborhood around δA within which the PDF of the solute molecule's position can be considered approximately constant at the value p_0; and (ii) the longest distance over which the trajectory of the solute molecule is approximately ballistic. In connection with (ii), if the solvent molecules were as massive as the solute molecule, this distance would be the mean free path of the solute molecule. But if, as we are assuming here, the solvent molecules are very small, so that a collision of one of them with the solute molecule will have only a very small effect on the velocity of the solute molecule, this distance will encompass many such collisions.

With λ the smaller of the lengths (i) and (ii), let δt be any time increment that satisfies $v_{\mathrm{max}} \delta t \leq \lambda$, i.e.,

$$\delta t \leq \lambda / v_{\mathrm{max}}, \tag{4.22}$$

where v_{max} is the largest possible speed of the solute molecule.[4] And let ω be a hemisphere of radius $v_{\mathrm{max}} \delta t$ whose center is at the center O of the surface element δA. We will denote the volume of this hemisphere also by ω. See Fig. 4.5. By construction, any solute molecule whose center can possibly reach the surface element δA within the next time δt must have its center somewhere inside the hemisphere ω, where by (i) the PDF of its position will have the value p_0, and by (ii) it will be moving ballistically.

With $d\omega$ denoting an infinitesimal subvolume of ω with polar coordinates (r, θ, ϕ) relative to the coordinate frame with origin O and polar axis normal to δA, we now make three observations: First, the probability that a randomly chosen solute molecule will have its center inside $d\omega$ is $p_0 \, d\omega$. Second, since the small surface element δA subtends at the location of $d\omega$ a solid angle $\delta A \cos \theta / r^2$, then the isotropy of the Maxwell–Boltzmann velocity distribution implies that the probability that the velocity vector of a solute molecule inside $d\omega$ will be aimed at the surface element δA will be equal to that solid angle divided by the total solid angle 4π, i.e., $\delta A \cos \theta / (4\pi r^2)$. Finally, with $f(v)$ denoting the PDF of the *speed* of a solute molecule, the probability that a solute molecule in $d\omega$ which is traveling toward the surface element δA will be going fast enough to reach δA within the next time interval δt is, by the addition law

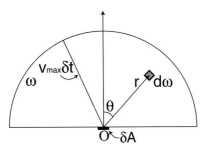

Fig. 4.5 *Diagram for calculating the probability of a collision in the next small time δt of a solute molecule with a small surface element δA. In this figure, δA is actually the area element that would be traced out by the* center *of the solute molecule as the molecule's boundary is traced over the physical surface element δA (which is not shown); thus, the collision of concern in this figure is a collision of the* center *of the solute molecule with the* artificial *area element δA. With δt chosen to be small enough to satisfy condition (4.22), where $v_{\max}$ is the maximum possible value of the solute molecule's velocity and λ is as defined in the text, the hemispheric volume ω will contain all present locations of the center of the molecule from which it could possibly reach δA within the next δt. We have also chosen δt small enough that ω will contain sufficiently few solvent molecules that the solute molecule will move inside ω approximately ballistically (in an approximately straight line at an approximately constant speed). All these conditions make it possible to express the desired collision probability as the multiple integral in Eq. (4.23).*

of probability, $\int_{r/\delta t}^{v_{\max}} f(v)dv$. Then by the laws of probability, {the probability that the center of the solute molecule will *collide* with the surface element δA during the next δt} can be computed as the *sum* over all volume elements $d\omega$ inside ω of the *product* of {the probability that the center of the solute molecule will be inside $d\omega$} times {the probability that it will be traveling toward δA} times {the probability that it will be moving fast enough to reach δA in time δt}:

$$p_{\delta t}^{\text{coll}} = \iiint_{\omega} [p_0 \, d\omega] \cdot \left[\frac{\delta A \cos\theta}{4\pi r^2}\right] \cdot \int_{v=r/\delta t}^{v_{\max}} f(v)dv. \tag{4.23}$$

Since

$$d\omega = |(dr)(rd\theta)(r\sin\theta d\phi)| = r^2 dr \, du \, d\phi,$$

where we have put $\cos\theta \equiv u$, Eq. (4.23) can be written more explicitly as

$$p_{\delta t}^{\text{coll}} = \int_{r=0}^{v_{\max}\delta t} dr \, r^2 p_0 \int_{u=0}^{1} du \left(\frac{\delta A \, u}{4\pi r^2}\right) \int_{\phi=0}^{2\pi} d\phi \int_{v=r/\delta t}^{v_{\max}} dv f(v)$$

$$= \frac{p_0 \delta A}{4\pi} \int_0^{v_{\max}\delta t} dr \int_0^1 du \, u \int_0^{2\pi} d\phi \int_{r/\delta t}^{v_{\max}} dv f(v).$$

The ϕ and u integrals here are easily performed, yielding factors of 2π and $1/2$ respectively, so this reduces to the double integral

$$p_{\delta t}^{\mathrm{coll}} = \frac{p_0 \delta A}{4} \int_0^{v_{\max}\delta t} dr \int_{r/\delta t}^{v_{\max}} dv f(v).$$

Switching the order of integration over r and v then gives (see Fig. 4.6)

$$p_{\delta t}^{\mathrm{coll}} = \frac{p_0 \delta A}{4} \int_0^{v_{\max}} dv \int_0^{v\delta t} dr f(v)$$

$$= \frac{p_0 \delta A}{4} \int_0^{v_{\max}} dv (v\delta t) f(v)$$

$$= \frac{p_0 \, \delta t \, \delta A}{4} \int_0^{v_{\max}} v f(v) \, dv \equiv \frac{p_0 \delta t \, \delta A}{4} \bar{v}.$$

Thus we have proved Eq. (4.21).

Notice that the two somewhat fuzzy parameters λ and $v_{\max}$ conveniently do not appear in our final formula for $p_{\delta t}^{\mathrm{coll}}$. But those two parameters *do* appear in the condition (4.22) that quantifies what "small δt" means. As for the value of $\bar{v}$ in our final formula for $p_{\delta t}^{\mathrm{coll}}$, it is easy to show that the average speed of a molecule of mass m in a Maxwell–Boltzmann velocity distribution at temperature T is (see Appendix 4C)

$$\bar{v} = \sqrt{\frac{8k_{\mathrm{B}}T}{\pi m}}. \tag{4.24}$$

Although Eq. (4.21) does not show us a clear path to an improved, more comprehensive theory of diffusion (we will embark on that journey in Chapter 8), it does provide us with a way of resolving the difficulty we encountered at the end of Chapter 3, where we tried to derive the stochastic rate of a diffusion-controlled bimolecular

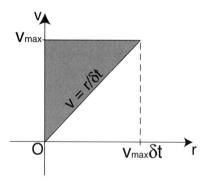

Fig. 4.6 *The shaded region shows the integration domain in vr-space that appears in the final stages of evaluating the multiple integral in Eq. (4.23). The figure shows how the integrations over first v (from r/δt to v_max) and then r (from 0 to v_maxδt) can instead be performed by integrating over first r (from 0 to v δt) and then v (from 0 to v_max).*

chemical reaction. A reminder of what that difficulty was, and the resolution of that difficulty and the conclusion of our derivation, are given in the next section.

4.8 The stochastic bimolecular chemical reaction rate: Part II

In Section 3.7, we set out to calculate the stochastic rate of the bimolecular reaction $S_1 + S_2 \rightarrow$ products when the molecules of the reactant species are solute molecules in a common bath of very many, much smaller solvent molecules. Our specific goal was to compute the probability $P_{\delta t}$ in Eq. (3.29) that this reaction will occur somewhere inside the system volume Ω in the next small time interval δt, given that we have a *dilute, macroscopically well-stirred* mixture of x_1 S_1 molecules and x_2 S_2 molecules at temperature T. We noted in Eq. (3.32) that this probability can be computed as

$$P_{\delta t} = p_{\delta t}\, x_1 x_2, \tag{4.25}$$

where $p_{\delta t}$ is the probability that a *single* randomly chosen S_1–S_2 molecular pair will react according to (3.28) in the next δt. And in Eq. (3.43), we concluded from the Einstein model of diffusion that the single-pair reaction probability $p_{\delta t}$ can be written

$$p_{\delta t} = 4\pi D_{12}\sigma_{12} \left(\Omega^{-1} - P_1(\sigma_{12}) \right) \delta t. \tag{4.26}$$

Here, D_{12} is the sum of the diffusion coefficients D_1 and D_2 of the two molecules; σ_{12} is their average center separation at collision, and hence the radius of the "action sphere" that surrounds the chosen S_2 molecule (see Fig. 3.1); and $P_1(\sigma_{12})$ is the value of the PDF of the center of the S_1 molecule at the surface of that action sphere. But our analysis stalled at that point, because we did not know what value to assign to $P_1(\sigma_{12})$.

The clarification in Section 4.6 of how a solute molecule moves on very small temporal and spatial scales affords a new perspective on this problem: In the *immediate vicinity* of the surface of the action sphere around the S_2 molecule, the S_1 molecule will be moving *ballistically*, as in a dilute gas. That implies that in the rest frame of the S_2 molecule, each rectilinear component of the velocity of an S_1 molecule that is very near the action sphere will be normally distributed with mean 0 and variance $k_B T/m_{12}$, where $m_{12} \equiv m_1 m_2/(m_1 + m_2)$.[5] Therefore, by Eq. (4.24) the S_1 molecule will be moving relative to the action sphere with average speed

$$\bar{v}_{12} = \sqrt{\frac{8 k_B T}{\pi m_{12}}}. \tag{4.27}$$

Equation (4.21) now allows us to assert that {the probability that the center of a randomly chosen S_1 molecule will strike a given infinitesimally small area element dA on the surface of the action sphere around the S_2 molecule in the next small time δt} is equal to $\frac{1}{4}P_1(\sigma_{12})\bar{v}_{12}\delta t dA$. Summing (integrating) this probability over all the area elements dA on the surface of the action sphere then gives, by the addition law of probability, the probability that the S_1 molecule will *collide* with the S_2 molecule in the next δt:

$$\tfrac{1}{4}P_1(\sigma_{12})\bar{v}_{12}\delta t \cdot 4\pi\sigma_{12}^2 = \pi\sigma_{12}^2 P_1(\sigma_{12})\bar{v}_{12}\delta t. \tag{4.28}$$

Now we multiply this *collision* probability with the probability q of a reaction *given* a collision, which was introduced in Section 3.7. That gives, by the multiplication law, the probability that the S_1 molecule will *react* with the S_2 molecule in the next δt:

$$p_{\delta t} = \pi\sigma_{12}^2 P_1(\sigma_{12})\bar{v}_{12}q\delta t. \tag{4.29}$$

We now have, in Eqs (4.29) and (4.26), *two equations* in the two unknowns $p_{\delta t}$ and $P_1(\sigma_{12})$. Equation (4.26) was derived from Einstein's theory of diffusion, which accurately describes the motion of the S_1 and S_2 molecules on larger-than-microscopic length and time scales. Equation (4.29) was derived from the kinetic theory of dilute gases, which accurately describes the motion of the S_1 and S_2 molecules microscopically close to the action sphere. Solving Eqs (4.26) and (4.29) simultaneously for the two unknowns $P_1(\sigma_{12})$ and $p_{\delta t}$ is algebraically straightforward, and yields

$$P_1(\sigma_{12}) = \frac{4D_{12}}{4D_{12} + \sigma_{12}\bar{v}_{12}q}\Omega^{-1}, \tag{4.30}$$

and

$$p_{\delta t} = \frac{4\pi\sigma_{12}^2 D_{12}\bar{v}_{12}q\Omega^{-1}}{4D_{12} + \sigma_{12}\bar{v}_{12}q}\delta t. \tag{4.31}$$

Equation (4.30) tells us that the PDF of the S_1 molecule at the surface of the action sphere is equal to the macroscopic value Ω^{-1} multiplied by the factor $4D_{12}/(4D_{12} + \sigma_{12}\bar{v}_{12}q)$; that factor, as expected, will always be some number between 0 and 1. The main result for us though is Eq. (4.31). When that formula is substituted into Eq. (4.25), we finally obtain the quantity we have been seeking: *The probability that reaction (3.28) will occur somewhere inside the system in the next small time δt is*

$$P_{\delta t} = cx_1 x_2 \cdot \delta t, \tag{4.32}$$

where c is given by

$$c \equiv \frac{4\pi\sigma_{12}^2 D_{12}\bar{v}_{12}q\Omega^{-1}}{4D_{12} + \sigma_{12}\bar{v}_{12}q}. \tag{4.33}$$

The factor $cx_1 x_2$ multiplying δt in Eq. (4.32) is called the *propensity function* of reaction (3.28). As its name suggests, the propensity function of a reaction channel is a measure of the immediate tendency of that reaction to occur somewhere inside Ω. The constant c is called the *specific reaction probability rate constant*. Since Eq. (4.32) can be viewed mathematically as the sum of $c\delta t$ over all $x_1 x_2$ distinct pairs of S_1–S_2 molecules inside Ω, then $c\delta t$ can be interpreted as the *probability that a typical S_1–S_2 molecular pair will react in the next δt*. The result that c is inversely proportional to the system volume reflects the physically obvious fact that two reactant molecules will have a harder time finding each other inside a larger volume.

Two important limiting cases of the result (4.33) arise at the extremes of the relative sizes of the two terms in its denominator: It is easy to see from Eq. (4.33) that:

$$\text{if } 4D_{12} \gg \sigma_{12}\bar{v}_{12}q, \text{ then } c \approx \pi\sigma_{12}^2\bar{v}_{12}q\Omega^{-1} \equiv c_{\text{ball}}; \tag{4.34a}$$

$$\text{if } 4D_{12} \ll \sigma_{12}\bar{v}_{12}q, \text{ then } c \approx 4\pi D_{12}\sigma_{12}\Omega^{-1} \equiv c_{\text{diff}}. \tag{4.34b}$$

The regime (4.34a) is called the *dilute gas regime*, because c_{ball} is the specific probability rate constant when the reactant molecules are ballistically moving dilute-gas molecules (as can be proved directly by a much simpler kinetic theory argument). In this regime, diffusion is so strong that the diffusional motion of two reactant molecules can bring them into close proximity faster than their small-scale ballistic motion can initiate a chemical reaction between them. This will happen in particular if the collision-conditioned reaction probability q is very small, so that many collisions are typically required in order for a chemical reaction to occur. The opposite regime (4.34b) is called the *Smoluchowski regime*, because c_{diff} is what arises from the Smoluchowski formula for $p_{\delta t}$ in Eq. (3.44). In this regime, diffusion is so weak that the small-scale ballistic motion of two reactant molecules can initiate a chemical reaction faster than their diffusional motion can bring them into close proximity; diffusion in that circumstance is the bottleneck to the chemical reaction process. But notice that although the vanishingly small q limit of Eq. (4.33) necessarily yields the dilute-gas result (4.34a), the limit $q \to 1$ of Eq. (4.33) does *not* necessarily yield the Smoluchowski result (4.34b).

By taking the reciprocal of Eq. (4.33), it is easy to show that

$$c^{-1} = c_{\text{diff}}^{-1} + c_{\text{ball}}^{-1}. \tag{4.35}$$

If we think of the *reciprocal* of the reaction rate constant as a "resistance" to the corresponding reaction (the smaller the rate constant the larger the resistance and vice versa), then Eq. (4.35) suggests a "two-resistors-in-series" heuristic from elementary electrical circuit theory: Diffusion first brings the two reactant molecules together via a "diffusional resistor" c_{diff}^{-1}, and then ballistic dynamics causes them to react via a "ballistic resistor" c_{ball}^{-1}. Since these two processes occur sequentially, i.e., in series, then the resistance c^{-1} of the overall process is given by the *sum* of the diffusional and ballistic resistors.

Since in practice it is usually not possible to determine the values of all the microphysical parameters in formula (4.33), the benefits of the foregoing derivation are somewhat indirect. At the very least, the derivation provides a justification for using a propensity function of the mathematical form cx_1x_2 for a bimolecular chemical reaction in a solution, even if the value of c is only empirically estimated. And formula (4.33) does show how c can be expected to change if any specific microphysical parameter is increased or decreased. Especially noteworthy is the implication of Eqs (4.34) that it is the ratio of the two specific quantities $4D_{12}$ and $\sigma_{12}\bar{v}_{12}q$ that ultimately determines whether a bimolecular chemical reaction proceeds in a ballistic mode or a diffusional mode; i.e., we now have a quantitative indicator of what "large" and "small" mean for the diffusion coefficients of chemically reacting species.

Notes to Chapter 4

[1]The analytical solution (4.4) for the perfectly reflecting boundary condition (4.3), which is also known as the homogeneous Neumann boundary condition, is obtained by first recognizing that Eq. (3.10) is the Green's function for the diffusion equation, and then applying the method of images. The analytical solution (4.8) for the perfectly absorbing boundary condition (4.7), also known as the homogeneous Dirichlet boundary condition, is obtained using the same procedure. For details of those derivations, see for example E. C. Zachmanoglou and Dale W. Thoe, *Introduction to Partial Differential Equations with Applications* (Dover, 1987).

[2]For two dimensions, we can replace the reflecting (absorbing) boundary *point* at x_r (x_a) by a reflecting (absorbing) boundary *line* through that point, a line perpendicular to the x-axis and extending from $y = -\infty$ to $y = +\infty$. Since the diffusional motions of the molecule in the x- and y-directions will then be independent of each other, we can use the unrestricted simulation formula (4.2) for the y-motion and the appropriate reflecting algorithm (4.5)/(4.6) or absorbing algorithm (4.5)/(4.9) for the x-motion. However, if the boundary line has one or more *openings* in it, this strategy will not work. That's because the generation of a value for x_t^{tent} that lies to the right of the boundary line can no longer be taken as a sure sign that the molecule was either reflected or absorbed by the boundary during (t_0, t), as the derivations in Appendices 4A and 4B assume; the solute molecule might have "legitimately" diffused to the right of the boundary through the opening. Nor will this strategy work if the boundary line is *curvilinear*, since in that case the x-location of the boundary will depend on the y-coordinate of the molecule at the unknown instant in (t_0, t) when it encountered the boundary. For a *reflecting corner* (x_r, y_r) whose two boundary legs extend away from the corner to infinity, the reflecting algorithm (4.5)/(4.6) can be used for each coordinate independently; that's because the x- and y-motions of the solute molecule will be independent of each other, even if the molecule gets reflected at one or both boundary legs during (t_0, t). Similarly, for an *absorbing corner* (x_a, y_a), the absorbing algorithm (4.5)/(4.9) can be used for each coordinate independently, because the x- and y-motions of the molecule will be independent of each other up until the instant one of the boundary legs absorbs the molecule and terminates its trajectory.

[3]A random variable X whose PDF has the form $(2a^2)\, x \exp(-a^2 x^2)$ is called a *Rayleigh* random variable, so Eq. (4.17) implies that, in two dimensions, R_t is a Rayleigh random variable.

[4]Although the Maxwell–Boltzmann distribution (see Appendix 4C) technically implies $v_{\max} = \infty$, in any real system with a finite total energy E_T, the solute molecule's speed can obviously be no larger than the value $v_{\max}$ that satisfies $\frac{1}{2} m v_{\max}^2 = E_T$. For speeds larger than that, the probability assigned by the Maxwell–Boltzmann distribution, although technically non-zero, will be so small as to be "effectively" zero.

[5]Since the S_1 and S_2 molecules have Maxwell–Boltzmann velocity distributions, each component of the velocity of an S_i molecule will be a statistically independent *normal* random variable with mean 0 and variance $k_B T m_i^{-1}$. Therefore, each

component of the velocity of an S_1 molecule *relative* to an S_2 molecule will be the *difference* between the two statistically independent normal random variables $\mathcal{N}\left(0, k_B T m_1^{-1}\right)$ and $\mathcal{N}\left(0, k_B T m_2^{-1}\right)$. From the theorem at Eq. (2.25) (taking $\alpha_1 = 1$ and $\alpha_2 = -1$), it follows that this difference will be a *normal* random variable with mean $0 - 0 = 0$ and variance

$$k_B T m_1^{-1} + k_B T m_2^{-1} = k_B T \left(m_1^{-1} + m_2^{-1}\right) \equiv k_B T m_{12}^{-1}.$$

Here, m_{12} is the so-called "reduced mass" $m_1 m_2/(m_1 + m_2)$. Thus, as seen from the rest frame of the S_2 molecule, the S_1 molecule moves as though it had mass m_{12}.

Appendix 4A: Proof of the reflecting boundary point simulation procedure

Here we will show that the simulation procedure of Eqs (4.5) and (4.6) produces an exact sampling of the PDF $P_X(x, t \mid x_0, t_0)$ when there is a single perfectly reflecting boundary at $x_r > x_0$. By construction, the point x_t^{tent} in Eq. (4.5) can be regarded as the common end-point of an infinite set S of possible trajectories over the time interval $[t_0, t]$ of a molecule diffusing *unrestrictedly* along the x-axis. This set of unrestricted trajectories consists of two subsets: subset S_0 contains all trajectories in S that never reached the point x_r in $[t_0, t]$; and subset S_1 contains all trajectories in S that reached the point x_r at least once in $[t_0, t]$.

For the trajectories in S_0, all of which will of course have $x_t^{\text{tent}} < x_r$, it cannot have made any difference to the diffusing molecule whether or not x_r was any kind of boundary point. Therefore, every trajectory in S_0 will also be a valid trajectory for the case in which x_r is actually a reflecting boundary point. The algorithm of Eqs (4.5) and (4.6) is thus a valid generating procedure for all trajectories in subset S_0.

Now consider any trajectory in subset S_1 of unrestricted trajectories in S that reached point x_r at least once in $[t_0, t]$. Note that such a trajectory might have either $x_t^{\text{tent}} < x_r$ or $x_t^{\text{tent}} \geq x_r$. Let $\bar{t} \in (t_0, t]$ be the instant of the trajectory's *first* arrival at x_r. Since the portion of that trajectory during the time interval $[t_0, \bar{t})$ does not contain the point x_r, it will be a valid trajectory over $[t_0, \bar{t})$ whether or not x_r is any kind of boundary point, and hence a valid trajectory for the case in which x_r is a reflecting boundary point. Now, the Markov nature of Einstein diffusion ensures that the remaining portion of this unrestricted trajectory from $\bar{t}$ to t (which we note can have any number of additional arrivals at x_r) can be regarded as a simulation of the solution $P_{\text{unres}}(x, t \mid x_r, \bar{t})$ of the Einstein diffusion equation (3.9) for the initial condition $X(\bar{t}) = x_r$ and *no* boundary points – a solution that can be immediately read off from Eq. (3.10):

$$P_{\text{unres}}(x, t \mid x_r, \bar{t}) = \frac{1}{\sqrt{4\pi D(t - \bar{t})}} \exp\left(-\frac{(x - x_r)^2}{4D(t - \bar{t})}\right) \quad (-\infty < x < \infty; \, t > \bar{t}).$$

$$(4A.1)$$

However, for the *restricted* case in which x_r is a perfectly reflecting boundary, the Einstein diffusion equation (3.9) for $P_{\text{res}}(x, t \mid x_r, \bar{t})$ must be solved for $-\infty < x \leq x_r$

and $t \geq \bar{t}$ subject to the reflecting boundary condition (3.26),

$$\frac{\partial P_{\text{res}}(x, t \,|\, x_{\text{r}}, \bar{t})}{\partial x}\bigg|_{x=x_{\text{r}}} = 0, \tag{4A.2}$$

and the initial condition

$$P_{\text{res}}(x, \bar{t} \,|\, x_{\text{r}}, \bar{t}) = 2\delta(x - x_{\text{r}}). \tag{4A.3}$$

In Eq. (4A.3), the factor 2 is required because the integral of $P_{\text{res}}(x, \bar{t} \,|\, x_{\text{r}}, \bar{t})$ from $x = -\infty$ to $x = x_{\text{r}}$ picks up only the left half of the delta-function peak at x_{r}. It is straightforward to show by direct substitution that the solution to the Einstein diffusion equation (3.9) for the boundary and initial conditions (4A.2) and (4A.3) is the above unrestricted solution with its $x > x_{\text{r}}$ part "folded over" onto its symmetric $x < x_{\text{r}}$ part (cf. Example 2 in Section 1.2):

$$P_{\text{res}}(x, t \,|\, x_{\text{r}}, \bar{t}) = \frac{2}{\sqrt{4\pi D(t - \bar{t})}} \exp\left(-\frac{(x - x_{\text{r}})^2}{4D(t - \bar{t})}\right) \quad (x \leq x_{\text{r}}; t > \bar{t}). \tag{4A.4}$$

Comparing this restricted solution (4A.4) to the unrestricted solution (4A.1), we see that the probability of any point $x_t < x_{\text{r}}$ in the restricted case will be exactly twice the probability of that point in the unrestricted case. But in the unrestricted case, the probability of any final point x_t will, by symmetry, be equal to the probability of its mirror image point about x_{r}. Therefore, the probability of any final point $x_t < x_{\text{r}}$ in the *restricted* case will be equal to the *sum* of the probabilities of that point and its mirror image point above x_{r} in the *unrestricted* case. A simple way to choose x_t according to that probability sum is to take x_t to be x^{tent} if $x^{\text{tent}} \leq x_{\text{r}}$, or its mirror image below x_{r} if $x^{\text{tent}} > x_{\text{r}}$. This selection procedure for any S_1 trajectory is exactly the selection procedure asserted in Eq. (4.6).

Appendix 4B: Proof of the absorbing boundary point simulation procedure

The key to proving that the simulation procedure described in connection with Eqs (4.5) and (4.9) produces an unbiased sample of $P_X(x, t \,|\, x_0, t_0)$ when there is a single perfectly absorbing boundary at $x_{\text{a}} > x_0$ is the following lemma.

Lemma: Define $\bar{p}(\tilde{x} \,|\, x, t; x_0, t_0)$ to be {the probability that a solute molecule, having diffused *unrestrictedly* from x_0 at time t_0 to x at time $t > t_0$, will have reached point $\tilde{x}$ at least once during that journey}. It is clear that if $\tilde{x}$ lies between x_0 and x, then this probability is 1. But otherwise, we have

$$\bar{p}(\tilde{x} \,|\, x, t; x_0, t_0) = \exp\left(-\frac{(\tilde{x} - x)(\tilde{x} - x_0)}{D(t - t_0)}\right) \quad \text{if} \begin{cases} \tilde{x} > x_0, x \\ \text{or} \\ \tilde{x} < x_0, x. \end{cases} \tag{4B.1}$$

This result is far from obvious. But it is plausible, inasmuch as it approaches 1 if $(t - t_0) \to \infty$, or if $x \to \tilde{x}$, or if $x_0 \to \tilde{x}$. Before proving this result, let us show how it establishes the procedure described in connection with Eqs (4.5) and (4.9).

By construction, the point x_t^{tent} in Eq. (4.5) is the end-point of an *unrestricted* journey of the solute molecule from x_0 at time t_0 to time $t > t_0$. Our main interest though is in a journey from x_0 at time t_0 to time $t > t_0$ when there is an *absorbing* boundary at $x_a > x_0$, a journey we will refer to as being *restricted*. For the generated point x_t^{tent} for the unrestricted journey, there are two possibilities: $x_t^{\text{tent}} \geq x_a$, and $x_t^{\text{tent}} < x_a$. For the first possibility we will have $x_0 < x_a \leq x_t^{\text{tent}}$, which means that the molecule, diffusing unrestrictedly, *must* have reached the point x_a at least once in its journey. Since the unrestricted and restricted motions of the molecule will be indistinguishable up until the instant the molecule first reaches point x_a, then the molecule on the *restricted* journey must have gotten absorbed by time t. This is the assertion made in the lines just above Eq. (4.9).

For the second possibility, we will have $x_a >$ both x_0 and x_t^{tent}. Then Eq. (4B.1) tells us that $\bar{p}(x_a \,|\, x_t^{\text{tent}}, t; x_0, t_0)$ gives the probability that the *unrestricted* journey from x_0 to x_t^{tent} will have visited x_a at least once. That implies that $\bar{p}(x_a \,|\, x_t^{\text{tent}}, t; x_0, t_0)$ gives the probability that the *restricted* journey will have terminated by time t. And that in turn implies that $1 - \bar{p}(x_a \,|\, x_t^{\text{tent}}, t; x_0, t_0)$ is the probability that the restricted journey will *not* have reached x_a by time t, and hence that x_t^{tent} is a valid sampling of the terminus of that journey. This is the either-or decision that is decided by the procedure described at Eq. (4.9), comparing $\bar{p}(x_a \,|\, x_t^{\text{tent}}, t; x_0, t_0)$ to a sample value of $\mathcal{U}(0, 1)$.

Proof of the lemma: According to the Einstein model of diffusion, if the solute molecule is at x_0 at time t_0 and diffuses *unrestrictedly* along the x-axis, then the probability that it will be in the interval $[x, x + dx)$ at time $t > t_0$ is $P(x, t \,|\, x_0, t_0)dx$, where

$$P(x, t \,|\, x_0, t_0) = \frac{1}{\sqrt{4\pi D(t - t_0)}} \exp\left(-\frac{(x - x_0)^2}{4D(t - t_0)} \right) \equiv f\left((x - x_0)^2, (t - t_0) \right). \quad (4B.2)$$

We now define two additional probabilities, both also for *unrestricted* diffusion:

$p_{\text{and}}(x, t; \tilde{x} \,|\, x_0, t_0)dx \equiv$ the probability that the solute molecule, at x_0 at time t_0, will be in the interval $[x, x + dx)$ at time $t > t_0$ *and* will have reached $\tilde{x}$ at least once during that journey.

$p_{\text{1st}}(t_1, \tilde{x} \,|\, x_0, t_0)dt_1 \equiv$ the probability that the molecule, at x_0 at time t_0, will *first* reach $\tilde{x}$ in the time interval $[t_1, t_1 + dt_1)$.

Observe that the probability $\bar{p}$ defined in the statement of the lemma is related to the probabilities $P\,dx$ and $p_{\text{and}}\,dx$, through the multiplication law:

$$p_{\text{and}}(x, t; \tilde{x} \,|\, x_0, t_0)\,dx = P(x, t \,|\, x_0, t_0)\,dx \cdot \bar{p}(\tilde{x} \,|\, x, t; x_0, t_0).$$

After cancelling the dx's and invoking the f notation in Eq. (4B.2), we can write this as

$$\bar{p}(\tilde{x} \mid x, t; x_0, t_0) = \frac{p_{\text{and}}(x, t; \tilde{x} \mid x_0, t_0)}{f((x - x_0)^2, (t - t_0))}. \tag{4B.3}$$

Thus, if we can derive an explicit formula for p_{and}, then Eq. (4B.3) will evidently give us an explicit formula for $\bar{p}$.

To deduce an explicit formula for p_{and}, we start by writing $p_{\text{and}} dx$ as the *sum* over all times $t_1 \in [t_0, t]$ of the *product* of {the probability that the molecule *first* reaches $\tilde{x}$ in the time interval $[t_1, t_1 + dt_1)$} times {the probability that a molecule, which is at $\tilde{x}$ at time t_1, will be in $[x, x + dx)$ at the later time t}:

$$p_{\text{and}}(x, t; \tilde{x} \mid x_0, t_0)dx = \int_{t_1 = t_0}^{t} [p_{1\text{st}}(t_1, \tilde{x} \mid x_0, t_0) \, dt_1] \cdot [P(x, t \mid \tilde{x}, t_1) \, dx].$$

Cancelling the dx's gives

$$p_{\text{and}}(x, t; \tilde{x} \mid x_0, t_0) = \int_{t_0}^{t} p_{1\text{st}}(t_1, \tilde{x} \mid x_0, t_0) P(x, t \mid \tilde{x}, t_1) dt_1. \tag{4B.4}$$

All equations we have written down thus far hold regardless of the relative ordering of x_0, x, and $\tilde{x}$. But in the particular case $x_0 < \tilde{x} < x$, it is obvious that the molecule *must* reach $\tilde{x}$ at least once in its journey from x_0 to x; therefore,

$$p_{\text{and}}(x, t; \tilde{x} \mid x_0, t_0) \, dx = P(x, t \mid x_0, t_0) \, dx \quad \text{if } x_0 < \tilde{x} < x. \tag{4B.5}$$

After cancelling the dx's and substituting on the left side from Eq. (4B.4), this becomes

$$\int_{t_0}^{t} p_{1\text{st}}(t_1, \tilde{x} \mid x_0, t_0) P(x, t \mid \tilde{x}, t_1) dt_1 = P(x, t \mid x_0, t_0) \quad \text{if } x_0 < \tilde{x} < x.$$

With the f notation for P introduced in Eq. (4B.2), this in turn can be written

$$\int_{t_0}^{t} p_{1\text{st}}(t_1, \tilde{x} \mid x_0, t_0) \, f((x - \tilde{x})^2, (t - t_1)) \, dt_1 = f((x - x_0)^2, (t - t_0)) \quad \text{if } x_0 < \tilde{x} < x.$$

In this equation, we now make the variable change $x \to x' \equiv 2\tilde{x} - x$. Since under this variable change, $(x - \tilde{x}) = -(x' - \tilde{x})$ and $(x - x_0) = (2\tilde{x} - x' - x_0)$, and also the condition $\tilde{x} < x$ becomes the condition $x' < \tilde{x}$, then the equation becomes

$$\int_{t_0}^{t_1} p_{1\text{st}}(t_1, \tilde{x} \mid x_0, t_0) \, f((x' - \tilde{x})^2, (t - t_1)) \, dt_1 = f((2\tilde{x} - x' - x_0)^2, (t - t_0))$$

$$\text{if } x_0 < \tilde{x} \text{ and } x' < \tilde{x}.$$

Now we relabel $x' \to x$, and then restore the P notation of Eq. (4B.2) on the left side:

$$\int_{t_0}^{t} p_{1\text{st}}(t_1, \tilde{x} \mid x_0, t_0) \, P(x, t \mid \tilde{x}, t_1) dt_1 = f((2\tilde{x} - x - x_0)^2, (t - t_0))$$

$$\text{if } x_0 < \tilde{x} \text{ and } x < \tilde{x}. \tag{4B.6}$$

Since the left side of Eq. (4B.6) is identical to the right side of Eq. (4B.4), then the other sides of those two equations must also be equal. This gives us our explicit formula for p_{and}:

$$p_{\text{and}}(x, t; \tilde{x} \mid x_0, t_0) = f\left((2\tilde{x} - x - x_0)^2, (t - t_0)\right) \quad \text{if } x_0, x < \tilde{x}. \tag{4B.7}$$

Substituting this into Eq. (4B.3) gives

$$\bar{p}(\tilde{x} \mid x, t; x_0, t_0) = \frac{f\left((2\tilde{x} - x - x_0)^2, (t - t_0)\right)}{f\left((x - x_0)^2, (t - t_0)\right)} \quad \text{if } x_0, x < \tilde{x}. \tag{4B.8}$$

Upon substituting for the f's the explicit expressions implied by Eq. (4B.2) and then simplifying the algebra, we obtain the asserted result (4B.1) for the case $x_0, x < \tilde{x}$.

Finally, an examination of the foregoing arguments will reveal, as symmetry considerations also imply, that if $\tilde{x}$ lies *below* instead of *above* both x_0 and x, then we will get exactly the same result.

Appendix 4C: The Maxwell–Boltzmann distribution

The Maxwell–Boltzmann velocity distribution asserts that each rectilinear component of the velocity of a molecule of mass m is effectively a statistically independent *normal* random variable with mean 0 and variance $\sigma^2 = k_B T/m$, where k_B is Boltzmann's constant and T is the absolute temperature. It follows from Eq. (2.4) that the joint PDF of the three rectilinear velocity components V_x, V_y, and V_z is given by:

$$P_{V_x, V_y, V_z}(v_x, v_y, v_z) = \left(2\pi\sigma^2\right)^{-3/2} e^{-v_x^2/2\sigma^2} e^{-v_y^2/2\sigma^2} e^{-v_z^2/2\sigma^2}, \tag{4C.1}$$

where of course each velocity component ranges over the entire real axis. The corresponding polar coordinate velocity variables

$$V_x \equiv V \sin\Theta \cos\Phi, \quad V_y \equiv V \sin\Theta \sin\Phi, \quad V_z \equiv V \cos\Theta,$$

where $0 \le V < \infty$, $0 \le \Theta \le \pi$, and $0 \le \Phi \le 2\pi$, will by Eq. (2.17) have the joint PDF

$$P_{V,\Theta,\Phi}(v, \theta, \phi) = P_{V_x, V_y, V_z}(v_x, v_y, v_z) \left| \frac{\partial(v_x, v_y, v_z)}{\partial(v, \theta, \phi)} \right|. \tag{4C.2}$$

Here, the Jacobian is for the transformation

$$v_x = v \sin\theta \cos\phi, \quad v_y = v \sin\theta \sin\phi, \quad v_z = v \cos\theta.$$

Evaluating that Jacobian gives $\left|v^2 \sin\theta\right|$, and that with Eq. (4C.1) gives

$$P_{V,\Theta,\Phi}(v, \theta, \phi) = \left(2\pi\sigma^2\right)^{-3/2} v^2 e^{-v^2/2\sigma^2} \left|\sin\theta\right|. \tag{4C.3}$$

Integrating this joint PDF first over $\phi \in [0, 2\pi]$ and then over $\theta \in [0, \pi]$ gives respective factors of 2π and 2, so the PDF of the *speed* V in three dimensions is

$$P_V(v) = \left(2\pi\sigma^2\right)^{-3/2} 4\pi v^2 e^{-v^2/2\sigma^2} \quad (0 \le v < \infty). \tag{4C.4}$$

Therefore, the *mean speed* is

$$\bar{v} \equiv \langle V \rangle = \int_0^\infty v P_V(v) dv = \int_0^\infty \left(2\pi\sigma^2\right)^{-3/2} 4\pi v^3 e^{-v^2/2\sigma^2} dv.$$

Evaluating this last integral finally gives for the mean speed in three dimensions,

$$\bar{v} = \sigma\sqrt{\frac{8}{\pi}} = \sqrt{\frac{8k_{\mathrm{B}}T}{\pi m}}. \tag{4C.5}$$

Since $\sqrt{8/\pi} \approx 1$, we see that the mean speed $\bar{v}$ is very nearly equal to the root-mean-square (rms) value of each rectilinear component of the velocity, namely,

$$\sqrt{\langle V_x^2 \rangle} = \sqrt{\langle V_y^2 \rangle} = \sqrt{\langle V_z^2 \rangle} = \sigma \equiv \sqrt{k_{\mathrm{B}}T/m}. \tag{4C.6}$$

5

The discrete-stochastic approach

In this chapter we introduce a model of diffusion in which the position of the solute molecule is described by a *discrete* stochastic variable, in contrast to the continuous stochastic variable of the Einstein model. Despite that difference, the intent of this new model is to give a description of diffusion that is for most practical purposes equivalent to the Einstein model. We will carefully examine the basis for that presumption of equivalence in order to clarify the limitations of the discrete-stochastic model. We will illustrate the model in the context of a real laboratory experiment, where the solute molecules are tiny polystyrene beads diffusing in water inside a specially fabricated microfluidic chip. The main motivation for this new approach is the new suite of mathematical methods it offers for quantitatively analyzing diffusion. In Chapter 6 we will present the mathematical framework for this approach, which features two master equations and their associated stochastic simulation algorithms.

5.1 Specification of the system

To keep the math simple, we will focus on a quasi-one-dimensional system: We assume the solute and solvent molecules are confined to a cylindrical volume Ω of length L and cross-sectional area A. We let this cylinder be coaxial with the x-axis, with its left face at $x = 0$ and its right face at $x = L$. We (mentally) subdivide Ω into M "cells" of equal length $l = L/M$ by means of division points at $x_i = i \cdot l$ for $i = 1, \ldots, M - 1$. With $x_0 \equiv 0$ and $x_M \equiv L$, we define "cell i" for $i = 1, \ldots, M$ to be the portion of Ω between x_{i-1} and x_i. See Fig. 5.1. The instantaneous state of the system is then defined by the M-tuple of integer random variables $\mathbf{N}(t \mid \mathbf{n}_0) \equiv \{N_1(t \mid \mathbf{n}_0), \ldots, N_M(t \mid \mathbf{n}_0)\}$, where

$$N_i(t \mid \mathbf{n}_0) \equiv \text{the number of solute molecules in cell } i \text{ at time } t, \text{ given the}$$
$$\text{cell populations } \mathbf{n}_0 \equiv (n_{10}, \ldots, n_{M0}) \text{ at time } 0 \ (i = 1, \ldots, M). \quad (5.1)$$

We will find that a key requirement underlying this approach to modeling diffusion is that within each cell the solute molecules must be approximately *uniformly distributed*. But this intra-cellular spatial homogeneity requirement is to be interpreted in the following special *stochastic* sense: The *probability* that an infinitesimal subinterval dx of cell i will contain the center of a solute molecule must be (at least approximately) independent of the *location* of dx inside $[x_{i-1}, x_i]$. Notice that this definition of intra-cellular spatial homogeneity does not depend on the number of solute molecules inside

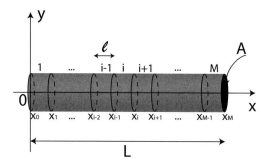

Fig. 5.1 *The system volume Ω envisioned in the simplest application of the discrete-stochastic approach to diffusion. It is a right cylinder of length L and cross-sectional area A, which is aligned with the x-axis in the interval $0 \leq x < L$. It is subdivided into M "cells" of equal length $l = L/M$ by the points $x_i = i \cdot l$ $(i = 0, \ldots, M)$, with cell i occupying the subinterval $x_{i-1} \leq x < x_i$. It is assumed to be filled with a mixture of solvent and solute molecules, and to be relatively dilute in the latter, so that the solute molecules move about practically independently of each other. Unless otherwise stated, all boundary surfaces of Ω are assumed to be perfectly reflecting. All the molecules move three-dimensionally, but it is only the x-coordinate of the solute molecules that is of interest to us.*

a cell; in particular, it is *not* necessary that there be a large number of solute molecules inside a cell in order for the molecules in that cell to be "uniformly distributed" in the sense that we are using that terminology here.

To see more clearly what this definition of intra-cellular spatial homogeneity requires, let the PDF q be defined so that $q(x)dx$ gives the probability that the infinitesimal section of Ω between x and $x + dx$ will contain the center of a solute molecule. And let x_c be the *center* of one of the cells of length l. Then the requirement that the solute molecules be approximately uniformly distributed inside that cell means that q should be approximately constant in the interval $[x_c - \frac{1}{2}l, x_c + \frac{1}{2}l)$. We will interpret this requirement as saying that the total variation Δq in q inside that interval should be a *small fraction* of $q(x_c)$. As is shown in Fig. 5.2, in any physically realistic situation, the small variation Δq in q over the length of the cell can be approximated by $q'(x_c) \cdot l$; thus, the *fractional variation* in q inside the cell is approximately

$$\frac{\Delta q}{q(x_c)} \approx \frac{q'(x_c)}{q(x_c)} \cdot l. \tag{5.2}$$

The first factor on the right in Eq. (5.2) should be finite in any physically reasonable situation, but will otherwise be beyond our control. However, we *can* control the value of the cell length l. And it is obvious from Eq. (5.2) that by taking l sufficiently small, we can make the fractional variation in q over the cell as small as we please.

So we see that the requirement that each cell be spatially uniform in the solute molecules can be met in any physically realistic situation by taking l small enough.

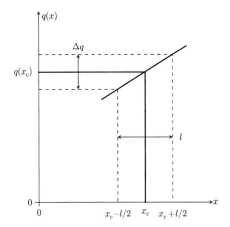

Fig. 5.2 *The discrete-stochastic approach requires the solute molecules to be approximately spatially uniform inside each cell. That does* not *require there to be many solute molecules in a cell; it requires only that the PDF $q(x)$ of a randomly chosen solute molecule in a cell should not vary appreciably over the length of the cell. As the sketch shows, the relative variation $\Delta q/q(x_c)$ of any physically reasonable PDF inside a cell of length l centered at x_c can generally be made as small as we wish by taking l sufficiently small. But as we will see later, we are prevented from taking l arbitrarily small by physical reasons which are ultimately traceable to the fact that the Einstein theory of diffusion imposes a lower limit on the size of any time increment.*

But we will discover shortly that other considerations prevent us from taking l *arbitrarily* small. These opposing requirements on l conspire to make satisfying the intra-cellular uniformity condition a major challenge to be met in order for this model of diffusion to be viable.

5.2   The key dynamical hypothesis

Any solute molecule in any cell might, as a result of collisions with the solvent molecules, suddenly find itself in an adjacent cell. In the discrete-stochastic model, we make the *assumption* that such diffusive transfers of solute molecules to adjacent cells happen independently of each other according to the following rule: There exists a constant $\kappa_l > 0$ such that

$$\kappa_l \delta t + o(\delta t) \equiv \text{the probability that a randomly chosen molecule}$$
$$\text{in cell } i \text{ will move to a particular adjacent cell}$$
$$\text{in the next macroscopically infinitesimal time } \delta t. \qquad (5.3)$$

Here, $o(\delta t)$ denotes a term of order > 1 in δt, i.e., a term that goes to zero with δt faster than δt.

Although Eq. (5.3) in effect defines κ_l, it also is an assertion as to how the collisions that a solute molecule has with solvent molecules physically affect the solute molecule. Indeed, Eq. (5.3) is analogous to Einstein's assumption in Eq. (3.1) of a function $\phi(\xi; \delta t)$, whose product with $d\xi$ gives the probability that collisions with solvent molecules during time δt will cause a solute molecule to be displaced along the x-axis by an amount between ξ and $\xi + d\xi$. But Eq. (5.3) evidently specifies the stochastic displacement in time δt in a more "coarse-grained" way than in Einstein's approach, because Eq. (5.3) only resolves the solute molecule's position to the cell length l. This effectively "discretizes" the molecule's position, replacing the continuum of position states $0 < x < L$ with a finite set of M position states. That can be computationally advantageous if M is not too large.

But there are several *assumptions* underlying hypothesis (5.3) that need to be recognized. First, the limit $\delta t \to 0$ is not allowed in its strictest sense, because δt here is a *macroscopic* infinitesimal; i.e., while δt is infinitesimally small on a macroscopic scale, it must be large enough that a solute molecule will experience many collisions with solvent molecules in that time. So δt in Eq. (5.3) is essentially the same as the δt in Einstein's theory: it has some *physically* imposed lower bound, which we have yet to specify quantitatively.

Second, hypothesis (5.3) assumes that the solute molecules inside cell i are uniformly distributed over the interval $[x_{i-1}, x_i]$, in the stochastic sense defined in the preceding section. To see why, suppose a randomly chosen solute molecule in cell i were more likely to be found in the left half of cell i than in the right half. Then the probability for that randomly chosen molecule to diffuse across the left boundary x_{i-1} in the next δt would be greater than the probability for it to diffuse across the right boundary x_i, and that is not allowed by hypothesis (5.3). Since, as was noted in connection with Eq. (5.2), intra-cellular spatial homogeneity requires l to be "small", it follows that hypothesis (5.3) is assuming that l is "not too large".

Finally, hypothesis (5.3) makes no allowance for jumps to *non-adjacent* cells in time δt; e.g., it does not allow a molecule in cell i to jump in the next δt to cell $i - 2$ or cell $i + 2$, even though that is physically possible. Since the likelihood of jumps to non-adjacent cells increases with both increasing δt and decreasing l, it follows that (5.3) is assuming that δt is "not too large" and l is "not too small".

To use the discrete-stochastic approach to model diffusion in all three Cartesian directions inside a cubic volume Ω of edge length L, we would subdivide Ω into cubic cells of edge length l, choosing l so that it evenly divides L. The probability that a solute molecule in some interior cell will leave that cell in the next δt through any of its six faces will then be $6\kappa_l \delta t$, by the addition law of probability. And when such an exit does occur, it will be equally likely through any cell face. But in this situation, yet another restriction on the cell length l comes into play: *l should be much larger than the diameter of a solute molecule*. There are two reasons for this requirement, one rooted in theory and the other in computational practicality. On the theory side, observe that hypothesis (5.3) makes the tacit assumption that the probability that a solute molecule will jump to a particular adjacent cell is independent of the number of solute molecules that are already in that cell. But that can be true only if the solute molecules that are already in that cell occlude a negligibly small fraction of the cell's volume. Indeed, in the extreme case where l is less than or equal to the radius of a solute molecule, a cell will be completely occluded by any solute molecule whose center lies inside that cell, and it will therefore be impossible for any other solute molecule to jump into that cell. From the standpoint of computational practicality, taking l so small that no more than one solute molecule at a time can have its center inside a cell means that there will be only two possible cell populations: 1 and 0. In that circumstance, we will effectively be tracking the detailed motion of every individual solute molecule in the system; thus we will have lost the computational advantages of coarse-graining the system, which is a principal rationale for using the discrete-stochastic model of diffusion.

We will discuss the restrictive assumptions on hypothesis (5.3) mentioned above in more quantitative detail later. But first we need to establish a physical rationale for that hypothesis by connecting it to the models of diffusion discussed in earlier chapters, namely the classical Fickian model of Chapter 1 and the Einstein model of Chapters 3 and 4. We will find that both connections lead to the same formula for κ_l in terms of the solute molecule's diffusion coefficient D and the cell length l.

5.3   Connection to the classical Fickian model

Suppose that there are currently n_i solute molecules in cell i for $i = 1, \ldots, M$ (see Fig. 5.1). Then the average number of solute molecules per unit volume in cell i will be n_i/Al. If the cell length l has been chosen small enough, n_i/Al should accurately approximate the average solute molecule density $\rho(x)$ at the midpoint $x = x_i - l/2$ of cell i. For any interior cell i we have $\rho(x_i - l/2) \doteq n_i/Al$ and $\rho(x_i + l/2) \doteq n_{i+1}/Al$, so Fick's Law (1.2) gives for the net flux of solute molecules at point x_i,

$$J(x_i) = -D \left. \frac{\partial \rho}{\partial x} \right|_{x=x_i} \doteq -D \left(\frac{(n_{i+1}/Al) - (n_i/Al)}{l} \right),$$

or

$$J(x_i) \doteq \frac{D}{Al^2} (n_i - n_{i+1}). \tag{5.4}$$

By the definition of the flux, the average net number of solute molecules crossing x_i in the positive x-direction in the next infinitesimal time dt will then be given by

$$J(x_i) \cdot A \, dt \doteq \frac{D}{l^2} n_i \, dt - \frac{D}{l^2} n_{i+1} \, dt. \tag{5.5}$$

The qualifier "net" here means that Eq. (5.5) gives {the average number of solute molecules crossing from cell i to cell $i + 1$ in time dt} *minus* {the average number of solute molecules crossing from cell $i + 1$ to i in time dt}. Associating those terms with the two terms on the right side of Eq. (5.5), we conclude that, quite generally,

$$\frac{D}{l^2} n_i \, dt = \text{the } \textit{average number} \text{ of the } n_i \text{ solute molecules inside cell } i$$
$$\text{that will cross a particular boundary of that cell in the next } dt. \tag{5.6a}$$

Now we take dt so small that it will be overwhelmingly unlikely for more than one of the n_i molecules in cell i to leave that cell in the next dt. Then in the next dt, the number of solute molecules in cell i that will cross the boundary in question will be either 1 or 0. Letting $p(n_i)$ denote the probability of the former, so that $1 - p(n_i)$ is the probability of the latter, we can compute the average in Eq. (5.6a) by using the usual weighted-sum rule (2.37):

$$0 \cdot (1 - p(n_i)) \; + \; 1 \cdot p(n_i) = p(n_i).$$

Thus, the *average* in Eq. (5.6a) is also a *probability*:

$$\frac{D}{l^2} n_i dt = \text{the } \textit{probability} \text{ that any one of the } n_i \text{ solute molecules in cell}$$
$$i \text{ will cross a particular boundary of that cell in the next } dt. \qquad (5.6b)$$

Now in simple diffusion, the solute molecules move about independently of each other. So the probability that *any* one of the n_i molecules in cell i will cross the boundary in question in the next dt will be equal to n_i times the probability that a *particular* one of them will do that.[1] We may thus conclude from Eq. (5.6b) that

$$\frac{D}{l^2} dt = \text{the probability that a } \textit{particular} \text{ solute molecule in cell}$$
$$i \text{ will cross a particular boundary of that cell in the next } dt. \qquad (5.7)$$

This prediction of Fick's Law will evidently agree with hypothesis (5.3) if and only if

$$\kappa_l = \frac{D}{l^2}. \qquad (5.8)$$

Equation (5.8) is the quantitative connection between the discrete-stochastic hypothesis (5.3) and the classical Fickian model of diffusion. But notice that, whereas hypothesis (5.3) makes allowance for some as yet unspecified lower bound on its time increment δt, the above argument allows, and in fact seems to insist, that its time increment dt can be arbitrarily small. This reflects the failure of the classical theory of diffusion to recognize that there is a physically imposed lower bound on time increments. In the next two sections, we will see that the same result (5.8) follows from the Einstein model, which however does recognize that there must be some lower bound on δt.

5.4 Connection to the Einstein model

There are several ways of establishing the connection between the discrete-stochastic model of diffusion and Einstein's model. In this section we will examine two of those ways. Both conclude that, subject to certain restrictions on δt and l, hypothesis (5.3) will be consistent with the Einstein model provided κ_l is given by formula (5.8). The first argument is disarmingly simple. The second is more involved, but it is also more versatile in that it can be adapted to other cellular geometries, and in some cases extended to situations that lie outside the domain of the Einstein model.

Derivation 1

The simpler argument focuses on a single solute molecule which is known to be in some interior cell (see Fig. 5.1), and it hinges on the following three implications of hypothesis (5.3): first, the probability that in the next δt the molecule will jump to the cell to its right, thereby augmenting its x-coordinate by $+l$, is $\kappa_l \delta t + o(\delta t)$; second, the probability that it will jump to its left, thereby augmenting its x-coordinate by $-l$, is also $\kappa_l \delta t + o(\delta t)$; and third, the probability that it does not jump at all, thereby augmenting its x-coordinate by 0, is $(1 - 2\kappa_l \delta t) + o(\delta t)$. Therefore, by Eq. (2.37), the *average* x-displacement of the molecule in the next δt is, to first order in δt,

$$\langle \Delta x_{\delta t} \rangle = (\kappa_l \delta t)(+l) + (\kappa_l \delta t)(-l) + (1 - 2\kappa_l \delta t)(0) = 0,$$

a very unsurprising result. More interesting is the *average squared* x-displacement of the molecule in the next δt. Again by Eq. (2.37), that average is

$$\left\langle (\Delta x_{\delta t})^2 \right\rangle = (\kappa_l \delta t)(+l)^2 + (\kappa_l \delta t)(-l)^2 + (1 - 2\kappa_l \delta t)(0)^2,$$

whence,

$$\left\langle (\Delta x_{\delta t})^2 \right\rangle = 2 \left(\kappa_l l^2 \right) \delta t, \tag{5.9}$$

again to first order in δt. Now recall that Einstein's definition of D in Eq. (3.5), as well as his result (3.12), implies that the mean-square x-displacement of a diffusing molecule in a "sufficiently large" time δt is $2D\delta t$. Agreement between that result and Eq. (5.9) can evidently be obtained if and only if we have $\kappa_l l^2 = D$, or $\kappa_l = D/l^2$. That is precisely the result in Eq. (5.8).

Derivation 2

For a more robust derivation, we need to recognize that, because the discrete-stochastic model is ambiguous as to exactly where a solute molecule in a cell starts and finishes its jump to an adjacent cell, that jump is not a sharply defined physical process; therefore, we will need to create a more specific microscopic model of that process. We will model the jump from cell i to cell $i + 1$ as the diffusional journey of a solute molecule from the middle of cell i, at point $x = x_i - \frac{1}{2}l$, to the middle of cell $i + 1$, at point $x = x_i + \frac{1}{2}l$ (see Fig. 5.3). Furthermore, because the jump probability in hypothesis (5.3) is presumed to be independent of time, in the sense that the probability that the molecule will jump in the next δt stays constant until the molecule actually jumps, we will use a *steady-state* model which captures the essential dynamics of that diffusional journey. Specifically, with $q(x)$ denoting the steady-state PDF of the solute molecule's position, and A the cross-sectional area of the cylindrical volume Ω, we will model this diffusional journey by imposing on $q(x)$ the following two requirements:

$$q(x_i - \tfrac{1}{2}l) = (Al)^{-1}, \tag{5.10a}$$

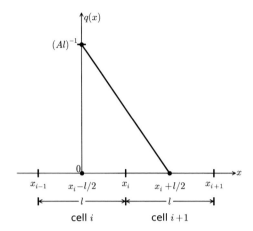

Fig. 5.3 *A steady-state diffusion model for computing κ_l. By buffering the solute molecule's position PDF $q(x)$ to the uniform-inside-cell-i value $(Al)^{-1}$ at $x = x_i - \frac{1}{2}l$, and to the value zero at $x = x_i + \frac{1}{2}l$, we obtain a steady-state probability flux from which we can compute a formula for κ_l.*

$$q(x_i + \tfrac{1}{2}\, l) = 0. \tag{5.10b}$$

Requirement (5.10a) reflects the assumption that, at the start of the solute molecule's journey, its PDF has the value of the randomly uniform distribution inside cell i, namely the reciprocal of the cell's volume. Requirement (5.10b) stipulates, by Eq. (3.27), that the destination point $x = x_i + \frac{1}{2}l$ is considered in our steady-state model to be a perfectly absorbing boundary point, thereby ensuring that the first arrival of the solute molecule at $x = x_i + \frac{1}{2}l$ will end its journey. Between the two points $x_i - \frac{1}{2}l$ and $x_i + \frac{1}{2}l$, the PDF $q(x)$ must of course satisfy the steady-state version of the Einstein diffusion equation (3.9), namely, $\partial^2 q(x)/\partial x^2 = 0$. The solution to that equation is a straight line, and to satisfy the two boundary conditions (5.10) that line must contain the two points $\left(x_i - \frac{1}{2}l, (Al)^{-1}\right)$ and $\left(x_i + \frac{1}{2}l, 0\right)$, as shown in Fig. 5.3 (cf. also Example 4 in Chapter 1). What is physically happening in our model is this: The single-molecule position probability is flowing out of the interval $[x_i - \frac{1}{2}l, x_i + \frac{1}{2}l]$ at the absorbing boundary $x = x_i + \frac{1}{2}l$, and the steady state is maintained by a flow of position probability into the interval at $x = x_i - \frac{1}{2}l$ at the same rate.

As discussed in Section 3.6, the rate of flow of position probability across any point x is given by the product of the cross-sectional area A times the local value of the single-molecule position probability flux, $\hat{J}_x(x) \equiv -D\,dq(x)/dx$. The latter can be seen from Fig. 5.3 to have the following value everywhere in the interval $[x_i - \frac{1}{2}l, x_i + \frac{1}{2}l]$:

$$\hat{J}_x(x) = -D\,\frac{q(x_i + \frac{1}{2}l) - q(x_i - \frac{1}{2}l)}{(x_i + \frac{1}{2}l) - (x_i - \frac{1}{2}l)} = -D\,\frac{0 - (Al)^{-1}}{l} = \frac{D}{A\,l^2}. \tag{5.11}$$

The result in Eq. (3.25a) then allows us to conclude that the probability that the solute molecule described by $q(x)$ will be absorbed at the surface of area A through $x = x_i + \frac{1}{2}l$ in the next small time δt is equal to $\hat{J}_x(x_i + \frac{1}{2}l) \cdot A \cdot \delta t$. Equating that *absorption* probability in the next δt to the probability $\kappa_l \delta t$ that the molecule will *jump* from cell i to cell $i+1$ in the next δt gives

$$\kappa_l \delta t = \hat{J}_x(x_i + \tfrac{1}{2}l) \cdot A \cdot \delta t = \left(\frac{D}{A\,l^2}\right) A\,\delta t = \left(\frac{D}{l^2}\right) \delta t. \qquad (5.12)$$

This evidently implies the result $\kappa_l = D/l^2$.

The foregoing two independent derivations make a convincing case that the Einstein model of diffusion supports the discrete-stochastic hypothesis (5.3) with $\kappa_l = D/l^2$. Notice that the probability $(D/l^2)\,\delta t$ for a solute molecule to diffusively move out of its current cell to a particular adjacent cell in the next δt gets larger as l gets smaller. That is reasonable, since a solute molecule should escape from a smaller cell more quickly than from a larger cell. The inverse-square dependence on l is not so obvious, although arguably that could have been guessed by dimensional reasoning: Eqs (3.9) and (4.11) both show that D has dimensions of $length^2\ time^{-1}$, while κ_l in Eq. (5.3) evidently has dimensions of $time^{-1}$.

Derivation for unequal cell lengths

The second of the above two derivations has the advantage that it can be extended in several useful ways. For example, suppose cells i and $i+1$ have unequal lengths

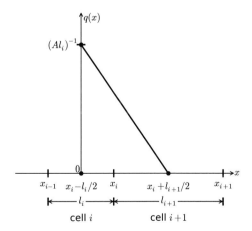

Fig. 5.4 *A steady-state diffusion model for computing $\kappa_{i \to i+1}$ when cells i and $i+1$ have unequal lengths l_i and l_{i+1}, respectively. By buffering the solute molecule's position PDF $q(x)$ to the uniform-inside-cell-i value $(Al_i)^{-1}$ at $x = x_i - \frac{1}{2}l_i$, and to the value zero at $x = x_i + \frac{1}{2}l_{i+1}$, we obtain a steady-state probability flux from which we can compute a formula for $\kappa_{i \to i+1}$. A straightforward reversal of the roles of cells i and $i+1$ yields a formula for $\kappa_{i+1 \to i}$ that is not equal to $\kappa_{i \to i+1}$.*

l_i and l_{i+1}, as shown in Fig. 5.4. To compute the probability $\kappa_{i \to i+1} \delta t$ that a solute molecule in cell i will jump to cell $i+1$ in the next δt, we need make only minor modifications to the reasoning that led us to Eq. (5.12). We model the jump as the steady-state diffusional journey of a solute molecule from the middle of cell i, which is now the point $x = x_i - \frac{1}{2}l_i$, to the middle of cell $i+1$, which is now the point $x = x_i + \frac{1}{2}l_{i+1}$. Since at the start of this journey the solute molecule is supposed to typify a uniform distribution inside cell i, the boundary conditions on $q(x)$ are now, instead of Eqs (5.10),

$$q(x_i - \tfrac{1}{2}\, l_i) = (Al_i)^{-1},$$

$$q(x_i + \tfrac{1}{2}\, l_{i+1}) = 0.$$

The straight-line solution of the steady-state diffusion equation now connects the points $\left(x_i - \frac{1}{2}l_i, (Al_i)^{-1}\right)$ and $\left(x_i + \frac{1}{2}l_{i+1}, 0\right)$, as shown in Fig. 5.4. So the single-molecule position probability flux everywhere in the interval $[x_i - \frac{1}{2}l_i, x_i + \frac{1}{2}l_{i+1}]$ is, instead of Eq. (5.11),

$$\hat{J}_x(x) = -D\,\frac{q(x_i + \tfrac{1}{2}l_{i+1}) - q(x_i - \tfrac{1}{2}l_i)}{(x_i + \tfrac{1}{2}l_{i+1}) - (x_i - \tfrac{1}{2}l_i)} = D\,\frac{(Al_i)^{-1}}{\tfrac{1}{2}(l_i + l_{i+1})}.$$

The probability that the solute molecule will be absorbed at the surface of area A through $x = x_i + \frac{1}{2}l_{i+1}$ in the next small time δt is, by Eq. (3.25a), $\hat{J}_x(x_i + \frac{1}{2}l_{i+1}) \cdot A \cdot \delta t$. Equating that probability to the probability that the molecule will jump from cell i to cell $i+1$ in the next δt gives

$$\kappa_{i \to i+1}\delta t = \hat{J}_x(x_i + \tfrac{1}{2}l_{i+1}) \cdot A \cdot \delta t.$$

Substituting into this equation the above formula for $\hat{J}_x$, and then cancelling the δt's, we conclude that

$$\kappa_{i \to i+1} = \frac{D}{l_i \cdot \tfrac{1}{2}(l_i + l_{i+1})}. \tag{5.13a}$$

In a similar way, one can show that the jump probability rate constant for the reverse jump, from cell $i+1$ to cell i, is

$$\kappa_{i+1 \to i} = \frac{D}{l_{i+1} \cdot \tfrac{1}{2}(l_i + l_{i+1})}. \tag{5.13b}$$

If $l_i = l_{i+1} = l$, Eqs (5.13a) and (5.13b) both reduce to the expected result, D/l^2. But if $l_i \neq l_{i+1}$, the two jump probability rate constants are not equal. They do, however, satisfy the ratio relation

$$\frac{\kappa_{i \to i+1}}{\kappa_{i+1 \to i}} = \frac{l_{i+1}}{l_i} = \frac{\Omega_{i+1}}{\Omega_i}, \tag{5.14}$$

where $\Omega_i \equiv Al_i$ is the volume of cell i. This relation is in fact required for proper behavior at equilibrium. At equilibrium, the PDF of the solute molecule will have the constant value Ω^{-1} throughout all the cells, so the probability of the molecule being inside cell i will be $\Omega^{-1}\Omega_i$. Therefore, the probability of a jump from cell i to cell $i+1$ in the next δt will be, by the multiplication law of probability, $(\Omega^{-1}\Omega_i) \cdot \kappa_{i \to i+1} \delta t$. To maintain equilibrium, that jump probability must be equal to the probability $(\Omega^{-1}\Omega_{i+1}) \cdot \kappa_{i+1 \to i} \delta t$ for a jump the other way, from cell $i+1$ to cell i. Equating those two probabilities yields the relation (5.14) between the κ's. And since the κ's do not depend on whether or not the system is in equilibrium (i.e., on whether or not the system has "forgotten" its initial state), then Eq. (5.14) must always hold.

Another variation on the derivation of Eq. (5.12) yields a formula for κ_l that improves on the accuracy of the D/l^2 formula by taking account of the fact that solute molecules move *ballistically* on very small spatial scales. We will derive that more general formula in Section 5.6. But before doing that, we need to examine more thoroughly the constraints on both the cell length l and the time increment δt that have been tacitly assumed by Eqs (5.3) and (5.8).

5.5   Constraints on l and δt

We remarked earlier that hypothesis (5.3) makes some implicit approximating assumptions that are not entirely obvious, but which have important consequences. We will now discuss those assumptions in more detail. We will see that they impose constraints on both l and δt, essentially requiring each of those two variables to be neither too large nor too small.

To begin with, we have the Einstein requirement, discussed in Chapter 3, that δt must be *large* enough that a solute molecule will usually experience many collisions with solvent molecules in time δt. That must be true in order for the resulting random displacements of the solute molecule in successive δt intervals to be statistically independent and identically distributed. But the assertion of (5.3) that $\kappa_l \delta t$ is a "probability" will obviously not make sense unless δt is also *small* enough that $\kappa_l \delta t \leq 1$. In fact, the implicit assumption in Eq. (5.3) that in time δt we need *not* be concerned with diffusive transfers to *non-adjacent cells* imposes the much stronger requirement,

$$\kappa_l \delta t \ll 1. \qquad (5.15a)$$

One way to see why this must be true is to observe that, if the probability of a diffusive jump to a non-adjacent cell in the next δt is to be "very very small", then the probability of a diffusive jump to an adjacent cell had better be "very small"—and the latter condition is what (5.15a) is asserting. Another way to see the necessity of condition (5.15a) is to reason as follows: In order for it to be very unlikely that a net diffusive displacement greater than l will occur in the next δt, the mean-square diffusive displacement in time δt had better be very much less than l^2. Since the mean-square displacement in time δt is $2D\delta t$, this requirement can be written

$$2D\delta t \ll l^2. \tag{5.15b}$$

Dividing this through by l^2 and then invoking Eq. (5.8) yields $2\kappa_l\delta t \ll 1$, which is essentially the same as condition (5.15a). We conclude that the Einstein requirement, that δt be large enough that the random displacements of a solute molecule in time δt will be independent and identically distributed, must be met *without* taking δt so large that it violates condition (5.15a), which ensures that jumps to non-adjacent cells in time δt will practically never happen.

Suppose we have a δt that satisfies the opposing requirements just described. Then Eq. (5.15b) evidently becomes a constraint on l, requiring that l be "not too small". This is not hard to understand: the smaller l is, the easier it will be for diffusive transfers to non-adjacent cells to occur in time δt, which we want to avoid. This restriction against values for l that are too small implies that the mathematical limit $l \to 0$ cannot be taken in its strictest sense. In particular, *the limit $l \to 0$ cannot be taken in the expectation of obtaining an exact theoretical description of the physical process of diffusion*, such as for instance to make exact the derivative approximation leading to Eq. (5.4). But l must not be too large either. For, as was discussed in connection with Eq. (5.2), if l is taken too large, the density of the solute molecules within a single cell will not be uniform.

One might also worry that the cell width l might be so large that, immediately after a solute molecule has jumped from cell $i - 1$ to cell i, it will be closer to cell $i - 1$ than to cell $i + 1$, and thus more likely to jump next to cell $i - 1$ than to cell $i + 1$. That kind of jumping asymmetry is not allowed by hypothesis (5.3). But as we will see in Chapter 11, this turns out *not* to be a problem: We should really regard $\kappa_l\delta t$ in (5.3) as the probability that a solute molecule which is "well inside" cell $i - 1$ at time t will be, at time $t + \delta t$, "well inside" cell i.

We conclude that the physical fidelity of hypothesis (5.3) depends on it being possible to choose a temporal scale on which δt is neither too large nor too small, and also a spatial scale on which l is neither too large nor too small. If, for whatever reason, such a "Goldilocks" space-time scale cannot be found for a given system, then the ability of the discrete-stochastic approach to describe the behavior of that system in a physically accurate way will be compromised. We will shed a little more light on these matters in the next section, where we will derive a quantitative lower bound on l when taking $\kappa_l = D/l^2$, and still more light in Section 9.4 where we will deduce quantitative lower bounds on both δt and l.

5.6   A more accurate formula for κ_l

We saw in the preceding two sections that if the cell length l in the linear cell array of Fig. 5.1 is not too small, Einstein's theory of diffusion supports the hypothesis that the probability that a solute molecule will jump from its present cell to a particular adjacent cell in the next small time δt is equal to $\kappa_l\delta t$, where $\kappa_l = D/l^2$. In this section, we will derive a refinement of that formula for κ_l which holds over a wider range of l-values, and which allows us to quantify the "not-too-small-l" restriction on the result $\kappa_l = D/l^2$.

The derivation of this more general formula for κ_l is a modification of the argument used in Section 5.4 to derive Eq. (5.12). That argument modeled the jump of a solute molecule from cell i to cell $i+1$ as the diffusional journey of the molecule from the middle of cell i at point $x = x_i - \frac{1}{2}l$ to the middle of cell $i+1$ at point $x = x_i + \frac{1}{2}l$, and more specifically as a *steady-state* diffusional journey in which the solute molecule's PDF $q(x)$ was held at the uniform-inside-a-cell value $(Al)^{-1}$ at the initial point, and at the perfectly absorbing value 0 at the final point; see Eqs (5.10). With those boundary conditions, the steady-state Einstein diffusion equation mandated for $q(x)$ the straight line from point $\left(x_i - \frac{1}{2}l, (Al)^{-1}\right)$ to point $\left(x_i + \frac{1}{2}l, 0\right)$, as shown in Fig. 5.3. From that solution, we computed the single-molecule position probability flux in Eq. (5.11), $\hat{J}_x(x) \equiv -D\,dq(x)/dx$. Then we computed, using Eq. (3.25a), the probability that the solute molecule will be absorbed at $x = x_i + \frac{1}{2}l$ in the next δt. By equating that absorption probability to the jump probability $\kappa_l \delta t$, we finally inferred the result $\kappa_l = D/l^2$ in Eq. (5.12).

But as we found in Sections 4.6 and 4.7, the Einstein theory of diffusion fails to take account of the fact that the motion of a solute molecule over very small distances is *ballistic* instead of diffusional. An important result in that connection is Eq. (4.21). That equation implies that the probability that the solute molecule described by the PDF $q(x)$ will collide with the cross-section A at $x = x_i + \frac{1}{2}l$ in the next small time δt is equal to $\frac{1}{4} q(x_i + \frac{1}{2}l)\, \bar{v}\, A\, \delta t$; here,

$$\bar{v} = \sqrt{\frac{8k_{\mathrm{B}}T}{\pi m}} \qquad (5.16)$$

is the average speed of the molecule (of mass m) if it were in a dilute gas at absolute temperature T. This probability that the solute molecule will collide with the cross-section at $x = x_i + \frac{1}{2}l$ in the next δt ought to be the same as the probability $\kappa_l \delta t$ that the molecule will make its "jump" to $x = x_i + \frac{1}{2}l$ in the next δt:

$$\kappa_l \delta t = \frac{1}{4} q(x_i + \tfrac{1}{2}l)\, \bar{v}\, A\, \delta t. \qquad (5.17)$$

But here we encounter a difficulty: the perfectly absorbing assumption $q(x_i + \frac{1}{2}l) = 0$ in Eq. (5.10b), when combined with Eq. (5.17), implies that $\kappa_l = 0$, and that contradicts the result $\kappa_l = D/l^2$ in Eq. (5.12).

The way out of this difficulty is to relax our assumption (5.10b) that $q(x_i + \frac{1}{2}l)$ vanishes, and assume instead that $q(x_i + \frac{1}{2}l)$ has *some constant but as yet unspecified positive value*, which however is still less than the value $q(x_i - \frac{1}{2}l) = (Al)^{-1}$ in Eq. (5.10a). Doing that would have the effect of making the point $x_i + \frac{1}{2}l$ *partially* absorbing instead of perfectly absorbing. And that seems reasonable; the point $x_i + \frac{1}{2}l$ is not in reality a sink that immediately removes forever a molecule from Ω, but merely an ordinary interior point from which the solute molecule could next move to the left just as easily as to the right.

With this modification to our steady-state model of the $i \to i+1$ cell jump, the $q(x)$ profile in Fig. 5.3 becomes the profile shown in Fig. 5.5. So instead of Eq. (5.11), we now have for the single-molecule position probability flux throughout the interval $[x_i - \frac{1}{2}l, x_i + \frac{1}{2}l]$,

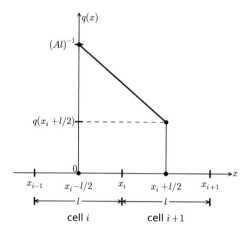

Fig. 5.5 *Modifying the model of Fig. 5.3 to obtain a formula for κ_l that is consistent with the fact that on sufficiently small spatial-temporal scales, a solute molecule moves ballistically, as in an ideal gas. We now allow $q(x_i + \frac{1}{2}l)$ to be some as yet unspecified positive value.*

$$\hat{J}_x(x) = -D\,\frac{q(x_i + \frac{1}{2}l) - q(x_i - \frac{1}{2}l)}{(x_i + \frac{1}{2}l) - (x_i - \frac{1}{2}l)} = D\,\frac{(Al)^{-1} - q(x_i + \frac{1}{2}l)}{l}. \tag{5.18}$$

As before, the result at Eq. (3.25a) implies that the probability that our solute molecule will be absorbed at the surface of area A through $x = x_i + \frac{1}{2}l$ in the next δt is $\hat{J}_x(x_i + \frac{1}{2}l) \cdot A \cdot \delta t$. When we equate that absorption probability to the probability $\kappa_l \delta t$ that the molecule will make the jump $i \to i+1$ in the next δt, we now obtain,

$$\kappa_l \delta t = D\,\frac{(Al)^{-1} - q(x_i + \frac{1}{2}l)}{l}\,A\,\delta t. \tag{5.19}$$

Equations (5.17) and (5.19) are *both* valid equations. The former arises from the ballistic motion of the solute molecule on small length scales, while the latter arises from the diffusional motion of the molecule on large length scales. Solving those two equations simultaneously for the two unknown quantities κ_l and $q(x_i + \frac{1}{2}l)$ is straightforward. The result is

$$q(x_i + \tfrac{1}{2}l) = \frac{4D}{4D + \bar{v}l}(Al)^{-1}, \tag{5.20}$$

$$\kappa_l = \frac{D\bar{v}/l}{4D + \bar{v}l}. \tag{5.21}$$

Equation (5.20) shows that $q(x_i + \frac{1}{2}l)$ is indeed always greater than zero, and indeed always less than $(Al)^{-1}$. But our main result here is Eq. (5.21). While that formula might seem strange at first sight, its plausibility becomes more apparent when we examine its approximate forms in the two cases $4D \ll \bar{v}l$ and $4D \gg \bar{v}l$: It is easy to see from Eq. (5.21) that

$$\kappa_l \approx \begin{cases} \dfrac{D}{l^2} & \text{if } 4D \ll \bar{v}l, \\[2mm] \dfrac{\bar{v}}{4l} & \text{if } 4D \gg \bar{v}l. \end{cases} \tag{5.22}$$

The upper part of Eq. (5.22) shows that, in order to take $\kappa_l = D/l^2$, the cell length must be "not too small" in the specific sense that

$$l \gg 4D/\bar{v}. \tag{5.23}$$

The lower part of Eq. (5.22) tells us that in the extreme opposite case $l \ll 4D/\bar{v}$, the length of the cell will be so small that the motion of the molecule inside a cell will be essentially ballistic. If we define $\kappa_l^{\text{diff}} \equiv D/l^2$ and $\kappa_l^{\text{ball}} \equiv \bar{v}/(4l)$, then it is easy to show from Eq. (5.21) that

$$\frac{1}{\kappa_l} = \frac{1}{\kappa_l^{\text{diff}}} + \frac{1}{\kappa_l^{\text{ball}}}. \tag{5.24}$$

If we view $1/\kappa$ as a "resistance" to the cell jump, then Eq. (5.24) can be viewed as a "resistors-in-series" heuristic analogous to Eq. (4.35).

The foregoing derivation of the ballistically refined version (5.21) of the formula $\kappa_l = D/l^2$ for cells of equal length l can be straightforwardly extended to derive ballistically refined versions of the formulas for $\kappa_{i \to i+1}$ and $\kappa_{i+1 \to i}$ in Eqs (5.13) for cells of unequal lengths l_i and l_{i+1}. The result in each case is the same formula (5.21), but with two replacements: the l in the *denominator* of Eq. (5.21) gets replaced by $\frac{1}{2}(l_i + l_{i+1})$; and the l in the *numerator* gets replaced by l_i for $\kappa_{i \to i+1}$, and l_{i+1} for $\kappa_{i+1 \to i}$. It is easy to verify that these ballistically refined formulas for $\kappa_{i \to i+1}$ and $\kappa_{i+1 \to i}$ still satisfy the ratio condition (5.14), as required for proper behavior at equilibrium.

Granting the greater range of validity of formula (5.21), it turns out that in most practical situations condition (5.23) will be satisfied, and the approximation $\kappa_l = D/l^2$ will be adequate. In what follows, we will assume that to be the case unless stated otherwise.

5.7 The discrete-stochastic model's version of Fick's Law

We found in Chapter 3 that Fick's Law, which plays such a central role in the classical model of diffusion in Chapter 1, emerges in Einstein's model of diffusion as simply the definition of the "probability flux" in Eq. (3.20). In this section we will elaborate on our derivation of formula (5.4) for the flux in an attempt to see more clearly how Fick's Law manifests itself in the discrete-stochastic model. We begin by defining the *net flux* $J_i(n_i, n_{i+1}; l, \delta t)$ of solute molecules in time δt in the positive i-direction at the boundary between cells i and $i+1$ by the following statement:

$$J_i(n_i, n_{i+1}; l, \delta t) \cdot A\delta t \equiv \text{the *net number* of solute molecules that will diffuse,}$$
in the next δt, from cell i to cell $i+1$, given that there are n_i molecules in cell i and n_{i+1} molecules in cell $i+1$. $\hspace{2em}$ (5.25)

Here, A is as usual the cross-sectional area of Ω. Notice that the quantity on the right side of Eq. (5.25) is an *integer random variable*; it is *not* an "average". But since the factor $A\delta t$ on the left side may not be an integer, then the net flux $J_i(n_i, n_{i+1}; l, \delta t)$ itself must be a *non-integer* random variable.

The definition (5.25) evidently makes the assumption that the net flux through the boundary between cells i and $i+1$ depends on the number of molecules in *only* cells i and $i+1$. Therefore, *definition (5.25) tacitly assumes that no solute molecule from any non-adjacent cell will cross that boundary in time δt.* As discussed earlier, this will be true only if δt is sufficiently small (and l is sufficiently large); therefore, δt in definition (5.25) is the same "macroscopic infinitesimal" that appears in hypothesis (5.3).

We might expect, on the basis of the deterministic version of Fick's Law, that the net flux $J_i(n_i, n_{i+1}; l, \delta t)$ would be independent of l and δt. But as we will see shortly, this is not entirely true.

To compute the quantity on the right side of Eq. (5.25), we introduce two new integer random variables $M_i^{\pm}(n_i, \delta t)$, which are defined as {the number of the n_i molecules in cell i that end up in cell $i \pm 1$ in the next δt}. (These new random variables $M_i^{\pm}$ are not to be confused with the total number of cells M.) The quantity on the right side of Eq. (5.25) is thus seen to be the difference

$$J_i(n_i, n_{i+1}; l, \delta t)A\delta t = M_i^+(n_i, \delta t) - M_{i+1}^-(n_{i+1}, \delta t).$$

The nature of the random variable $M_i^+(n_i, \delta t)$ can be deduced, at least to lowest order in δt, simply by taking note of the following fact: According to our hypothesis (5.3), each of the n_i solute molecules in cell i in effect "tosses a weighted coin" in order to decide, independently of the other solute molecules, whether or not it will move to cell $i+1$ in the next δt. With probability $\kappa_l \delta t$ the molecule *will* move to cell $i+1$ in the next δt, and with probability $1 - \kappa_l \delta t$ it *won't*, at least to lowest order in δt. This immediately implies, as we discussed in connection with the definition of the binomial random variable in Eq. (2.38), that the number $M_i^+(n_i, \delta t)$ of the n_i molecules in cell i that will move to cell $i+1$ in the next δt is the *binomial* random variable $\mathcal{B}(\kappa_l \delta t, n_i)$. Similarly, $M_{i+1}^-(n_{i+1}, \delta t)$, the number of solute molecules in cell $i+1$ that will move to cell i in the next δt, is the binomial random variable $\mathcal{B}(\kappa_l \delta t, n_{i+1})$. Therefore, the above formula for $J_i(n_i, n_{i+1}; l, \delta t)A\delta t$ becomes

$$J_i(n_i, n_{i+1}; l, \delta t)A\delta t = \mathcal{B}(\kappa_l \delta t, n_i) - \mathcal{B}(\kappa_l \delta t, n_{i+1}) + o(\delta t). \tag{5.26}$$

Now we take the mean of Eq. (5.26), with the help of Eqs (2.21a) and (2.39). That gives

$$\langle J_i(n_i, n_{i+1}; l, \delta t)A\delta t \rangle = \langle \mathcal{B}(\kappa_l \delta t, n_i) \rangle - \langle \mathcal{B}(\kappa_l \delta t, n_{i+1}) \rangle + o(\delta t)$$
$$= n_i \kappa_l \delta t - n_{i+1} \kappa_l \delta t + o(\delta t).$$

Thus we conclude that the *mean* net number of molecules moving from cell i to cell $i+1$ in the next δt is, to lowest order in δt,

$$\langle J_i(n_i, n_{i+1}; l, \delta t)A\delta t \rangle = (n_i - n_{i+1})\,\kappa_l \delta t + o(\delta t). \tag{5.27}$$

Similarly, taking the *variance* of Eq. (5.26) with the help of Eqs (2.21b) and (2.39) gives

$$\text{var} \left\{ J_i(n_i, n_{i+1}; l, \delta t) A \delta t \right\} = \text{var} \left\{ \mathcal{B}(\kappa_l \delta t, n_i) \right\} + \text{var} \left\{ \mathcal{B}(\kappa_l \delta t, n_{i+1}) \right\} + o(\delta t)$$

$$= \kappa_l \delta t (1 - \kappa_l \delta t) n_i + \kappa_l \delta t (1 - \kappa_l \delta t) n_{i+1} + o(\delta t).$$

This, to lowest order in δt, is

$$\text{var} \left\{ J_i(n_i, n_{i+1}; l, \delta t) A \delta t \right\} = (n_i + n_{i+1}) \kappa_l \delta t + o(\delta t). \tag{5.28}$$

Equation (5.27) gives the *average net number* of solute molecules moving from cell i to cell $i + 1$ in the next δt, and Eq. (5.28) quantifies the fluctuations about that average by giving the associated variance. Together, those two formulas imply that we should expect to see a *net* transfer of, on average, $(n_i - n_{i+1}) \kappa_l \delta t$ solute molecules from cell i to cell $i + 1$, *and* we should expect to see *fluctuations* about that average of order $\pm \sqrt{(n_i + n_{i+1}) \kappa_l \delta t}$. Notice that these fluctuations might well be substantial. For example, if $n_i \approx n_{i+1} \approx n$, then the mean net number transferred will, according to Eq. (5.27), be ≈ 0, but the fluctuations will be of order $\pm \sqrt{2 n \kappa_l \delta t}$, which will be large if n is large. Furthermore, when $n_i < n_{i+1}$, the fluctuations could cause the "up-gradient" transfers from cell i to cell $i + 1$ in δt to outnumber the "down-gradient" transfers from cell $i + 1$ to cell i, even though *on average* the latter will dominate. In any case, since the net number of molecules crossing from cell i to cell $i + 1$ in time δt is in principle experimentally measureable, Eqs (5.27) and (5.28) are amenable to direct experimental testing.

Equations (5.27) and (5.28) together constitute a discrete-stochastic generalization of Fick's Law, at least within the context of the discrete-stochastic model. To better appreciate this point, first divide Eq. (5.27) through by $A\delta t$ to get

$$\langle J_i(n_i, n_{i+1}; l, \delta t) \rangle = \kappa_l \frac{(n_i - n_{i+1})}{A} + \frac{o(\delta t)}{\delta t}.$$

Next, eliminate κ_l using formula (5.8), and finally let $\rho_i \equiv n_i / Al$ denote the *average molecular density* of solute molecules in cell i. The result is

$$\langle J_i(n_i, n_{i+1}; l, \delta t) \rangle = -D \left(\frac{\rho_{i+1} - \rho_i}{l} \right) + \frac{o(\delta t)}{\delta t}. \tag{5.29}$$

The quantity in parentheses on the right of Eq. (5.29) is the discretized gradient of the average molecular density at the interface between cells i and $i + 1$; thus, Eq. (5.29) has essentially the form of Fick's Law (1.2). Furthermore, Eq. (5.29) shows that if l and δt are both taken *small enough* that we can make the replacements $(\rho_{i+1} - \rho_i)/l \rightarrow \partial \rho / \partial x$ and $o(\delta t)/\delta t \rightarrow 0$, then $\langle J_i(n_i, n_{i+1}; l, \delta t) \rangle$ will be *independent of both l and δt*. But that convenient and intuitively attractive property of the *mean* of $J_i(n_i, n_{i+1}; l, \delta t)$ does *not* extend to its *variance*.

To prove this last statement, divide Eq. (5.28) through by $(A\delta t)^2$ and use the fact, from Eq. (2.19b), that $\text{var}\{\alpha X\} \equiv \alpha^2 \text{var}\{X\}$:

$$\text{var}\left\{J_i(n_i, n_{i+1}; l, \delta t)\right\} = \kappa_l \frac{(n_i + n_{i+1})}{A^2 \delta t} + \frac{o(\delta t)}{(\delta t)^2}.$$

With $\kappa_l = D/l^2$ and $\rho_i \equiv n_i/Al$, this becomes

$$\text{var}\left\{J_i(n_i, n_{i+1}; l, \delta t)\right\} = D\frac{(\rho_i + \rho_{i+1})}{Al\delta t} + \frac{o(\delta t)}{(\delta t)^2}. \tag{5.30}$$

Equation (5.30) is obviously *not* well behaved in either of the limits $l \to 0$ or $\delta t \to 0$. The *non-zero lower bounds* on l and δt discussed in Section 5.5 thus serve the useful purpose of preventing $\text{var}\{J_i(n_i, n_{i+1}; l, \delta t)\}$ from becoming undefined. This untoward behavior of $\text{var}\{J_i(n_i, n_{i+1}; l, \delta t)\}$ can be traced to the fact that the flow of matter described by the flux is not the continuum flow envisioned in the classical approach to diffusion, but rather a flow of discrete packets of matter, namely molecules. Equation (5.30) thus describes the breakdown of the traditional "rate" view of diffusional flux on very small spatial and temporal scales.

5.8   Does the concentration gradient "cause" diffusion?

For the one-dimensional diffusion problem we have been considering, Fick's Law (1.2) implies that in an infinitesimal time dt there will be a net transfer of, on average,

$$J_x dt = -D\frac{\partial\rho}{\partial x}dt \tag{5.31}$$

molecules in the x-direction across a unit area normal to the x-axis. In a different vein, a famous law of physics makes a statement that can be phrased in a mathematically similar way: Newton's second law $F_x = m(dv_x/dt)$ implies that, if a particle of mass m moves on the x-axis in a conservative force field $F_x \equiv -\partial U/\partial x$, where U is the potential energy, then in an infinitesimal time dt the x-component of the velocity of the particle will change by the amount

$$dv_x = -\frac{1}{m}\frac{\partial U}{dx}dt. \tag{5.32}$$

The change in the particle's velocity described by Eq. (5.32) is generally viewed as being *caused by* the force $-\partial U/\partial x$. The mathematically similar form of Eq. (5.31) thus tempts us to similarly regard the average net flow on the left side of Eq. (5.31) as being "caused by" the concentration gradient $-\partial\rho/\partial x$ on the right side.

But a moment's reflection will reveal that the notion that the concentration gradient is the *cause* of diffusion, and the concomitant notion that in the *absence* of a concentration gradient there will be *no* diffusion, are physically erroneous. To see why, notice that the number of molecules transferred in time dt in Eq. (5.31) is qualified by the two words "average" and "net". The first implies that *variations* in that number should be expected; the second implies that that number is the number

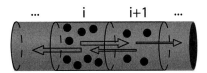

Fig. 5.6 *How the flux originates. Here cell i contains twice as many solute molecules as cell i + 1. We know that any individual solute molecule will be equally likely to jump next to the right or to the left, independently of what the other solute molecules do. It follows that the average flux of solute molecules out of the left and right boundaries of cell i (the two thick arrows) will be twice as large as the average flux out of the left and right boundaries of cell i + 1 (the two thin arrows). Therefore, the* net *flux across the boundary between cells i and i + 1 (the signed sum of the two arrows in the middle) will be to the right—in the direction of decreasing solute molecule concentration. But this average net flow in the direction of decreasing solute molecule concentration is evidently* not *the result of the concentration gradient actively encouraging individual solute molecules to move down the gradient. It is simply an incidental consequence of the independent isotropic jumping tendencies of the individual molecules coupled with the population imbalance between the two cells.*

of molecules transported in the positive x-direction *minus* the number transported in the negative x-direction. These qualifications mean that diffusing molecules are not affected by the concentration gradient $-\partial\rho/\partial x$ in the same inexorable way that a mass particle is affected by the potential energy gradient $-\partial U/\partial x$.

Physically, we know that the diffusional movement of a solute molecule is caused *solely* by collisional impacts of that molecule with surrounding solvent molecules. And since in "simple" diffusion the solute molecules move about independently of each other, then *the motions of the solute molecules are not influenced in any way by their concentration gradient.* It is true that formulas (5.27) and (5.29) both imply that the direction of the net average flow of solute molecules is always from regions of higher concentration to regions of lower concentration. But, as a careful consideration of Fig. 5.6 will make clear, this average trend is merely an incidental consequence of the independent motions of the solute molecules in the presence of a concentration gradient: Any given solute molecule is equally likely to jump to the left or to the right in the next δt, regardless of the presence or direction of a concentration gradient. If there *is* a concentration gradient at some point, say a *decreasing* gradient as in Fig. 5.6 which simply reflects the fact that there are more molecules to the left of that point than to the right, then *on average* there will be more molecules diffusing left-to-right than right-to-left. But clearly, the concentration gradient is not what is "causing" the diffusional motion of the solute molecules. And the diffusional motion of the solute molecules is not going to cease if the concentration gradient goes away.

5.9 A microfluidics diffusion experiment

We will now describe a laboratory experiment that was performed on a real physical system of the kind we have been discussing. The quasi-one-dimensional volume Ω of this system is housed in a microfluidic chip, which was designed and constructed using techniques of soft lithography and chip fabrication. In this system, the solvent molecules are water molecules, and the role of the solute molecules is played by green-fluorescent polystyrene microspheres, or beads, of radius 0.145 μm. The system is kept at room temperature ($T = 300$ K). Since polystyrene has a density of 1.05 g/cm^3, the beads "just sink" in water. The experiment affords a way to test, directly or indirectly, many assumptions and results of the theory developed thus far, including the classical diffusion equation, the kappa hypothesis (5.3), the relation $D = \kappa_l l^2$, and the generalized Fick's Law equations (5.27) and (5.28). We will later use the system described here as a basis for performing some stochastic simulations in Chapters 6 and 9. Comparison of those simulated results with the laboratory results described here will further test and illustrate the theory.[2]

The specially fabricated microfluidic chip used in this experiment consists of two layers. The lower layer, where the diffusional motion of the beads takes place, is called the "flow layer". The upper layer is called the "control layer". As shown schematically in Fig. 5.7a, the lower flow layer is a horizontal channel of width 100 μm. The upper control layer consists of three vertical channels or "valves", which are used to control the flow of the sample in the lower flow layer. Each control valve expands in a balloon-like manner when pressurized, and that expansion causes the cross–section of the valve to change from a flat rectangular shape to a circular shape. The consequent intrusion of the expanded valve into the lower flow channel forms a barrier there. In order for this action to completely close the flow channel, the flow channel is fabricated so that its cross–section is a segment of a circular disk of maximum depth 10 μm with its flat side facing down, as shown in Fig. 5.7b. Two "outer" control valves are spaced 400 μm apart; when pressurized to their closed state, they form the two ends of the system volume Ω. A "middle" control valve functions as a removable partition, which can be inserted to divide the system in half by pressurizing it to its closed state, or removed by deflating it to its open state.

The experiment was prepared by first closing the middle control valve and opening the two outer control valves, and then filling the left half of the chamber with pure water and the right half with a water suspension of approximately 200 beads. The two outer control valves were then closed to form the cylindrical volume Ω of length $L = 400$ μm and cross–section shown in Fig. 5.7b. After the beads in the right half of Ω had equilibrated to a randomly uniform distribution, which *on average* is the cell population profile shown by the heavy curve in Fig. 5.8b, the middle control valve was opened at time $t = 0$, and the fluorescent beads were then observed through an inverted fluorescence microscope as they diffused freely throughout Ω.[3]

The parameters of the experiment were carefully chosen to make the system dilute in the solute molecules: The total number of beads was such that the average distance between nearest neighbors in the initial right half of Ω was approximately 8 μm, which is 27 times the bead diameter. At that separation, electrostatic and hydrodynamic forces between the beads are negligible.[4] Therefore, to a good approximation, the beads should move about in the flow layer independently of one another. The diluteness of the beads also ensures that beads will only rarely occlude each other when viewed through the microscope—an important consideration for accurately observing the diffusional motion of each bead.

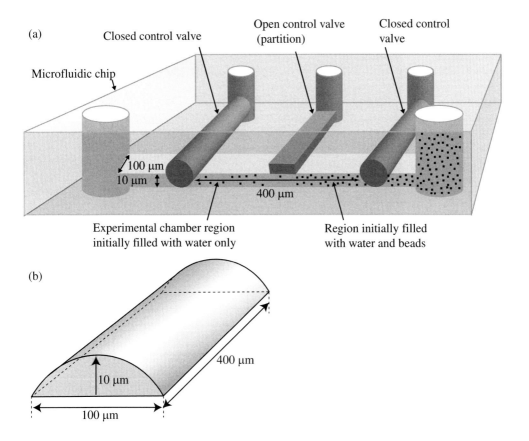

(a)

Closed control valve

Open control valve
(partition)

Closed control
valve

Microfluidic chip

100 μm
10 μm

400 μm

Experimental chamber region
initially filled with water only

Region initially filled
with water and beads

(b)

400 μm

10 μm

100 μm

Fig. 5.7 *Set-up for the microfluidics diffusion experiment. (**a**) The lower level of the microfluidic chip houses the experimental chamber, in which green-fluorescent polystyrene beads of radius 0.145 μm diffuse in a sea of water molecules. The upper level of the chip houses three control valves, the outer two defining (when closed as shown) the ends of the system volume Ω, and the middle one (shown open) serving as a removable partition between the left and right halves of Ω. The large vertical chamber on the right serves as a bead reservoir. The system is prepared by first closing the center valve, and then opening the left and right valves to fill the left half of the chamber with water and the right half of the chamber with a solution of beads in water. The left and right valves are then closed, and the beads in the right half of the chamber are allowed to become randomly uniform there. Then, at time $t = 0$, the center valve is opened, and the beads are allowed to diffuse freely throughout the portion of the experimental chamber between the closed left and right control valves. The motion of the beads is tracked in time by a computer controlled microscope-camera-recording apparatus that records the cell location of every bead every $\Delta t = 10$ seconds. (**b**) The actual shape of the system volume Ω. The curved upper surface allows for more effective functioning of the control valves.* (Reprinted with permission from E. Seitaridou et al. (2007), Journal of Physical Chemistry B **111**:2288–2292. © 2007 American Chemical Society.)

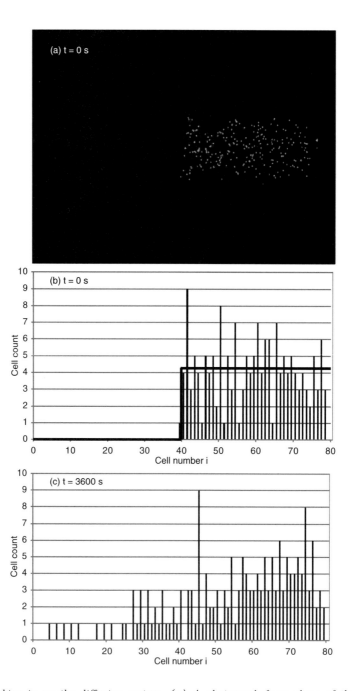

Fig. 5.8 *Looking in on the diffusing system.* **(a)** *A photograph from above of the fluorescent beads in the flow chamber just before the middle control valve is opened at time 0.* **(b)** *A typical histogram of the cell populations at time 0. The superimposed heavy curve shows the theoretically expected average of many repeated observations of such initial cell counts.* **(c)** *A typical histogram of the cell populations one hour later. (Reprinted with permission from E. Seitaridou et al. (2007),* Journal of Physical Chemistry B **111***:2288–2292.* © *2007 American Chemical Society.)*

Data acquisition was accomplished by taking snapshots with a camera mounted on the microscope. The snapshots were taken at successive time intervals of $\Delta t = 10\,\text{s}$ following the removal of the partition at time $t = 0$. Since there was a possibility that some beads might be momentarily overlapping or out of focus, each observation actually consisted of three rapid snapshots by the camera. The three images were then compared and combined, and eventually replaced with a single image that most clearly located each bead at that nominal instant. The data, therefore, consisted of a set of N_{im} images, which correspond to $(N_{\text{im}} - 1)$ 10-second time intervals, and which essentially provide a "motion picture" of the unfolding diffusion process. An example of a $t = 0$ snapshot is shown in Fig. 5.8a.

The immediate objective of the experiment was to measure the quantities

$N_i(t_k) \equiv$ the number of beads in cell i at observation time $t_k \equiv (k-1) \cdot \Delta t$,

$$(i = 1, \ldots, M; \ k = 1, \ldots, N_{\text{im}}). \quad (5.33)$$

These measured values can be regarded as sample values of the random variables $N_i(t \mid \mathbf{n}_0)$ in Eq. (5.1), except that here the initial state $\mathbf{n}_0$ is randomly sampled from the step-function distribution indicated by the heavy curve in Fig. 5.8b. To obtain the values (5.33), each of the N_{im} images was imported into a numerical computing environment for image processing, and a computerized centroid-tracking algorithm was then used to determine the position of each of the 200 beads. Next, the 400 μm by 100 μm flow chamber was (mathematically) partitioned along its long axis (the x-axis) into $M = 80$ cells of equal length $l = 5$ μm, and the number of beads in each cell was determined for each image. The histogram in Fig. 5.8b shows, for a typical run of the experiment, the initial ($t_1 = 0$) cell population profile $\{N_1(0), \ldots, N_{80}(0)\}$. Figure 5.8c shows what the population profile looked like one hour later.

In Fig. 5.9, the jagged curve shows the cell population profile 90 minutes after the partition was removed. The quantity actually plotted here is $N_i(5400\,\text{s})/l$, the number of beads per unit length in cell i. Since $l \ll L$, this quantity is essentially the density of the beads in cell i. According to theory, that density, when averaged over many runs to time $t = 5400\,\text{s}$, should approximate $\rho(x_i, 5400\,\text{s})$, where $\rho(x, t)$ is the function in the classical diffusion equation (1.4), and $x_i \equiv (i - \tfrac{1}{2})l$ locates the center of cell i. Now, we found in Example 5 of Section 1.2 that the solution of the diffusion equation (1.4) on the interval $0 \le x \le L$ for the reflecting boundary conditions (1.16) and the initial condition (1.17), which conditions describe our problem here, is given by Eq. (1.30), which we will rewrite here:

$$\rho(x, t) = \frac{\rho_0}{2} + \sum_{n=1(\text{odd})}^{\infty} (-1)^{\frac{n+1}{2}} \frac{2\rho_0}{n\pi} \cos\left(\frac{n\pi x}{L}\right) e^{-\left(\frac{n\pi}{L}\right)^2 Dt}. \quad (5.34)$$

Note that of the three parameters ρ_0, L, and D appearing in this formula, only D is experimentally unknown. Superimposed on the histogram in Fig. 5.9 are *two smooth curves*, which to the resolution of the figure exactly overlay each other. One curve is the best least-squares fit of the function (5.34) to the single-run jagged curve as obtained using D as the sole fitting parameter. The other curve is a plot of Eq. (5.34)

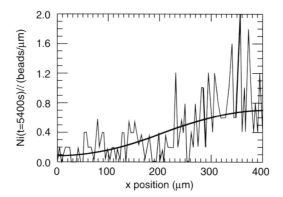

Fig. 5.9 *Comparing experimental results with theoretical predictions for the bead density. The jagged curve shows the cell population profile 90 minutes after the center partition was removed. The quantity actually plotted here is the number of beads in cell i divided by l, which is essentially the density of the beads in cell i. According to theory, that density, when averaged over many runs to time $t = 5400\,\mathrm{s}$, should approximate $\rho(x_i, 5400\,\mathrm{s})$, where $\rho(x,t)$ is the function (5.34) that we obtained in Example 5 of Chapter 1 by solving the classical diffusion equation for this set-up. The smooth curve is the best least-squares fit of the function (5.34) to the jagged curve, obtained using D as the only fitting parameter. That smooth curve is exactly overlaid, to within the resolution of the figure, by a second curve that plots Eq. (5.34) using $D = 1.51\ \mu m^2/s$, the value of the diffusion coefficient of the beads predicted by the Einstein–Stokes formula (which we will discuss in Section 9.7). That exact overlay implies experimental confirmation of the underlying theory.*

with $D = 1.51\ \mu m^2/s$, which is the value predicted by the Einstein–Stokes formula (which will be discussed in Sec. 9.7).[5] The excellent agreement between these two curves gives experimental confirmation of the underlying theory.[6]

The experiment also sought to test some of the *statistical* aspects of the stochastic version of Fick's Law in Eqs (5.27) and (5.28). To describe how that was done, let A denote the cross-sectional area of the volume Ω, and let AJ_i denote the *total flow* of beads across the boundary between cells i and $i+1$ at time t_k; i.e., the product of AJ_i with Δt gives, by definition,

$$AJ_i(N_i(t_k), N_{i+1}(t_k); l, \Delta t)\Delta t \equiv \text{the net number of beads that diffuse, in}$$
$$\text{time } [t_k, t_k + \Delta t), \text{ from cell } i \text{ to cell } i+1$$
$$(i = 1, \ldots, M-1; \ k = 1, \ldots, N_{\mathrm{im}}). \quad (5.35)$$

The index i here does not run to M, since there is no transfer of beads across the right boundary of cell M. This net flow AJ_i is of course just the net flux J_i defined in Eq. (5.25) multiplied by A, *provided Δt can be regarded as a macroscopic infinitesimal δt*. This constraint on Δt is implicit in the definition (5.35), via its tacit assumption that the net number of beads crossing the boundary between cells i and $i+1$ in time $[t_k, t_k + \Delta t)$ depends on the bead populations of *only* cells i and $i+1$ at time t_k.

That's because only when Δt is sufficiently small can we neglect the flow across that boundary of beads originating in such non-adjacent cells as cells $i-1$ and $i+2$.

Assuming that Δt is small enough to qualify as a "δt", it follows upon dividing Eq. (5.27) through by δt, and Eq. (5.28) through by $(\delta t)^2$, that the *mean* and *variance* of AJ_i are given by

$$\langle AJ_i(N_i(t_k), N_{i+1}(t_k); l, \Delta t) \rangle = \kappa_l \left(N_i(t_k) - N_{i+1}(t_k) \right) + \frac{o(\Delta t)}{\Delta t} \qquad (5.36a)$$

and

$$\mathrm{var}\left\{ AJ_i(N_i(t_k), N_{i+1}(t_k); l, \Delta t) \right\} = \left(\frac{\kappa_l}{\Delta t} \right) \left(N_i(t_k) + N_{i+1}(t_k) \right) + \frac{o(\Delta t)}{(\Delta t)^2}, \qquad (5.36b)$$

where as usual $\kappa_l = D/l^2$.

Now, according to Eq. (5.15a), the macroscopic infinitesimal δt should satisfy

$$\delta t \ll \frac{1}{\kappa_l} = \frac{l^2}{D} = 16.6\,\mathrm{s}, \qquad (5.37)$$

where in the last step we have used the values $l = 5\ \mu\mathrm{m}$ and $D = 1.51\ \mu\mathrm{m}^2/\mathrm{s}$ appropriate to this experiment. But this condition is only *weakly* satisfied by the observational time interval $\Delta t = 10\,\mathrm{s}$ used in the experiment. To see why that might cause a problem with regard to formulas (5.36), notice that with $\kappa_l = 1/16.6 = 0.060\,\mathrm{s}^{-1}$, the probability for a bead to jump from its present cell to a particular adjacent cell in the next $\Delta t = 10\,\mathrm{s}$ will be $\kappa_l \Delta t = 0.6$. That probability is so large that we cannot safely ignore the possibility of some bead diffusing to a *non-adjacent* cell in the next $\Delta t = 10\,\mathrm{s}$, a circumstance that is not allowed for in the derivation of the flux formulas (5.36). But since $\Delta t = 10\,\mathrm{s}$ is at least *less* than the limit in (5.36), it will be interesting to see how well the theoretical predictions (5.36) fare here.

The possibility of experimentally testing Eqs (5.36) arises from the fact that the data set (5.33) *implicitly* contains many sample values of the random variables $AJ_i(N_i(t_k), N_{i+1}(t_k))$. (We will suppress for now the dependency of AJ_i on Δt and l.) This is a consequence of the definitions (5.33) of N_i and (5.35) of AJ_i. Taken together, those two definitions imply that

$$N_i(t_k + \Delta t) - N_i(t_k) = AJ_{i-1}\left(N_{i-1}(t_k), N_i(t_k) \right) \Delta t - AJ_i\left(N_i(t_k), N_{i+1}(t_k) \right) \Delta t$$

$$(i = 1, \ldots, M-1;\ k = 1, \ldots, N_{\mathrm{im}} - 1). \qquad (5.38)$$

This is simply the statement that {the increase in the number of beads inside cell i between times t_k and $t_k + \Delta t \equiv t_{k+1}$} must be equal to {the net number of beads that *enter* cell i through its *left* boundary in that time} *minus* {the net number of beads that *leave* cell i through its *right* boundary in that time}. Since there can be no flow of beads through the left boundary of cell 1, then in the $i = 1$ version of Eq. (5.38) it is to be understood that

$$AJ_0 \equiv 0. \qquad (5.39a)$$

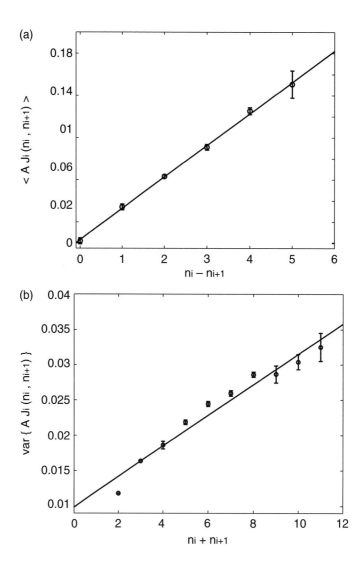

Fig. 5.10 *Comparing experimental results with theoretical predictions for the bead flux. With A denoting the cross-sectional area of the volume Ω, and n_i the current number of beads in cell i, we examine here the mean and variance of $AJ_i(n_i, n_{i+1})$, the average net number of beads that move from cell i to cell $i+1$ during the next $\Delta t = 10$ s.* **(a)** *A plot of the mean of $AJ_i(n_i, n_{i+1})$ versus $(n_i - n_{i-1})$, with the circles and error bars showing experimentally measured values, and the line showing the best least-squares straight-line fit to those values. The experimental results evidently support the theoretical prediction of Eq. (5.36a) that the mean of the net flow of beads between two adjacent cells is proportional to the difference between their bead populations.* **(b)** *A plot of the variance of $AJ_i(n_i, n_{i+1})$ versus $(n_i + n_{i-1})$, with the circles and error bars showing experimentally measured values, and the line showing the best least-squares straight-line fit to those values. The theoretical prediction of Eq. (5.36b), that the variance of the net flow of beads between two adjacent cells is proportional to the sum of their bead populations, is roughly confirmed by the data. We will revisit these experimental results in Chapter 6 (see Figs 6.8 though 6.10). (Reprinted with permission from E. Seitaridou et al. (2007), Journal of Physical Chemistry B* **111**:2288–2292. © *2007 American Chemical Society.)*

Solving Eq. (5.38) for AJ_i yields

$$AJ_i\left(N_i(t_k), N_{i+1}(t_k)\right) = AJ_{i-1}\left(N_{i-1}(t_k), N_i(t_k)\right) - \frac{N_i(t_{k+1}) - N_i(t_k)}{\Delta t}$$

$$(i = 1, \ldots, M-1; \; k = 1, \ldots, N_{\text{im}} - 1). \quad (5.39b)$$

Equations (5.39) evidently form a *recursion relation* from which all the AJ_i values can be computed, in the order $i = 1, 2, \ldots, M-1$, from the measured N_i values.

The flux analysis therefore proceeds as follows: From the matrix of experimentally measured cell population values

$$n_{i,k} \equiv N_i(t_k) \quad (i = 1, \ldots, M; k = 1, \ldots, N_{\text{im}}), \quad (5.40a)$$

we first compute, from Eqs (5.39) the matrix of total flux values

$$f_{i,k} \equiv AJ_i\left(N_i(t_k), N_{i+1}(t_k)\right) \quad (i = 1, \ldots, M-1; k = 1, \ldots, N_{\text{im}} - 1). \quad (5.40b)$$

Then, to test the theoretical formula (5.36a) for the *mean* of AJ_i, we let d (for "difference") be some fixed integer, and we scan the matrix $\{n_{i,k}\}$ to find all pairs $(n_{i,k}, n_{i+1,k})$ satisfying $n_{i,k} - n_{i+1,k} = d$. For each such pair, we record the corresponding value of $f_{i,k}$. Then we compute the *average* of those $f_{i,k}$ values, $\langle f_d \rangle$, and we plot $\langle f_d \rangle$-versus-d. Figure 5.10a shows the plotted results obtained in this experiment for d ranging over the 11 integers between -5 and 5. In this plot, the value of $f_{i,k}$ for each *negative* d-value was multiplied by -1 and then averaged with the $f_{i,k}$'s from the corresponding positive d-values. The respectable linear-in-d fit, shown by the line, confirms the theoretical prediction of Eq. (5.36a) that the *mean* of the net flux between two cells is proportional to the *difference* between their bead populations.

The procedure for testing formula (5.36b) for the *variance* of AJ_i is similar: Letting s (for "sum") be some fixed positive integer, we scan the matrix $\{n_{i,k}\}$ to find all pairs $(n_{i,k}, n_{i+1,k})$ satisfying $n_{i,k} + n_{i+1,k} = s$. For each such pair we record the corresponding value of $f_{i,k}$. Then we compute the *variance* of those $f_{i,k}$ values, var$\{f_s\}$, and we plot var$\{f_s\}$-versus-s. Figure 5.10b shows the plotted results obtained in the experiment for s ranging over the 10 integers between 2 and 11. The linear-in-s fit indicated by the line is not quite as good as the straight–line fit in Fig. 5.10a. Whether this can be attributed to the fact that Δt did not fully satisfy condition (5.37) is not clear at this juncture. Nevertheless, Fig. 5.10b broadly confirms the theoretical prediction of Eq. (5.36b) that the *variance* of the net flow of beads between two cells is proportional to the *sum* of their bead populations.

Additional insight into these experimental flux results will emerge from the numerical simulation studies that will be described in Section 6.11 (in the discussion of Figs 6.8 through 6.10).

Notes to Chapter 5

[1]We are invoking here the *addition* law of probability for *mutually exclusive* events: Adding n_i equal probabilities just multiplies that probability by n_i. The events in

question here are mutually exclusive because dt has been taken so small that it is virtually impossible for more than one solute molecule to leave cell i in the next dt.

[2]Details of the experimental set-up and procedures can be found in the paper "Measuring flux distributions for diffusion in the small-numbers limit", *Journal of Physical Chemistry B* **111**:2288–2292 (2007), by E. Seitaridou, M. Inamdar, R. Phillips, K. Ghosh, and K. Dill. This paper also describes how the population and flux data obtained in the experiment can be used to investigate several other intriguing aspects of diffusional motion, such as for instance measuring the fraction of the beads that move "against" the prevailing concentration gradient.

[3]One might wonder if the sudden deflation of the middle control valve, whose width inside the flow channel is much larger than the cell width l, introduces convective mixing of the beads. That this should *not* happen was concluded theoretically on the basis of the Navier–Stokes equation, and confirmed experimentally via direct observation through the microscope. It turns out that the sudden deflation of the middle control valve approximates rather well the sudden disappearance of a thin partition between cells 40 and 41.

[4]A detailed argument justifying this claim can be found in the paper cited in Note 2, but in summary: Electrostatic forces arise between the beads from a small negative surface charge on each bead, which the bead manufacturer estimates to be about $0.05\,C/m^2$. Using the Poisson–Boltzmann equation, it can be shown that a sphere of diameter $0.29\,\mu m$ with a surface charge density of $0.05\,C/m^2$ will appear neutral at a distance of approximately 1 *nanometer* above its surface; therefore, at the micrometer separations between the beads typical of this experiment, the electrostatic forces between the beads should be insignificant. As for hydrodynamic interactions between the beads, the hydrodynamic correction to the diffusion coefficient D predicted by the Einstein–Stokes equation turns out to be only about 2% for a sphere of diameter $0.29\,\mu m$ for separations between beads of $8\,\mu m$.

[5]In all numerical evaluations of the function (5.34), the infinite sum there was truncated at 200 terms (i.e., $n = 399$). That truncation gave an accurate representation of the known step-function form of $\rho(x, t = 0)$, namely the heavy curve in Fig. 5.8b, so it was reasoned that it should be more than adequate for computing the subsequent smoother curves.

[6]In the experiment, this least-squares curve-fitting procedure was performed every half hour for 5 hours, giving 10 slightly different values for the diffusion coefficient D. Averaging those values gave the result $D = (1.3 \pm 0.27)\mu m^2/s$, which is consistent with the value $1.51\,\mu m^2/s$ predicted by the Einstein–Stokes formula.

6

Master equations and simulation algorithms for the discrete-stochastic approach

Hypothesis (5.3), which defines the dynamics of the discrete-stochastic model of diffusion, implies that the system's state $\mathbf{N}(t|\mathbf{n}_0) \equiv \{N_1(t|\mathbf{n}_0), \dots, N_M(t|\mathbf{n}_0)\}$ defined in Eq. (5.1) evolves in time as what is technically called a jump Markov process. In general, the time evolution of such a process is governed by what is called a master equation. In this case, the assumption that the solute molecules move about independently of each other gives rise to *two* master equations, one for a single solute molecule, and one for the entire collection of solute molecules. In this chapter, we will derive both of those master equations and examine some of their consequences. We will also derive the two stochastic simulation algorithms that complement these master equations. We will illustrate those simulation algorithms by applying them to the microfluidic diffusion experiment discussed in Section 5.9. We will show how such simulations can be used to further test the stochastic version of Fick's Law that was developed in Section 5.7.

6.1   The single-molecule diffusion master equation

The single-molecule master equation focuses on the probability function $p(i, t|i_0)$, which is defined as the probability that a given solute molecule, which was known to be in cell i_0 at time 0, will at the later time $t > 0$ be found in cell i. The laws of probability, together with the hypothesis (5.3), require that this function satisfies

$$p(i, t + \delta t|i_0) = \varepsilon_{i1} p(i - 1, t|i_0) \cdot \kappa_l \delta t + \varepsilon_{iM} p(i + 1, t|i_0) \cdot \kappa_l \delta t$$
$$+ p(i, t|i_0) \cdot (1 - \varepsilon_{i1} \kappa_l \delta t - \varepsilon_{iM} \kappa_l \delta t) + o(\delta t). \tag{6.1}$$

Here we have defined

$$\varepsilon_{ij} \equiv \begin{cases} 0, & \text{if } i = j \\ 1, & \text{if } i \neq j. \end{cases} \tag{6.2}$$

In Eq. (6.1), the first term on the right is the probability that the molecule will be in cell $i - 1$ at time t and then will jump to cell i in the next δt; that probability will of course be 0 if $i = 1$. The second term is the probability that the molecule will be in cell $i + 1$ at time t, and then will jump to cell i in the next δt; that probability will be 0 if $i = M$. The third term is the probability that the molecule will be in cell i at time t and then will *not* jump across either boundary of cell i in the next δt. Finally, the term $o(\delta t)$ accounts for higher order terms in the aforementioned probabilities due to the term $o(\delta t)$ in Eq. (5.3), as well as for probabilities of other pathways to state i at time $t + \delta t$ that entail more than one jump in the time interval $[t, t + \delta t)$. Subtracting $p(i, t|i_0)$ from both sides of Eq. (6.1), dividing through by δt, and then taking the limit $\delta t \to 0^+$, we obtain the *single-molecule diffusion master equation*:

$$\frac{dp(i, t|i_0)}{dt} = \kappa_l \left[\varepsilon_{i1} p(i - 1, t|i_0) - (\varepsilon_{i1} + \varepsilon_{iM}) p(i, t|i_0) + \varepsilon_{iM} p(i + 1, t|i_0) \right]$$

$$(t \geq t_0; i = 1, \ldots, M). \quad (6.3)$$

Notice that the limit $\delta t \to 0^+$ is not a "true" mathematical limit, because δt must always be large enough that the solute molecule will typically experience many collisions with solvent molecules in time δt. The legitimacy of Eq. (6.3) thus hinges on, among other things, whether it is possible to find a range of δt-values that, on the one hand, is large enough to encompass many collisions with the solvent molecules, and on the other hand is small enough to give a good approximation to the derivative on the left and kill the $o(\delta t)/\delta t$ terms on the right.

6.2 Relation to the Einstein model of diffusion

In Section 5.4, we found that consistency between the discrete-stochastic model of diffusion and the Einstein model of diffusion requires that we take $\kappa_l = D/l^2$. We will now use that result and the single-molecule master equation (6.3) to establish a little more definitively the connection between these two models of diffusion.

With $\kappa_l = D/l^2$, we can write Eq. (6.3) for any *interior* cell i $(2 \leq i \leq M - 1)$ as

$$\frac{dp(i, t|i_0)}{dt} = \frac{D}{l^2} [p(i - 1, t|i_0) - 2p(i, t|i_0) + p(i + 1, t|i_0)]. \quad (6.4)$$

Labeling cell i by its *midpoint* x, we take the cell length l small enough that the probability of the solute molecule being inside cell i at time t can be written in terms of the PDF $P_X(x, t|x_0, 0)$ of its position $X(t)$ in accordance with the definition (3.7), treating l as a differential dx:

$$p(i, t|i_0) \doteq P_X(x, t|x_0, 0) \cdot l. \quad (6.5)$$

In effect, we are again insisting that the cell length l be small enough that $P_X(x, t|x_0, 0)$ is nearly constant inside each cell. Dividing Eq. (6.4) through by l then gives

$$\frac{\partial P_X(x, t|x_0, 0)}{\partial t} \doteq D \left[\frac{P_X(x - l, t|x_0, 0) - 2P_X(x, t|x_0, 0) + P_X(x + l, t|x_0, 0)}{l^2} \right].$$

Since the quantity in brackets is the standard discrete approximation to the second derivative of $P_X(x, t|x_0, 0)$ with respect to x, we conclude that

$$\frac{\partial P_X(x, t|x_0, 0)}{\partial t} \doteq D \frac{\partial^2 P_X(x, t|x_0, 0)}{\partial x^2}. \tag{6.6}$$

The foregoing argument proves that if κ_l is defined by Eq. (5.3) and l is taken sufficiently small, then the master equation (6.3) reduces to Einstein's time-evolution equation (3.9) for the solute molecule's position. This implies that, to the extent that l is small enough, the solute molecule dynamics arising from the discrete-stochastic hypothesis (5.3) will be the same as the solute molecule dynamics prescribed by Einstein's model of diffusion. This argument also constitutes yet another derivation of the relation (5.8).

But what does it mean to require l to be "small enough" here? Any such restriction on the size of l requires comparison with some other physically relevant length. The only length that is thus far available to us in this problem is *the spatial scale on which we are observing the motion of the solute molecule*, which we will denote by Δx_{obs}. If we are observing the solute molecule over a period of time t_{obs}, then a reasonable measure of Δx_{obs} should be the root-mean-square displacement of the solute molecule in that time:

$$\Delta x_{\text{obs}} \equiv \sqrt{2Dt_{\text{obs}}}. \tag{6.7}$$

Therefore, the requirement that l in the argument leading from Eq. (6.3) to Eq. (6.6) be "small" must mean that l is small compared to this length:

$$l \ll \sqrt{2Dt_{\text{obs}}}. \tag{6.8a}$$

But this cannot be regarded as a universal upper bound on l, since it depends on the arbitrary observation time. Instead, we should regard condition (6.8a) as telling us how small l must be if the plot of the *discrete* state function $p(i, t_{\text{obs}}|i_0)$ is to *look like* the plot of the *continuous* state function $P_X(x, t_{\text{obs}}|x_0, 0)$. For example, if the strong inequality in condition (6.8a) went the other way, then it is likely that the solute molecule would never leave its initial cell by time t_{obs}. In that case, $p(i, t_{\text{obs}}|i_0)$ would, during our observation time, be nearly 1 for $i = i_0$ and nearly 0 for all other i. Although that extremely coarse-grained result would not be "wrong", it would not reveal the finer resolution of the diffusion process that is afforded by Einstein's PDF $P_X(x, t_{\text{obs}}|x_0, 0)$.

Another way of expressing condition (6.8a) can be obtained by first squaring it, to get $l^2 \ll 2Dt_{\text{obs}}$, and then replacing D by $\kappa_l l^2$. That gives

$$\frac{1}{2\kappa_l} \ll t_{\text{obs}}. \tag{6.8b}$$

We will see shortly (in Section 6.4) that this condition can be physically interpreted as saying that the *average dwell time in any interior cell* should be very small compared to the *total* time of observation. Or from a slightly different perspective, condition

(6.8b) says that the diffusing molecule will experience *many cell transitions* during the total time of observation.

A caveat on the foregoing considerations is the lower bound on l that is implied by condition (5.15b), $l^2 \gg 2D\delta t$. That lower bound arose from the fact that if l is taken too small, then transitions of the solute molecule to *non-adjacent* cells during time δt will no longer be infrequent enough to ignore, and such transitions were obviously not accounted for in our derivation of the master equation (6.3). So if, in the process of trying to satisfy condition (6.8a), we inadvertently take l so small that the condition $l^2 \gg 2D\delta t$ is violated, then even though the discrete-stochastic approach and the Einstein approach will agree with each other, *neither* will be *physically* accurate.

6.3 Solutions to the single-molecule master equation

The single-molecule master equation (6.3) is actually a set of M coupled, linear, ordinary differential equations for $i = 1, \ldots, M$. As outlined in Appendix 6A, Eq. (6.3) can be solved analytically for arbitrary M using matrix methods. But the algebraic complexity of the solution increases rapidly with M. As can be verified by direct differentiation, the solutions for the two cases $M = 2$ and $M = 3$ are as follows:

$$
\left.
\begin{aligned}
p(2,t|1) = p(1,t|2) = \tfrac{1}{2}\left(1 - e^{-2\kappa_l t}\right) \\
p(1,t|1) = p(2,t|2) = \tfrac{1}{2}\left(1 + e^{-2\kappa_l t}\right)
\end{aligned}
\right\} \quad (M = 2), \tag{6.9}
$$

$$
\left.
\begin{aligned}
p(2,t|1) = p(1,t|2) = p(2,t|3) = p(3,t|2) = \tfrac{1}{3}\left(1 - e^{-3\kappa_l t}\right) \\
p(3,t|1) = p(1,t|3) = \tfrac{1}{3} - \tfrac{1}{2}e^{-\kappa_l t} + \tfrac{1}{6}e^{-3\kappa_l t} \\
p(1,t|1) = p(3,t|3) = \tfrac{1}{3} + \tfrac{1}{2}e^{-\kappa_l t} + \tfrac{1}{6}e^{-3\kappa_l t} \\
p(2,t|2) = \tfrac{1}{3} + \tfrac{2}{3}e^{-3\kappa_l t}
\end{aligned}
\right\} \quad (M = 3). \tag{6.10}
$$

Plots of these curves are shown in Fig. 6.1.

It is instructive to examine the solutions (6.9) and (6.10) in the "small-t" case $t = \delta t \ll 1/\kappa_l$: Taylor expansions of the exponentials in the formulas for the adjacent-cell transition probabilities $p(i \pm 1, \delta t|i)$ reveal that, to lowest order in $\kappa_l \delta t$, *all* are equal to $\kappa_l \delta t$ regardless of the values of i or M. And that is exactly what we should expect on the basis of our hypothesis (5.3). For $M = 3$, the probabilities $p(3, \delta t|1) = p(1, \delta t|3)$ for diffusive transfers to next-nearest neighbor cells have a lowest order approximation of $\tfrac{1}{2}(\kappa_l \delta t)^2$. That is consistent with the implicit assumption in hypothesis (5.3) that we can ignore the possibility of transitions to non-adjacent cells whenever δt is "sufficiently small". However, the fact that the above formulas for $p(3, t|1) = p(1, t|3)$ approach 1/3 as $t \to \infty$ shows that for time intervals δt that are *not* sufficiently small, transitions to non-adjacent cells cannot be ignored.

An important property of $p(i, t|i_0)$ which is illustrated by the $M = 2$ and $M = 3$ formulas (6.9) and (6.10) is that

$$
\lim_{t \to \infty} p(i, t|i_0) = \frac{1}{M}. \tag{6.11}
$$

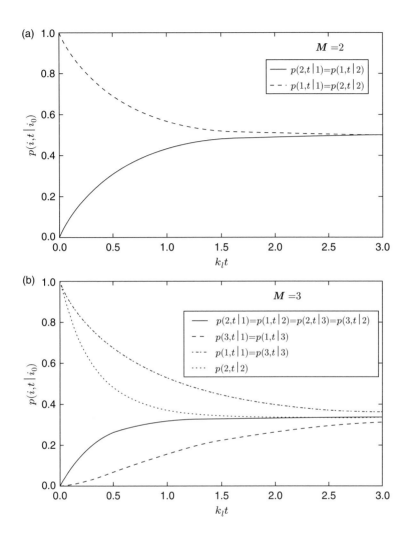

Fig. 6.1 *Plots of the solutions to the single-molecule diffusion master equation (6.3) for two-cell and three-cell systems.* (**a**) *Solution for $M = 2$ cells, as given in Eqs (6.9).* (**b**) *Solution for $M = 3$ cells, as given in Eqs (6.10). Notice that the probability of a solute molecule diffusing to an adjacent cell by a given time t, shown as the solid curves in these plots, depends on M. But that M-dependence goes away as $\kappa_l t \to 0$, in which limit all adjacent cell transition probabilities approximate to $\kappa_l t$ in accordance with hypothesis (5.3). Similarly, the probability of a solute molecule moving to a next-nearest neighbor cell in time t, shown by the dashed curve in (b) for $M = 3$, will change as M is increased beyond 3. But for $\kappa_l t$ sufficiently small, that probability can be well approximated for all M by $\frac{1}{2}(\kappa_l t)^2$.*

This result, which can be proved for all M from the general matrix-form solution (see Appendix 6A), says that after a sufficiently long time, the solute molecule is equally likely to be found in any cell. It can also be seen from Eqs (6.9) and (6.10) that this spatially homogeneous limit can be considered attained for most practical purposes whenever $\kappa_l t \gg 1$, i.e., whenever t is comfortably larger than $1/\kappa_l$. Thus, $1/\kappa_l$ can be regarded as the *relaxation time* of the diffusing molecule.

An interesting property that is exhibited by the explicit formulas for $M = 2$ and 3 is that the probability for the molecule to diffuse from cell i to cell j in time t is the same as the probability for the molecule to diffuse from cell j to cell i in time t:

$$p(j, t|i) = p(i, t|j). \qquad (6.12)$$

The validity of this result for all M can be proved from the general matrix-form solution (see Appendix 6A). This property is not surprising for $M = 2$, since in that case physical symmetry implies that $p(2, t|1) = p(1, t|2)$. Similarly, for $M = 3$ physical symmetry implies that $p(3, t|1) = p(1, t|3)$, which is another instance of Eq. (6.12). But for $M = 3$, physical symmetry also implies that $p(2, t|1) = p(2, t|3)$, and that relation is *not* an instance of Eq. (6.12). Even more surprisingly, physical symmetry does *not* imply for $M = 3$ the two (6.12) relations $p(2, t|1) = p(1, t|2)$ and $p(2, t|3) = p(3, t|2)$, even though an inspection of Eqs (6.10) shows that both equalities hold. So from a physical standpoint, Eq. (6.12) is a non-trivial result.

For all allowable $i \to i \pm n$ transitions, $p(i \pm n, \delta t|i)$ will, for sufficiently small $\kappa_l \delta t$, be approximately equal to some constant times $(\kappa_l \delta t)^n$. However, for $M > 3$ and *unrestricted* values of t, it is *not* true that the probabilities for all n-jump transitions are equal to each other. In the case $M = 4$ for example, while it *is* true that $p(2, t|1) \equiv p(1, t|2)$ and $p(3, t|2) \equiv p(2, t|3)$ as asserted by Eq. (6.12), it is *not* true that $p(2, t|1)$ and $p(3, t|2)$ are equal, nor that $p(1, t|2)$ and $p(2, t|3)$ are equal. The message here is that for any fixed $M \geq 4$ and t *not* an infinitesimal, $p(i \pm n, t|i)$ will depend on i as well as n. This makes it very difficult to develop general-purpose formulas for accurately simulating the cell changes of a solute molecule over time steps that are not small compared to $1/\kappa_l$.

6.4 Simulating the discrete-stochastic motion of a single solute molecule

We have just seen that, for arbitrary M, it is challenging to obtain exact analytical solutions to the single-molecule master equation (6.3). But it turns out to be easy to make exact *numerical simulations* of the trajectory of a molecule that moves about according to that master equation—i.e., according to hypothesis (5.3). What is needed to do that is a valid computational procedure that answers two questions: Given that the solute molecule is in cell i at time t, (i) at what time $t + \tau$ will it jump away from that cell, and (ii) to which of the adjacent cells will it jump?

To derive an algorithm that will enable us to answer these two questions, we start by assuming that our solute molecule is in cell i at time t. By hypothesis (5.3), the probability that the molecule will exit cell i, through either its left or right boundary, during the next effectively infinitesimal time $\delta t = dt$ is

$$\varepsilon_{i1}\kappa_l dt + \varepsilon_{iM}\kappa_l dt = (\varepsilon_{i1} + \varepsilon_{iM})\kappa_l dt. \tag{6.13}$$

Let $p_{0i}(\tau)$ be {the probability that a solute molecule in cell i will *not* leave that cell in the next *finite* time interval τ}. The multiplication and addition laws of probability require $p_{0i}(\tau)$ to satisfy

$$p_{0i}(\tau + d\tau) = p_{0i}(\tau) \cdot (1 - (\varepsilon_{i1} + \varepsilon_{iM})\kappa_l d\tau);$$

because, the first factor on the right is the probability that the molecule will not leave cell i in the time interval $[0, \tau)$, and the second factor is the *subsequent* probability that the molecule will not leave cell i in the time interval $[\tau, \tau + d\tau)$. This equation evidently implies the differential equation $dp_{0i}/d\tau = -(\varepsilon_{i1} + \varepsilon_{iM})\kappa_l p_{0i}$. The solution to that differential equation for the required initial condition $p_{0i}(0) = 1$ is

$$p_{0i}(\tau) = \exp\left(-(\varepsilon_{i1} + \varepsilon_{iM})\kappa_l \tau\right). \tag{6.14}$$

Now let the random variable T_i denote the time until the solute molecule jumps away from its present cell i; i.e., T_i is the *dwell time* of a solute molecule in cell i. Then the probability that $T_i \in [\tau, \tau + d\tau)$ can be written as the *product* of {the probability $p_{0i}(\tau)$ that the molecule will *not* leave cell i during $[0, \tau)$} times {the probability $(\varepsilon_{i1} + \varepsilon_{iM})\kappa_l d\tau$ that the molecule *will* leave cell i in the next $d\tau$}:

$$\text{Prob}\{T_i \in [\tau, \tau + d\tau)\} = \exp\left(-(\varepsilon_{i1} + \varepsilon_{iM})\kappa_l \tau\right) \cdot (\varepsilon_{i1} + \varepsilon_{iM})\kappa_l d\tau.$$

Observe that the *left* side of this last equation is by definition the product of $d\tau$ times the PDF of the random variable T_i, while the *right* side is by Eq. (2.3) the product of $d\tau$ times the PDF of the exponential random variable $\mathcal{E}\left((\varepsilon_{i1} + \varepsilon_{iM})\kappa_l\right)$. Thus we conclude that the dwell time T_i of the solute molecule in cell i is the exponential random variable $\mathcal{E}\left((\varepsilon_{i1} + \varepsilon_{iM})\kappa_l\right)$. A sample value of that random variable can be constructed from a unit-interval uniform random number u by using formula (2.35); i.e., we can generate a sample value of the dwell time of the solute molecule in cell i by taking

$$\tau = \frac{1}{(\varepsilon_{i1} + \varepsilon_{iM})\kappa_l} \ln\left(\frac{1}{1-u}\right) \quad (i = 1, \ldots, M). \tag{6.15}$$

We note in passing that, since the exponential random variable $\mathcal{E}\left((\varepsilon_{i1} + \varepsilon_{iM})\kappa_l\right)$ has mean $[(\varepsilon_{i1} + \varepsilon_{iM})\kappa_l]^{-1}$, then the quantity $(2\kappa_l)^{-1}$ on the left side of Eq. (6.8b) is indeed, as was claimed, the average dwell time of the solute molecule in a typical (i.e., interior) cell.

When the molecule in cell i does jump away, it will jump to cell 2 if $i = 1$, to cell $M - 1$ if $i = M$, and to either cell $i - 1$ or $i + 1$ with equal probability if $2 \le i \le M - 1$. Taking all this into account, we have the following procedure for numerically simulating the movement of a single solute molecule in the discrete-stochastic approach:

1° With the molecule in cell i at time t, generate a unit-interval uniform random number u and compute the time step τ to the next jump according to Eq. (6.15).

2° Choose the *next* cell index j on the basis of the present cell index i as follows:
  - if $i = 1$, take $j = 2$;
  - if $i = M$, take $j = M - 1$;
  - if $2 \leq i \leq M - 1$, generate another unit-interval uniform random number u', and take $j = i - 1$ if $u' \leq 1/2$, or $j = i + 1$ otherwise.

3° Effect the next jump by replacing $t \leftarrow t + \tau$ and $i \leftarrow j$.

4° Output (t, i) if desired. Then either return to 1°, or else end the simulation.

Notice how easily this simulation procedure takes account of the assumed *perfectly reflecting* boundary conditions at $x = 0$ and $x = L$: according to step 2°, the only allowed jump from cell $i = 1$ is to cell $j = 2$, and the only allowed jump from cell $i = M$ is to cell $j = M - 1$. But suppose instead that the boundary at $x = 0$ were *perfectly absorbing*. Then we would need only make the following minor change in the algorithm: In step 2° we would treat the case $i = 1$ in the same way as the interior cell case $2 \leq i \leq M - 1$, except that if the u' draw gives for the next cell $j = 0$, which would imply a transition through the left boundary of cell 1, we would simply "absorb" the molecule and terminate the simulation. Similar relatively minor changes can accommodate boundaries that are *partially* reflecting or absorbing; e.g., if an absorbing boundary at $x = 0$ were only 20% efficient, then upon generating a down-going jump from cell 1 we would absorb the molecule only if *another* unit-interval uniform random number u'' were found to be less than 0.2.

6.5   Some examples of single-molecule simulations

To illustrate the foregoing single-molecule stochastic simulation algorithm, let the cylindrical volume Ω pictured in Fig. 5.1 have its left face at $x = 0$ and its right face at $x = 400$ μm, as in the experiment described in Section 5.9. And let Ω be subdivided into $M = 80$ cells, each of length $l = 5$ μm, so that cell 1 is the interval $0 \leq x < 5$ μm and cell 80 is the interval $395 \leq x < 400$ μm. At time 0, let a particular solute molecule—in this case a microscopic polystyrene bead as described in Section 5.9—be inside cell $i_0 = 40$, which occupies the interval $195 \leq x < 200$ μm. For reasons that will be given later in Section 9.7, we take the diffusion coefficient of the bead in water at room temperature to be $D = 1.51$ μm^2 s^{-1}, a value that is also consistent with the experimental findings in Section 5.9. It then follows from formula (5.8) that

$$\kappa_l = \frac{D}{l^2} = \frac{1.51}{5^2} \approx 0.0604 \, \text{s}^{-1}. \tag{6.16}$$

We want to determine by numerical simulation the cell occupation probabilities at time $t = 900$ s; i.e., we want to construct a plot of the function $p(i, 900|40)$ versus i for $i = 1, \ldots, 80$. But before doing that, let us see if our chosen cell size satisfies condition (6.8a). The expected rms displacement of the bead in time $t_{\text{obs}} = 900$ s is

$$\sqrt{2Dt_{\text{obs}}} = \sqrt{2 \cdot 1.51 \cdot 900} \approx 52 \, \text{μm}.$$

Since this is comfortably larger than our cell size $l = 5$ μm, condition (6.8a) is indeed satisfied. So for this choice of l, a plot of the function $p(i, t|i_0)$ ought to give a decent approximation to a plot of the function $P_X(x, t|x_0, 0)$. Another perspective on this point is provided by the equivalent condition (6.8b): Since in this case the mean dwell time $1/(2\kappa_l)$ in any interior cell is approximately 8.3 s, then in 900 s the bead will make on average $900\,\text{s}/8.3\,\text{s} = 109$ diffusional cell transfers. That should give us enough non-zero cell data points to allow a comparison to be made between the two functions $p(i, t|i_0)$ and $P_X(x, t|x_0, 0)$.

Our plan for estimating $p(i, 900|40)$ is as follows: We will make a series of $K = 10^6$ independent "runs", each using the four-step simulation procedure given in the preceding section. Each run will start at time 0 in cell $i_0 = 40$, and terminate when the time variable t in step 3° first goes beyond 900. At run termination, we record the cell number of the bead *at* time 900, which will be the cell that the bead was in *just before* the final step that advanced the time variable *beyond* 900. After K runs, we will make a normalized frequency histogram of the K final cell numbers; i.e., we will plot the *fraction* of the K runs that ended up in each cell i. In the limit of infinitely large K, those fractions will by definition be the probabilities $p(i, t|i_0)$.

Figure 6.2 shows a histogram and a curve together on one graph: The histogram is the normalized frequency histogram of the $K = 10^6$ runs obtained by the procedure just described, and is thus our simulation estimate of the function $p(i, 900|40)$ versus i. The heavy dots in the figure is a graph of

$$P_X(x, 900; 197.5, 0) \cdot 5 = \frac{5}{\sqrt{4\pi(1.51) \cdot 900}} \exp\left(-\frac{(x - 197.5)^2}{4(1.51) \cdot 900}\right). \tag{6.17}$$

In this equation, which was obtained by substituting our experimental values into Eq. (3.10), the distance x of the *center* of cell i from the left boundary of Ω is related to the cell number i by $x = 5 \cdot i - 2.5$. This plot of Eq. (6.17) shows what the Einstein prediction (3.10) of the cell population probabilities at time $t = 900$ s would be *if there were no boundaries* at $x = 0$ and $x = 400$. Since those boundaries are almost never reached in these runs to $t = 900$ s, Eq. (6.17) should accurately represent the Einstein prediction for this problem. The vertical dashed line in the figure shows the time-0 location of the bead in cell 40. The excellent agreement shown in Fig. 6.2 confirms that our stochastic simulation procedure, and hence also the discrete-stochastic model of diffusion defined by hypothesis (5.3), is consistent with the Einstein model of diffusion.

The normal shape of $p(i, t|i_0)$ for the problem just considered could of course have been predicted without resorting to numerical simulation. The power of the stochastic simulation technique is that it can obtain results of acceptable accuracy for a great many *variations* on this problem that would be difficult if not impossible to obtain analytically. Two examples illustrating this are shown in Figs 6.3 and 6.4.

Figure 6.3 shows the stochastic simulation estimate of $p(i, 3000|20)$ for the case in which the bead is initially in cell 20 (dashed vertical line) and diffuses until time $t = 3000$ s. Apart from the different initial condition and the longer run time, the simulation runs for this figure were carried out in exactly the same way as for Fig. 6.2. Notice that in this case, both boundaries of Ω make their presence felt.

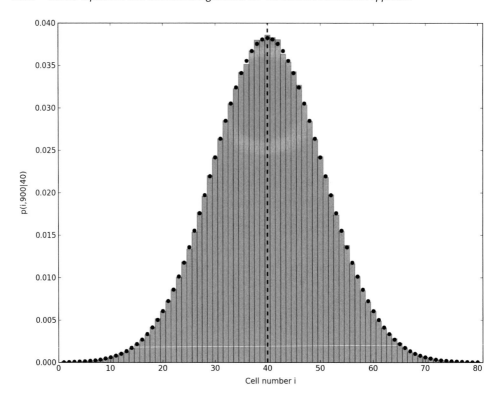

Fig. 6.2 *A single-molecule stochastic simulation study. Parameter values are taken from the microfluidics diffusion experiment in Section 5.9, where the solute molecule is a polystyrene bead of radius 0.145 μm, the solvent molecules are water, and the diffusion coefficient of the bead is $D = 1.51\ \mu m^2\ s^{-1}$. The system volume Ω, shown cartooned in Fig. 5.1, has length $L = 400\ \mu m$, and is subdivided into $M = 80$ cells, each of length $l = 5\ \mu m$. For that cell length, κ_l takes the value in Eq. (6.16). At time $t = 0$ the bead is assumed to be inside cell $i_0 = 40$ (the dashed vertical line). The gray area shows the normalized frequency histogram of the cell location of the bead at time $t = 900\,s$ obtained by averaging the results of 10^6 runs of the simulation algorithm of Section 6.4. Superimposed on the histogram as the solid dots is a plot of the function (6.17), which is the prediction of the Einstein diffusion equation for unrestricted diffusion. The unrestricted approximation is reasonable in this case, since the bead is evidently very unlikely to reach the reflecting boundaries at cells 1 and 80 by time 900 s. The excellent agreement of the simulation results with the prediction of the Einstein theory confirms the correctness of not only the single-molecule stochastic simulation algorithm, but also the central hypothesis (5.3) of the discrete-stochastic approach.*

Figure 6.4 is for the same problem as the one in Fig. 6.3, but with a twist. Here we have placed a *partially absorbing trap* at the boundary between cells 10 and 11 (arrow): On any transition *from* cell 11 *to* cell 10, there is a 10% chance that the trap will "absorb" the bead, essentially removing it from Ω. This modification would greatly complicate an analytical calculation of the function $p(i, 3000|20)$. But a numerical

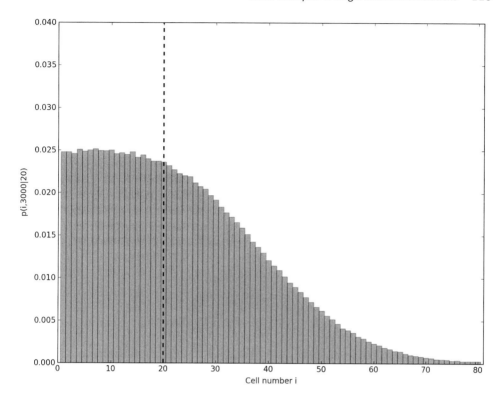

Fig. 6.3 *A single-molecule stochastic simulation study of the effect of a reflecting boundary. The set-up is the same as in Fig. 6.2, except that now the bead is initially in cell 20 (the dashed vertical line), and each of the 10^6 simulations goes to $t = 3000$ s. Notice that the zero slopes of the PDF of the bead at both reflecting boundaries, as required by condition (3.26), arise naturally from the single-molecule stochastic simulation algorithm.*

estimate by stochastic simulation requires only the insertion of one new step in the simulation procedure of Section 6.4: Immediately after performing step $2°$, if $i = 11$ and $j = 10$ we will with probability 0.10 terminate the run, and place the bead in a special "cell 0". Unlike the graphs in the preceding two figures, the area under the histogram in Fig. 6.4 is less than unity. That's because it is now possible that the bead will be absorbed before time 3000. Indeed, the number of times the bead ended up in cell 0 divided by the total number of runs K constitutes a direct estimate of the probability p_{abs} that a bead, if released from cell 20 at time 0, will be absorbed by the trap before time 3000 s. Of the $K = 10^6$ runs made for Fig. 6.4, a total of $n_K = 288\,108$ turned out to be absorbing. As explained in Eq. (6B.12) in Appendix 6B, this implies a "95% confidence" estimate of $p_{\text{abs}} = 0.2881 \pm 0.0009$. The area under the curve in Fig. 6.4 is less than 1 by this amount.

    Notice that in both Figs 6.3 and 6.4, the estimated function $p(i, t | i_0)$ has *zero slope* at the two perfectly reflecting boundaries of Ω. This is of course theoretically

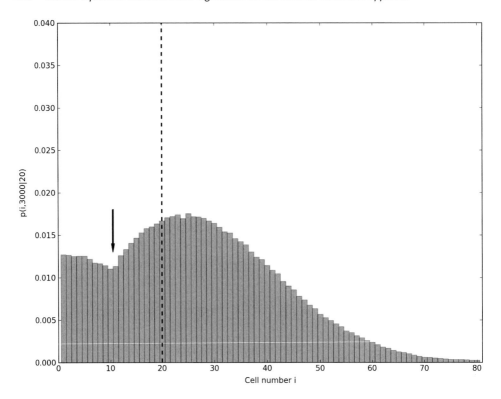

Fig. 6.4 *The same set-up as in Fig. 6.3, but with an added partially absorbing trap. The trap is at the boundary between cells 10 and 11 (arrow), and it operates as follows: On any transition from cell 11 to cell 10, there is a 10% chance that the trap will "absorb" the bead. But a transition of the bead from cell 10 to cell 11 is unhindered. The fact that the area under this histogram is only 71% of the (unit) area under the histograms in Figs 6.3 and 6.2 reflects the fact that with probability 0.29 the bead will have gotten absorbed by the trap by time $t = 3000$ s. The modifications to the single-molecule stochastic simulation algorithm that are needed to take account of this partially absorbing trap are simple (see text). But computing this solution analytically by solving the diffusion equation with suitable boundary conditions would be quite challenging.*

expected from Eq. (3.26). But here these zero slopes arose naturally in the simulations, essentially from the j-selection procedure in step 2°; we did not have to overtly impose this condition in executing the simulations, as we would have had to do if we had been solving the classical diffusion equation.

The stochastic simulation method of numerically estimating the solution of the master equation (6.3) allows a wide variety of other questions to be answered by simply adding a few lines of "monitoring code" to the simulation algorithm. For example, in the simulation of Fig. 6.4, one might want to know the probability that the bead will reach the boundary at $x = 0$, and then later be absorbed, all before time 3000 s. What recommends the stochastic simulation approach is its ability to provide acceptably accurate answers to such analytically difficult questions with relative ease. The curves in Figs 6.2–6.4 can be made smoother (and more accurate) by taking the number of runs K larger. To decrease the size of the fluctuations by half requires increasing K, and hence also the computer execution time, by a factor of 4. The computer execution time for each of the runs in Figs 6.3 and 6.4 was roughly twice that for the run in Fig. 6.2.

6.6　The many-molecule diffusion master equation

A second master equation that can be derived based on hypothesis (5.3) deals with the entire ensemble of solute molecules. The focus now is on the *joint* PDF of the cell population variables (5.1), namely,

$$P(\mathbf{n}, t|\mathbf{n}_0) \equiv \mathrm{Prob}\left\{\mathbf{N}(t) = \mathbf{n}|\mathbf{N}(0) = \mathbf{n}_0\right\}, \tag{6.18}$$

where $\mathbf{n} \equiv (n_1, \dots, n_M)$ and $\mathbf{n}_0 \equiv (n_{01}, \dots, n_{0M})$. By the fundamental hypothesis (5.3), this function must satisfy

$$P(\mathbf{n}, t + \delta t|\mathbf{n}_0) = \sum_{i=1}^{M} \left\{\varepsilon_{i1} P(n_{i-1} - 1, n_i + 1, \tilde{\mathbf{n}}, t|\mathbf{n}_0) \cdot (n_i + 1)\kappa_l \delta t\right\}$$

$$+ \sum_{i=1}^{M} \left\{\varepsilon_{iM} P(n_i + 1, n_{i+1} - 1, \tilde{\mathbf{n}}, t|\mathbf{n}_0) \cdot (n_i + 1)\kappa_l \delta t\right\}$$

$$+ P(\mathbf{n}, t|\mathbf{n}_0) \cdot \left[1 - \sum_{i=1}^{M}(\varepsilon_{i1} n_i \kappa_l \delta t + \varepsilon_{iM} n_i \kappa_l \delta t)\right] + o(\delta t), \tag{6.19}$$

where $\tilde{\mathbf{n}}$ in the argument of the function P stands for all components of $\mathbf{n}$ not explicitly listed. In this equation, the first term on the right is the probability of the system getting to state $\mathbf{n}$ at time $t + \delta t$ via one $i \to i - 1$ jump in $[t, t + \delta t)$; the second term is the probability of getting there via one $i \to i + 1$ jump in $[t, t + \delta t)$; the third term is the probability of getting there with no jump in $[t, t + \delta t)$; and the last term $o(\delta t)$ accounts for higher order corrections to the aforementioned probabilities, plus the probabilities of routes to state $\mathbf{n}$ at time $t + \delta t$ that involve more than one jump in $[t, t + \delta t)$. Subtracting $P(\mathbf{n}, t|\mathbf{n}_0)$ from both sides, dividing through by δt, and then taking the limit of "very small" δt, we obtain the *many-molecule diffusion master equation*:

$$\frac{dP(\mathbf{n}, t|\mathbf{n}_0)}{dt} = \kappa_l \sum_{i=1}^{M} \left\{\varepsilon_{i1}(n_i + 1)P(n_{i-1} - 1, n_i + 1, \tilde{\mathbf{n}}, t|\mathbf{n}_0)\right\}$$

$$+ \kappa_l \sum_{i=1}^{M} \left\{\varepsilon_{iM}(n_i + 1)P(n_i + 1, n_{i+1} - 1, \tilde{\mathbf{n}}, t|\mathbf{n}_0)\right\}$$

$$- \kappa_l \sum_{i=1}^{M} \left\{(\varepsilon_{i1} + \varepsilon_{iM})n_i P(\mathbf{n}, t|\mathbf{n}_0)\right\}. \tag{6.20}$$

As in the derivation of the single-molecule master equation (6.3), this derivation implicitly assumes that it is possible to satisfy some potentially conflicting requirements on δt: On the one hand, δt must be *small* enough to meaningfully approximate the time derivative on the left side, and to kill off the $o(\delta t)/\delta t$ terms on the right that arise from multiple cell jumps in time δt. And on the other hand, δt must be *large* enough that each solute molecule experiences many collisions with solvent molecules in time δt.

Like the single-molecule master equation (6.3), the many-molecule master equation (6.20) is actually a set of coupled ordinary differential equations. But Eq. (6.20) gives us one equation for each possible *combination* of the cell population values $(n_1, \ldots, n_M)$, so there are very many more equations in the set (6.20) than in the set (6.3). And as we will see in the following section for the simple case $M = 2$, exact analytical solutions to Eq. (6.20) for some initial conditions can be practically impossible to compute.

But it is not difficult to deduce the asymptotic or equilibrium solution, $P(\mathbf{n}, \infty | \mathbf{n}_0)$. The simplest way of doing that is to use the asymptotic result (6.11) of the *single-molecule* master equation. That result, which is proved for all $M \geq 2$ in Appendix 6A, tells us that after a sufficiently long time, each solute molecule, regardless of where it was initially, will be found in any particular cell i with probability M^{-1}. Thus, we imagine the N solute molecules inside Ω to be divided into M subgroups, with the ith subgroup containing $n_i \geq 0$ molecules, and $\sum_{i=1}^{M} n_i = N$. Since the solute molecules move about independently of each other (this is assumed by the many-molecule master equation), then by the multiplication law of probability, the probability that every molecule in subgroup i will be found at time $t = \infty$ to be in cell i for all $i = 1, \ldots, M$ is

$$(M^{-1})^{n_1} \cdot (M^{-1})^{n_2} \cdot \cdots \cdot (M^{-1})^{n_M} = (M^{-1})^{(n_1 + \cdots + n_M)} = M^{-N}.$$

But there are $N!/(n_1! \cdots n_M!)$ distinct ways of dividing N molecules into M groups of sizes $n_1, \ldots, n_M$; therefore, it follows from the addition law of probability that the net probability of finding the configuration $(n_1, \ldots, n_M)$ at $t = \infty$ is equal to that number of ways times the above probability:

$$P(n_1, \ldots, n_M, \infty | \mathbf{n}_0) = \frac{N! \, M^{-N}}{n_1! n_2! \cdots n_M!} \quad \left(n_i \geq 0, \sum_{i=1}^{M} n_i = N \right). \qquad (6.21)$$

The joint asymptotic PDF (6.21) defines a "multinomial" distribution. Using that PDF, we can compute various "marginal" asymptotic PDFs by summing over variables that are not of interest. However, it is actually easier to compute such marginal asymptotic PDFs by applying the logic we used above to deduce Eq. (6.21). For example, the probability that at $t = \infty$ a *particular* set of n_i molecules will be in cell i, and the remaining $N - n_i$ molecules will not, is $(M^{-1})^{n_i} \cdot (1 - M^{-1})^{N-n_i}$. Summing that probability over all $N!/[n_i!(N - n_i)!]$ distinct ways of dividing N molecules into two groups of n_i and $N - n_i$ molecules, we find that the probability that cell i will contain exactly n_i solute molecules at equilibrium, irrespective of the other cell populations, is

$$P_i(n_i, \infty) = \frac{N!}{n_i!(N - n_i)!} \left(\frac{1}{M}\right)^{n_i} \left(1 - \frac{1}{M}\right)^{N-n_i} \quad (0 \le n_i \le N). \tag{6.22a}$$

This we recognize as the PDF of the binomial random variable $\mathcal{B}(1/M, N)$. Analogous reasoning allows us to conclude that the equilibrium probability that cell i contains exactly n_i molecules *and* cell j contains exactly n_j molecules, irrespective of the other cell populations, is

$$P_{i,j}(n_i, n_j, \infty) = \frac{N!}{n_i! n_j! (N - n_i - n_j)!} \left(\frac{1}{M}\right)^{n_i} \left(\frac{1}{M}\right)^{n_j} \left(1 - \frac{2}{M}\right)^{N-n_i-n_j}$$

$$(0 \le n_i, n_j \le N). \tag{6.22b}$$

Here, the first factor is the number of distinct ways of dividing the N molecules into three groups of n_i, n_j, and $(N - n_i - n_j)$ molecules. The second factor is the probability that each of the n_i molecules in the first group goes to cell i. The third factor is the probability that each of the n_j molecules in the second group goes to cell j. And the last factor is the probability that each of the $(N - n_i - n_j)$ molecules in the third group does *not* go to either cell i or cell j; note in that connection that the probability that a molecule in the third group *does* go to either cell i or cell j is $M^{-1} + M^{-1} = 2M^{-1}$. The probability rule $\mathrm{Prob}(1\&2)/\mathrm{Prob}(1) = \mathrm{Prob}(2|1)$ now allows us to conclude that Eq. (6.22b) divided by Eq. (6.22a) will give the asymptotic probability that, *given* there are n_i molecules in cell i, cell $j \ne i$ will contain n_j molecules. That division yields, after a good deal of algebraic rearrangement,

$$P_{j|i}(n_j, \infty | n_i) = \frac{(N - n_i)!}{n_j! (N - n_i - n_j)!} \left(\frac{1}{M - 1}\right)^{n_j} \left(1 - \frac{1}{M - 1}\right)^{N-n_i-n_j}$$

$$(j \ne i; \ 0 \le n_j \le N - n_i). \tag{6.22c}$$

This will be recognized as the PDF of the binomial distribution $\mathcal{B}(1/(M - 1), N - n_i)$. We can understand this formula as resulting from randomly placing exactly n_j molecules out of $N - n_i$ in a particular one of the $M - 1$ cells that do not include cell i, a cell which by assumption already has its quota of n_i molecules.

6.7 The case $M = 2$: an exact solution of a different kind

Although exact time-dependent solutions to the many-molecule master equation (6.20) are difficult to obtain, an unconventional but nonetheless useful form of the exact solution for $M = 2$ can be obtained by reasoning as follows: Refining slightly the notation introduced in Eq. (5.1), let

$$N_j(t | n_{10}, n_{20}) \equiv \text{the number of solute molecules in cell } j \text{ at time } t \ge 0,$$
$$\text{given that at time } t = 0 \text{ there were } n_{10} \text{ molecules in}$$
$$\text{cell 1 and } n_{20} \text{ molecules in cell 2 } (j = 1, 2; M = 2). \tag{6.23}$$

These two integer-valued random variables are of course those for which $P(\mathbf{n}, t|\mathbf{n}_0)$ in Eq. (6.20) is the joint PDF for $M = 2$. With an eye to obtaining explicit formulas for these random variables, we introduce the following four auxiliary random variables:

$$N_{i \to j}(t|n_{i0}) \equiv \text{ the number of the } n_{i0} \text{ solute molecules in cell } i \text{ at} \atop \text{time 0 that will be in cell } j \text{ at time } t \quad (i, j = 1, 2; M = 2). \quad (6.24)$$

These auxiliary random variables can be determined from the solution (6.9) of the *single*-molecule master equation (6.3) for $M = 2$ by reasoning as follows: Each of the n_{i0} solute molecules in cell i at time 0 in effect "tosses a weighted coin" to decide whether it will be in cell j at time t. With probability $p(j, t|i)$ in Eqs (6.9) the molecule will be in cell j at time t, and with probability $1 - p(j, t|i)$ it won't. Therefore, as discussed in connection with Eq. (2.38), the probability that exactly n of those n_{i0} molecules will wind up in cell j at time t is the *binomial* random variable with parameters $p(j, t|i)$ and n_{i0}:

$$N_{i \to j}(t|n_{i0}) = \mathcal{B}\left(p(j, t|i), n_{i0}\right).$$

So with the exact formulas for $p(j, t|i)$ in Eqs (6.9), we have in the two-cell case:

$$N_{1 \to 1}(t|n_{10}) = \mathcal{B}\left(\tfrac{1}{2}\left(1 + e^{-2\kappa_l t}\right), n_{10}\right), \quad (6.25a)$$

$$N_{1 \to 2}(t|n_{10}) = \mathcal{B}\left(\tfrac{1}{2}\left(1 - e^{-2\kappa_l t}\right), n_{10}\right), \quad (6.25b)$$

$$N_{2 \to 1}(t|n_{20}) = \mathcal{B}\left(\tfrac{1}{2}\left(1 - e^{-2\kappa_l t}\right), n_{20}\right), \quad (6.25c)$$

$$N_{2 \to 2}(t|n_{20}) = \mathcal{B}\left(\tfrac{1}{2}\left(1 + e^{-2\kappa_l t}\right), n_{20}\right). \quad (6.25d)$$

Now we simply observe that $N_1(t|n_{10}, n_{20})$, as defined in Eq. (6.23), must be equal to the *sum* of {the number of solute molecules initially in cell 1 that wind up in cell 1 at time t} plus {the number of solute molecules initially in cell 2 that wind up in cell 1 at time t}:

$$N_1(t|n_{10}, n_{20}) = N_{1 \to 1}(t|n_{10}) + N_{2 \to 1}(t|n_{20}).$$

Therefore, from Eqs (6.25a) and (6.25c), we have

$$N_1(t|n_{10}, n_{20}) = \mathcal{B}\left(\tfrac{1}{2}\left(1 + e^{-2\kappa_l t}\right), n_{10}\right) + \mathcal{B}\left(\tfrac{1}{2}\left(1 - e^{-2\kappa_l t}\right), n_{20}\right) \quad (M = 2). \quad (6.26a)$$

With $N_1(t|n_{10}, n_{20})$ thus determined, conservation of the number of solute molecules implies that $N_2(t|n_{10}, n_{20})$, now viewed as the number of solute molecules in cell 2 *given* that there are $N_1(t|n_{10}, n_{20})$ solute molecules in cell 1, can be calculated as

$$N_2(t|n_{10}, n_{20}) = n_{10} + n_{20} - N_1(t|n_{10}, n_{20}) \quad (M = 2). \quad (6.26b)$$

Alternatively, we could have first sampled $N_2(t|n_{10}, n_{20})$ "unconditionally" as the sum of the two random variables in Eqs (6.25b) and (6.25d), analogous to what was done in Eq. (6.26a), and then computed $N_1(t|n_{10}, n_{20})$ from the conservation relation (6.26b). Note that since N_1 and N_2 are related by Eq. (6.26b), then fixing the value of one automatically fixes the value of the other; therefore, if we computed *both* N_1 and N_2 using Eqs (6.25), the conservation relation (6.26b) would usually not be satisfied.

Since formulas (6.26) fully characterize the two random variables whose PDF is the function $P(\mathbf{n}, t|\mathbf{n_0})$ for $M = 2$, then Eqs (6.26) constitute an exact solution of the master equation (6.20) for $M = 2$. The impossibility of obtaining a *tractable formula* for $P(\mathbf{n}, t|\mathbf{n_0})$ in this case is apparently due to the fact that the PDF of the sum of two statistically independent binomial random variables with different "p" parameters does not have a simple analytical form. But Eqs (6.26) are actually quite useful just as they are, because computer programs exist for efficiently sampling the binomial random variable $\mathcal{B}(p, N)$. Consequently, exact values for $N_1(t|n_{10}, n_{20})$ and $N_2(t|n_{10}, n_{20})$ for *any* $t > 0$ can be obtained simply by sampling the two statistically independent binomial random variables on the right side of Eq. (6.26a).

From the formulas (2.39) for the mean and variance of the binomial random variable, together with the result (2.21) that the mean (variance) of the *sum* of two statistically independent random variables is equal to the sum of their means (variances), it is easy to show from Eq. (6.26a) that

$$\langle N_1(t|n_{10}, n_{20}) \rangle = \frac{n_{10} + n_{20}}{2} + \frac{n_{10} - n_{20}}{2} e^{-2\kappa_l t}, \tag{6.27a}$$

$$\text{var}\{N_1(t|n_{10}, n_{20})\} = \frac{n_{10} + n_{20}}{4} \left(1 - e^{-4\kappa_l t}\right). \tag{6.27b}$$

Similar formulas of course hold for $N_2(t|n_{10}, n_{20})$; they can be obtained most easily by just interchanging indices $1 \leftrightarrow 2$.

Another interesting consequence of the result (6.26) has to do with the net number of molecules moving from cell 1 to cell 2 in any time t. That quantity is evidently equal to

$$N_{1 \to 2}^{\text{net}}(t|n_{10}, n_{20}) \equiv N_{1 \to 2}(t|n_{10}) - N_{2 \to 1}(t|n_{20}). \tag{6.28}$$

The mean and variance of this random variable can be straightforwardly computed using Eqs (6.25b), (6.25c), (2.21), and (2.39). The result is

$$\left\langle N_{1 \to 2}^{\text{net}}(t|n_{10}, n_{20}) \right\rangle = \frac{n_{10} - n_{20}}{2} \left(1 - e^{-2\kappa_l t}\right), \tag{6.29a}$$

$$\text{var}\left\{N_{1 \to 2}^{\text{net}}(t|n_{10}, n_{20})\right\} = \frac{n_{10} + n_{20}}{4} \left(1 - e^{-4\kappa_l t}\right). \tag{6.29b}$$

If $t \ll \kappa_l^{-1}$, we can expand the exponentials here in powers of $\kappa_l t$ to obtain

$$\left\langle N_{1 \to 2}^{\text{net}}(t|n_{10}, n_{20}) \right\rangle = (n_{10} - n_{20}) \kappa_l t + o(\kappa_l t),$$

$$\text{var}\left\{N_{1 \to 2}^{\text{net}}(t|n_{10}, n_{20})\right\} = (n_{10} + n_{20}) \kappa_l t + o(\kappa_l t).$$

These formulas are the $M = 2$ versions of Eqs (5.27) and (5.28). Thus, Eqs (6.29) can be viewed as the exact *finite-t* versions, for $M = 2$, of the stochastic Fick's Law formulas (5.27) and (5.28).

The method just described for "solving" the many-molecule diffusion master equation (6.20) can be extended to values of M greater than 2; however, the analysis, and the results, quickly become very much more complicated.

6.8   The moments of the cell populations: recovering the diffusion equation

As we have noted, it is usually too difficult to solve the many-molecule master equation (6.20) analytically for $P(\mathbf{n}, t|\mathbf{n}_0)$. But that equation can be used to derive a set of time-evolution equations for the *moments* of $P(\mathbf{n}, t|\mathbf{n}_0)$. Of particular interest are the *first* and *second* moments:

$$\langle N_i(t|\mathbf{n}_0)\rangle \equiv \sum_{\text{all } \mathbf{n}} n_i P(\mathbf{n}, t|\mathbf{n}_0) \quad (i = 1, \dots, M), \tag{6.30a}$$

$$\langle N_i(t|\mathbf{n}_0)N_j(t|\mathbf{n}_0)\rangle \equiv \sum_{\text{all } \mathbf{n}} n_i n_j P(\mathbf{n}, t|\mathbf{n}_0) \quad (i, j = 1, \dots, M). \tag{6.30b}$$

In Appendix 6C, we derive from the definition (6.30a) and the master equation (6.20) the following time-evolution equation for the *first* moments:

$$\frac{d\langle N_i(t|\mathbf{n}_0)\rangle}{dt} = \kappa_l \left(\varepsilon_{i1} \langle N_{i-1}(t|\mathbf{n}_0)\rangle - (\varepsilon_{i1} + \varepsilon_{iM}) \langle N_i(t|\mathbf{n}_0)\rangle + \varepsilon_{iM} \langle N_{i+1}(t|\mathbf{n}_0)\rangle \right)$$

$$(i = 1, \dots, M). \tag{6.31}$$

Physically, one can interpret the first and third terms on the right side of this equation as the average rates at which molecules are diffusing *into* cell i from the respective cells $i - 1$ and $i + 1$, and the middle term as the average rate at which molecules are diffusing *out of* cell i to those two adjacent cells.

Equation (6.31) is evidently a *closed* set of M linearly coupled ordinary differential equations. It is therefore possible to solve them exactly for the mean cell populations $\langle N_1(t|\mathbf{n}_0)\rangle, \dots, \langle N_M(t|\mathbf{n}_0)\rangle$. But the difficulty of obtaining those solutions, and the complexity of the solutions themselves, increases rapidly with increasing M. The reason why becomes clear once we recognize that *for M "very large", Eq. (6.31) is essentially the classical diffusion equation.* To see this, first use Eq. (5.8) to write Eq. (6.31) for any interior cell i $(2 \leq i \leq M - 1)$ as

$$\frac{d\langle N_i(t|\mathbf{n}_0)\rangle}{dt} = D \frac{\langle N_{i-1}(t|\mathbf{n}_0)\rangle - 2\langle N_i(t|\mathbf{n}_0)\rangle + \langle N_{i+1}(t|\mathbf{n}_0)\rangle}{l^2}. \tag{6.32a}$$

Since the volume of each cell is the product of its length l and cross-sectional area A, then $\langle N_i(t|\mathbf{n}_0)\rangle/Al \equiv \rho_i(t|\mathbf{n}_0)$ is the *average molecular population density* of solute molecules in cell i at time t. Therefore, dividing this equation through by Al gives

$$\frac{d\rho_i(t|\mathbf{n}_0)}{dt} = D \left(\frac{\rho_{i-1}(t|\mathbf{n}_0) - 2\rho_i(t|\mathbf{n}_0) + \rho_{i+1}(t|\mathbf{n}_0)}{l^2} \right). \tag{6.32b}$$

If l is sufficiently small, or equivalently if $M = L/l$ is sufficiently large, the quantity on the right in parentheses becomes the second derivative of the average population density at the location of cell i, and Eq. (6.32b) reduces to the classical diffusion equation (1.4).

But the discrete-stochastic approach offers the possibility of going beyond the classical diffusion equation by quantifying the fluctuations in the cell populations about their averages. For example, by using the same techniques that we used in Appendix 6C to derive the time-evolution equations (6.31) for the first moments of the cell populations, we can derive time-evolution equations for the second moments (6.30b). And if those formulas for the derivatives of the second moments are then inserted into the time derivatives of the two identities

$$\text{var}\left\{N_i(t|\mathbf{n_0})\right\} \equiv \left\langle N_i^2(t|\mathbf{n_0})\right\rangle - \left\langle N_i(t|\mathbf{n_0})\right\rangle^2, \tag{6.33a}$$

$$\text{cov}\left\{N_i(t|\mathbf{n_0}), N_j(t|\mathbf{n_0})\right\} \equiv \left\langle N_i(t|\mathbf{n_0})N_j(t|\mathbf{n_0})\right\rangle - \left\langle N_i(t|\mathbf{n_0})\right\rangle\left\langle N_j(t|\mathbf{n_0})\right\rangle, \tag{6.33b}$$

the result is a set of time-evolution equations for all the variances and covariances of the cell populations. That set of equations turns out to be *closed*, in that it contains only the variances, the covariances, and the (presumed known) solutions $\langle N_i(t|\mathbf{n_0})\rangle$ to Eqs (6.31). Therefore, we can, with sufficient effort, obtain exact analytical expressions for all the variances and covariances of the cell populations, thereby "stochastically generalizing" the mean behavior given by the solution of Eq. (6.31). But solving the second moment equations turns out to be much harder than solving the first moment equations (6.31). One reason is that, whereas the coupled differential equations (6.31) for the first moments number M, which is typically a large number, the coupled differential equations for the variances and the covariances number $\frac{1}{2}M^2$, which is typically a *very* large number.

The difficulties of analytically extracting quantitative information from the many-molecule master equation (6.20) prompt us to look for other ways of obtaining that information. Again, one versatile numerical approach is stochastic simulation.

6.9 Simulating the discrete-stochastic motion of an ensemble of solute molecules

Like the single-molecule diffusion master equation (6.3), the many-molecule diffusion master equation (6.20) has an associated stochastic simulation algorithm. This algorithm is an exact computational procedure for answering three questions: Given the solute molecule cell populations $n_1, \ldots, n_M$ at the present time t, (i) at what time $t + \tau$ will the next diffusive transfer inside Ω occur, (ii) what will be the originating cell j of that transfer, and (iii) what will be the destination cell k of that transfer?

To derive an algorithm that answers these questions, we begin by observing that hypothesis (5.3), together with the addition law of probability, implies that $n_i \cdot \kappa_l dt$ is the probability that *some* one of the n_i solute molecules in cell i will jump to a particular adjacent cell in the next effectively infinitesimal time interval $\delta t = dt$. It follows from a second application of the addition law that the probability that *some* solute molecule inside Ω will diffuse across *some* cell boundary in the next dt is given by

$$n_1(\kappa_l dt) + \sum_{i=2}^{M-1} 2n_i(\kappa_l dt) + n_M(\kappa_l dt) \equiv (2n_{\text{tot}} - n_1 - n_M)\,\kappa_l dt. \tag{6.34}$$

Here the factor of 2 on the left side arises from the fact that each interior cell has two adjacent cells, and we have put $n_{\text{tot}} \equiv \sum_{i=1}^{M} n_i$ for the total number of solute molecules in the system. Equation (6.34) tells us that the *next event* in the system, i.e., the next diffusive transfer of *some* solute molecule out of its current cell, will occur in the next dt with probability αdt, where $\alpha = (2n_{\text{tot}} - n_1 - n_M)\,\kappa_l$. Recall that, when we derived the single-molecule diffusion simulation algorithm in Section 6.4, our key premise was the fact, expressed in Eq. (6.13), that the next cell jump of a *particular* solute molecule will occur in the next dt with probability cdt, where $c = (\varepsilon_{i1} + \varepsilon_{iM})\kappa_l$ and ε_{ij} is 0 if $i = j$ and 1 otherwise. So, by applying the same reasoning we used to derive from Eq. (6.13) the result in Eq. (6.14), namely that $\exp(-c\tau)$ with $c = (\varepsilon_{i1} + \varepsilon_{iM})\kappa_l$ gives the probability that a time τ will elapse without a *particular* solute molecule leaving its current cell i, we may conclude that $\exp(-\alpha\tau)$ with $\alpha = (2n_{\text{tot}} - n_1 - n_M)\,\kappa_l$ is the probability that a time τ will elapse without *any* solute molecule in Ω leaving its current cell. From this result, we may infer that the probability that the *next event* in the system will occur in the time interval $[t + \tau, t + \tau + d\tau)$ *and* will be a solute molecule *leaving* cell j is

$$p(\tau, j)\,d\tau = \exp(-\alpha\tau) \times (\varepsilon_{j1} + \varepsilon_{jM})\,n_j \kappa_l d\tau; \tag{6.35}$$

because, the first factor on the right is the probability that nothing will happen in the time interval $[t, t + \tau)$; and the second factor is the *subsequent* probability that some one of the n_j solute molecules in cell j will exit that cell, through either of the cell boundaries, in the time interval $[t + \tau,\ t + \tau + d\tau)$. We conclude that the *joint PDF* of {the time τ to the next diffusive transfer event} and {the index j of the originating cell of that transfer} is

$$p(\tau, j) = \alpha \exp(-\alpha\tau) \cdot \frac{(\varepsilon_{j1} + \varepsilon_{jM})\,n_j \kappa_l}{\alpha}, \tag{6.36}$$

where $\alpha = (2n_{\text{tot}} - n_1 - n_M)\,\kappa_l$. Notice that the first factor on the right is the properly normalized PDF of the exponential random variable $\mathcal{E}(\alpha)$ [see Eq. (2.3)], and it depends on τ but not j, while the second factor depends on j but not τ. We may conclude from this that the time τ to the next diffusive transfer event inside Ω and the index j of the originating cell in that transfer are statistically independent random variables, with τ being the *exponential* random variable $\mathcal{E}(\alpha)$, and j being the *integer* random variable with probability mass

$$p_j = \frac{(\varepsilon_{j1} + \varepsilon_{jM})\,n_j \kappa_l}{\alpha} \equiv \frac{(\varepsilon_{j1} + \varepsilon_{jM})\,n_j}{2n_{\text{tot}} - n_1 - n_M}.$$

Exact algorithms for generating sample values of τ according to $\mathcal{E}(\alpha)$ and sample values of j according to p_j can be read off from Eqs (2.35) and (2.40)—see steps 2° and 3° below. As for the *destination* cell of the next diffusive transfer, it will be the only adjacent cell if $j = 1$ or M, and either of the two adjacent cells $j - 1$ or $j + 1$

with equal probability if $2 \leq j \leq M-1$. We thus have the following *exact* algorithm for numerically simulating the time evolution of a *collection* of solute molecules in the discrete-stochastic approach:

$1°$ With n_i molecules in cell i at time t, and $n_{\text{tot}} \equiv \sum_{i=1}^{M} n_i$ assumed constant throughout the simulation, compute $a_1 = \kappa_l n_1$, $a_M = \kappa_l n_M$, and for all $i \in [2, M-1]$, $a_i = 2\kappa_l n_i$. Then compute $a_0 \equiv \sum_{i=1}^{M} a_i = \kappa_l \left(2n_{\text{tot}} - n_1 - n_M \right)$.

$2°$ Where u_1 is a unit-interval uniform random number, compute the time step to the next diffusive transfer event as $\tau = (1/a_0) \ln \left(1/(1 - u_1) \right)$.

$3°$ Where u_2 is another unit-interval uniform random number, compute the index j of the cell *from* which the diffusive transfer will originate as the *smallest* integer in $[1, M]$ for which $\sum_{i=1}^{j} a_i \geq u_2 a_0$.

$4°$ Effect the diffusive transfer by replacing $t \leftarrow t + \tau$; $n_j \leftarrow n_j - 1$; and
   - if $j = 1$, $n_2 \leftarrow n_2 + 1$;
   - if $j = M$, $n_{M-1} \leftarrow n_{M-1} + 1$;
   - if $1 < j < M$, where u_3 is another unit-interval uniform random number, if $u_3 < \frac{1}{2}$ then $n_{j-1} \leftarrow n_{j-1} + 1$, otherwise $n_{j+1} \leftarrow n_{j+1} + 1$.

$5°$ Output $(t; n_1, \ldots, n_M)$ if desired. Either return to $1°$, or end the simulation.

Notice that although $a_1, \ldots, a_M$ and a_0 all must be computed in step $1°$ at the beginning of a run, after each diffusive transfer only two of the $a_1, \ldots, a_M$ values will need to be recomputed; furthermore, a_0 will need to be recomputed only if n_1 or n_M gets changed in the transfer. Attention to such computational efficiencies will be rewarded for large M and n_{tot}, because the many-molecule simulation algorithm is inherently more computationally intensive than the single-molecule simulation algorithm of Section 6.4. There are two reasons for that: First, the average τ in step $2°$ here is of order $(1/a_0) \sim (2\kappa_l n_{\text{tot}})^{-1}$, which is smaller by a factor of $1/n_{\text{tot}}$ than the average τ in the single-molecule simulation algorithm. And second, the computation of j in step $3°$ here is an order-M operation, whereas it is an order-1 operation in the single-molecule simulation algorithm.

6.10 Some examples of many-molecule simulations

To illustrate the many-molecule simulation algorithm, we will as in Section 6.5 base our modeling on the diffusion system described in Section 5.9. Our volume Ω is thus a long cylinder with its left face at $x = 0$ and its right face at $x = 400$ μm, and we imagine it to be subdivided into $M = 80$ cells, each of length $l = 5$ μm. The solvent molecules are water molecules, and the role of the solute molecules is played by 200 polystyrene beads of diameter 0.29 μm, for which the constant κ_l has the value in Eq. (6.16).

In the laboratory experiment described in Section 5.9, an impermeable partition was in effect placed between cells 40 and 41, dividing Ω into a left half (cells 1 through 40) and a right half (cells 41 through 80). Approximately 200 polystyrene beads were introduced into the right half and allowed to equilibrate to a "uniform" distribution there. Then, at time $t = 0$, the partition was removed, and the beads were allowed to diffuse freely throughout the entire volume Ω.

Since $P(\mathbf{n}, t|\mathbf{n}_0)$ is the joint PDF of 80 random variables, namely the cell populations $N_1(t|\mathbf{n}_0), \ldots, N_{80}(t|\mathbf{n}_0)$, the state space $\{\mathbf{n}\}$ of this system is an 80-dimensional integer lattice, with each axis extending from 0 to 200 (the total number of beads). The number of accessible states on this lattice is thus astronomically large, so estimating the solution $P(\mathbf{n}, t|\mathbf{n}_0)$ of the many-molecule master equation (6.20), from either experimental data or simulated data, is almost beyond feasibility. We will have to be satisfied with a much less detailed picture of the system's behavior, and in that regard we have essentially two options: Either (i) we can look at the populations of all the states at a particular time, or (ii) we can look at the population of a particular state as a function of time. In this section we will illustrate how both of these strategies can be pursued using the many-molecule simulation algorithm described in the preceding section.

For a type (i) simulation experiment, we will make a series of K "runs" of the many-molecule simulation algorithm. Each run will start at time $t = 0$ s with the 200 beads distributed uniformly over the right half of Ω, and will terminate when t first exceeds some chosen time $T > 0$. At the termination of each run k $(k = 1, \ldots, K)$, we will record the cell populations $n_1^{(k)}, \ldots, n_M^{(k)}$ at time T; these will be the cell populations *just before* the change caused by the final time step that took t beyond T. From this data, we will then compute the following 160 averages:

$$\overline{N_i(T)} = \frac{1}{K} \sum_{k=1}^{K} n_i^{(k)}, \quad \overline{N_i^2(T)} = \frac{1}{K} \sum_{k=1}^{K} \left(n_i^{(k)} \right)^2 \quad (i = 1, \ldots, 80). \tag{6.37}$$

These averages are our simulation estimates of the first and second moments, $\langle N_i(T) \rangle$ and $\langle N_i^2(T) \rangle$. Notice that to obtain these averages, the individual $n_i^{(k)}$ values need not be retained after each run; we need retain only the *cumulating sums* of $n_i^{(k)}$ and $\left(n_i^{(k)} \right)^2$, which after the Kth run we will divide by K to get the averages in Eqs (6.37). We will then plot, as a function of the cell number i, $\overline{N_i(T)}$ along with the associated "one-standard-deviation envelope"

$$\overline{N_i(T)} \pm \text{sdev} \{N_i(T)\}, \tag{6.38}$$

where

$$\text{sdev} \{N_i(T)\} \approx \sqrt{\overline{N_i^2(T)} - \overline{N_i(T)}^2}. \tag{6.39}$$

In addition—just to illustrate how it can be done—we will compute a typical covariance at time T, say between the populations of cells 30 and 34:

$$\text{cov} \{N_{30}(T), N_{34}(T)\} = \overline{N_{30}(T) \cdot N_{34}(T)} - \overline{N_{30}(T)} \cdot \overline{N_{34}(T)}. \tag{6.40}$$

This evidently requires tallying, in addition to the 160 averages in Eq. (6.37), the average

$$\overline{N_{30}(T) \cdot N_{34}(T)} = \frac{1}{K} \sum_{k=1}^{K} n_{30}^{(k)} n_{34}^{(k)}. \qquad (6.41)$$

For the initial distribution of the beads, which is stipulated to be "uniform" over cells 41 through 80, it would be simplest to allocate five beads to each of those 40 cells. But to conform more closely to the actual initial conditions in the experiments of Section 5.9, where all that could be assured at time 0 was that each bead is equally likely to be in any of cells 41 through 80, we instead proceed as follows: First we generate an *initial cell number* ϕ_m for each of the 200 beads by using the generating rule (2.41) to generate 200 random integers uniformly between 41 and 80 inclusively:

$$\phi_m = [41 + 40 \cdot u_m] \quad (m = 1, \dots, 200). \qquad (6.42)$$

Here $u_1, \dots, u_{200}$ are independent samples of $\mathcal{U}(0, 1)$, and $[x]$ denotes "the greatest integer in x". Then we take n_{i0} $(i = 1, \dots, 80)$ to be the number of those ϕ_m-values that are equal to i. The histogram in Fig. 6.5a shows the result of carrying out this initialization procedure for a single run. The superimposed curve shows the average of infinitely many histograms generated in this way. Note that the histogram is statistically similar to the experimental initial cell profile in Fig. 5.8b.

Figure 6.5b shows what the initial distribution in Fig. 6.5a evolved to, via the many-molecule simulation algorithm, at time $t = 3600\,\mathrm{s} = 1\,\mathrm{hr}$. During that time in this particular run, the beads executed 86 038 diffusional transitions between cells. This histogram evidently compares favorably to the histogram for the one-hour experimental run in Fig. 5.8c. Finally, Fig. 6.5c shows what the distribution in Fig. 6.5b evolved to at time $t = 36\,000\,\mathrm{s} = 10\,\mathrm{hr}$. By that time the beads have spread out to cover the entire volume in a nearly uniform way (dashed line), although even at this long time the right half of Ω is still slightly more heavily populated than the left half.

In Fig. 6.6, the large dots at the individual cell locations show the *mean* cell populations at time $t = 5400\,\mathrm{s} = 1.5\,\mathrm{hr}$, obtained by averaging over $K = 10^4$ independent runs. This curve agrees well with the solution to the classical diffusion equation in Fig. 5.9 if allowance is made for the fact that what is actually plotted in Fig. 5.9 are the cell populations divided by the cell length (5 µm). The small dots above and below the large dots in Fig. 6.6 show the corresponding one-standard-deviation envelope of Eq. (6.38). Note that this is new information that is not provided by the classical diffusion equation. The lightly dotted horizontal line is the theoretically expected average cell population profile for $t \to \infty$. Also shown in this figure is the cell population profile of a typical *single* run to time 90 minutes; it compares well with the (differently scaled) experimental histogram in Fig. 5.9. Notice that while the one-standard-deviation envelope characterizes the *scale* of the fluctuations in the cell populations about their averages, it is *not* a hard bound on those fluctuations. Interestingly, the computed value of the standard deviation of each cell population here differs from the square root of the corresponding mean by such a small amount that a plot of $\overline{N_i} \pm \sqrt{N_i}$ would be practically indistinguishable from the exact one-standard-deviation envelope shown. We will discuss why that is so later in this section.

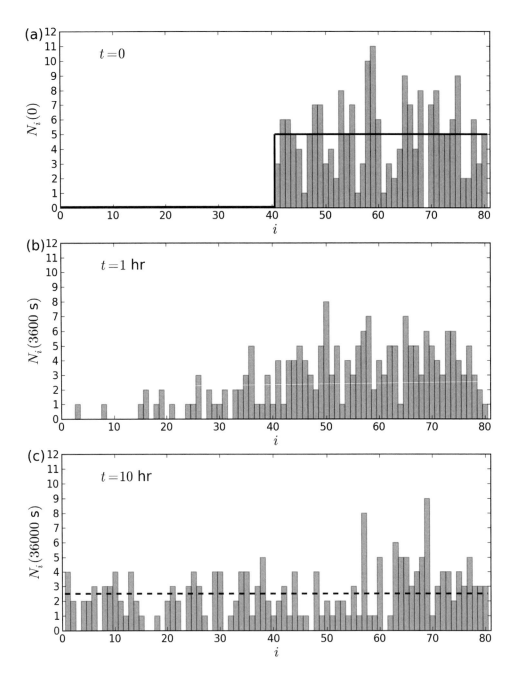

Fig. 6.5 *Computer simulation of the microfluidics diffusion experiment in Section 5.9. (a) A typical cell population profile of the initial state, in which 200 beads are distributed over cells 41 through 80 using formula (6.42); the solid curve shows the average of infinitely many such initial distributions. Notice that the deviations of the histogram from that solid curve are consistent with what was observed experimentally in Fig. 5.8b. (b) The cell populations one hour after the barrier between cells 40 and 41 was removed, obtained using the many-molecule simulation algorithm of Section 6.9. This population profile compares reasonably with the experimental result in Fig. 5.8c. (c) Extending the many-molecule simulation in (b) to 10 hours resulted in this population profile. The dashed line shows the PDF of a uniform distribution over cells 1 through 80. Evidence of the initial state is arguably still present.*

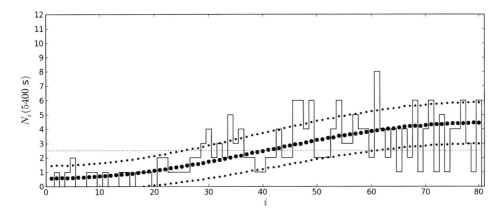

Fig. 6.6 *Computer simulation of the microfluidics diffusion experiment in Section 5.9. The histogram is an extension of the single 60-minute simulation run in Fig. 6.5b to 90 minutes (5400 s). An average over 10^4 such runs yielded the curve formed by the large dots at the individual cell locations, which thus shows the mean cell populations at this time. The light horizontal line at cell count 2.5 shows what we would expect this mean curve to be for $t \to \infty$. With only little additional tabulating, we were also able to extract from the 10^4 runs the one-standard deviation envelope about the mean; it is shown as the two smaller-dotted curves. All of these stochastic simulation results are consistent with the experimental and theoretical results reported in Fig. 5.9, allowing for the fact that what is plotted in Fig. 5.9 are the cell populations divided by the cell length (5 μm).*

In making the $K = 10^4$ runs for Fig. 6.6, we also computed, in addition to the moments in Eqs (6.37), the cross-moment in Eq. (6.41). That resulted in an estimate of -0.0072 for the covariance (6.40) of the bead populations in cells 30 and 34 at time $T = 5400$ s. That covariance becomes, after dividing by the product of the standard deviations of the populations in cells 30 and 34, a *correlation* of -0.0040. This near-zero value for the correlation, on a scale where 1 means perfectly correlated and -1 means perfectly anti-correlated, implies that the populations of cells 30 and 34 are *nearly uncorrelated* at 90 minutes. The explanation for the slight negative value of the correlation is that there is a fixed number of beads, and therefore if one cell contains more than the usual number of beads there will be fewer than usual beads available for the other cell. In the extreme case of only $M = 2$ cells, the correlation of the populations in cells 1 and 2 would be -1.

The second way of examining this system via stochastic simulation is to track the population of a *particular* cell over some specified interval of time. Figure 6.7 illustrates this strategy, focusing on cell 35. The step-curve in this figure shows the bead population of cell 35 as a function of time during a *single* simulation run from $t = 0$ s to $t = 900$ s, using as before a randomly uniform distribution of the 200 beads over the right half of Ω at time $t = 0$ s. During this 15-minute run, the population of cell 35 was observed to change 288 times. Notice that the jumps in this trajectory show *precisely* when each change in the population of cell 35 occurred; therefore, this trajectory accurately depicts the population of cell 35 at *all* times during the run, and not just at some discrete set of observation times.

The center dotted curve in Fig. 6.7 shows the *average* population of cell 35, as computed from $K = 10^4$ runs, and the two dotted curves above and below show the corresponding one-standard-deviation envelope. Computing these three curves required some additional measures that deserve comment: Since the time steps τ that carry the system from one diffusive transfer to the next are stochastically generated by the many-molecule simulation algorithm, those steps will be not only irregularly spaced within a single run, but also different from one run to the next. So to compute the *averaged* curves in Fig. 6.7, we first recorded *for each run* the population of cell 35 at regular 15-second time intervals, i.e., at times $t_0 = 0\,\text{s}$, $t_1 = 15\,\text{s}, \dots, t_{60} = 900\,\text{s}$. To do that, we monitored the time variable t in the simulation algorithm, and at the τ-step that *first* advances t past any t_i we recorded for $n_{35}(t_i)$ the population *just before* that step. At the end of the K runs, we computed for each t_i $(i = 1, \dots, 60)$ the average of the K $n_{35}(t_i)$ values, and also the average of the K $n_{35}^2(t_i)$ values. Again, to minimize data storage requirements, these averages were computed from cumulative sums of the $n_{35}(t_i)$-values and the $n_{35}^2(t_i)$-values. From those averages, we then obtained the mean and one-standard-deviation bounds in Eqs (6.37)–(6.39). The three smooth curves connecting the dots in Fig. 6.7 are least-square polynomial fits. Comparing the one-standard deviation envelope in Fig. 6.7 with the single-run

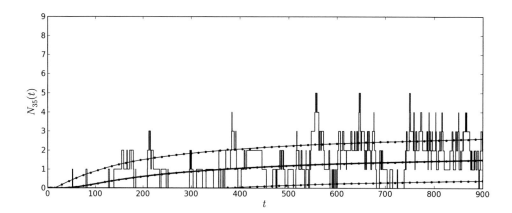

Fig. 6.7 *Computer simulation of the microfluidics diffusion experiment in Section 5.9. Here we focus on the bead population of a single cell, namely cell 35, as a function of time. The histogram is the temporal profile of the cell 35 population during the course of a typical 900 s simulation run of the many-molecule simulation algorithm in Section 6.9. As before, the simulation was started from an initially uniform distribution of 200 beads over cells 41 through 80, which was generated using formula (6.42). The center string of dots shows the average population of cell 35 at specific points in time as obtained from 10^4 independent simulation runs; the curve through those dots has been fitted. The two curves above and below the center curve show the corresponding one-standard deviation envelope, as computed from the simulation results.*

trajectory, we see once again that that envelope characterizes the *typical* size of the fluctuations about the mean, but is *not* a hard bound on those fluctuations.

As in the case of the one-standard deviation envelope in Fig. 6.6, the envelope in Fig. 6.7 turns out to be practically indistinguishable from a plot of $mean \pm \sqrt{mean}$. Some analytical insight on why the standard deviation of a cell population should be so well approximated by the square root of the mean of the cell population *in this particular system* can be gained by considering the $t \to \infty$ case, where each of the 200 beads has probability $\frac{1}{80}$ of being in any cell. In that case the number of beads in any given cell, irrespective of the numbers of beads in any of the other cells, will be the binomial random variable $\mathcal{B}(p, N)$ with $p = \frac{1}{80}$ and $N = 200$, as discussed in connection with Eq. (2.38). It then follows from Eqs (2.39) that the standard deviation of the asymptotic cell population is

$$\sqrt{Np(1-p)} = \sqrt{\left(200 \cdot \tfrac{1}{80}\right)\left(1 - \tfrac{1}{80}\right)} = \sqrt{2.5} \times \sqrt{1 - \tfrac{1}{80}}.$$

This is obviously very close to $\sqrt{2.5}$, the square root of the mean. Although this asymptotic result makes plausible our finding that, at least for these parameter values, the standard deviation in each cell population is approximately equal to the corresponding mean, it does *not* explain why that approximation is also good at *finite* times. But there is another way to look at this matter: Since our finding for finite times was obtained using a simulation procedure that is rigorously equivalent to the many-molecule master equation, then it can be regarded as a solid result *without* any further substantiation. In other words, by using the stochastic simulation approach, we have learned something interesting and useful about this system that we were not able to divine by purely analytical methods.

It would have required little extra effort, in the course of making the K simulation runs for Fig. 6.7, to have also computed any "two-time covariance" $\text{cov}\left\{N_i(t_1), N_j(t_2)\right\}$. All that would have been needed for that is a computation of the average over the K runs of $n_i(t_1)n_j(t_2)$. And confidence intervals for that estimate could be obtained by additionally computing the average of $(n_i(t_1)n_j(t_2))^2$, and then following the procedure described in Appendix 6B.

Generally speaking, the stochastic simulation approach has two major limitations: (i) the large amount of computer time that is sometimes needed to estimate dynamical averages with high accuracy; and (ii) the non-analytical, strictly numerical form of its results. But those limitations are balanced by the ability of the stochastic simulation approach to directly quantify the predictions of the discrete-stochastic hypothesis (5.3) for virtually *any* aspect of the system's dynamical behavior that can be observed experimentally. This is a valuable ability in light of the fact that many experimentally interesting aspects of the system's behavior simply cannot be derived from the many-molecule master equation by presently known methods of analysis. But since our simulation algorithms have been rigorously derived from the same hypothesis (5.3) that was used to derive the master equation, then the numerical results of our simulations are just as definitive as analytically derived formulas for the purpose of making comparisons with experiment.

6.11 A simulation study of Fick's Law

For a final demonstration of simulation strategies in the discrete-stochastic model of diffusion, we will replicate the experimental test carried out in Section 5.9 of Fick's Law, more specifically formulas (5.27) and (5.28) for the mean and variance of the net flux between two adjacent cells as defined in Eq. (5.25). The theory behind this test, and the test procedure itself, are described in the discussion in Section 5.9, and we will now briefly review that discussion.

Starting from the initial distribution of 200 beads shown in Fig. 5.8b or Fig. 6.5a, a *single* run (experimental or simulated) is made to time $t = 21\,600$ s (6 hours). During this run, observations of all cell populations are made at regularly spaced time intervals of $\Delta t = 10$ s; thus, the data collected (experimental or simulated) are the cell populations $N_i(t_k)$ $(i = 1, \ldots, M = 80)$ at times $t_k \equiv (k-1) \cdot \Delta t$ $(k = 1, \ldots, N_{\text{im}})$. Here, N_{im} is the total number of observations (photographic images), which for $\Delta t = 10$ s turns out to be 2161. These observed cell populations are then used to compute the net total fluxes $AJ_i(N_i(t_k), N_{i+1}(t_k); l, \Delta t)$ using the recursion relations (5.39). Recall that $AJ_i(N_i(t_k), N_{i+1}(t_k); l, \Delta t)$ is defined so that its product with Δt gives the *net* number of beads transferred in time $[t_k, t_k + \Delta t)$ from cell i to cell $i+1$. The tacit assumption that this number depends on the time t_k populations of *only* cells i and $i+1$ implies, however, that Δt should be *small* enough that bead jumps between non-adjacent cells in time Δt practically never occur. The requirement for that to be true is given by Eq. (5.15a) as $\Delta t \ll 1/\kappa_l$. For such a Δt, the theory based on hypothesis (5.3) implies that the *mean* of $AJ_i(N_i(t_k), N_{i+1}(t_k); l, \Delta t)$ should satisfy Eq. (5.36a), and its *variance* should satisfy Eq. (5.36b). It is these two theoretical predictions that we want to test by simulation.

Consider first Eq. (5.36a). It predicts that the *mean* of $AJ_i(N_i(t_k), N_{i+1}(t_k); l, \Delta t)$ should be proportional to the *difference* $[N_i(t_k) - N_{i+1}(t_k)]$, and further that the proportionality constant should be κ_l. To see if our *simulated* data support that prediction, we proceed as follows: First we scan the matrix of observed values $N_i(t_k)$ to find all pairs $(n_{i,k}, n_{i+1,k})$ satisfying $n_{i,k} - n_{i+1,k} = d$. For each such pair, we record the corresponding value of $AJ_i(N_i(t_k), N_{i+1}(t_k); l, \Delta t) \equiv f_{i,k}$, which was computed from the data via the recursion relations (5.39). Then we compute the *average* of all the $f_{i,k}$ values associated with the difference d, $\langle f_d \rangle$, and we plot $\langle f_d \rangle$ versus d. Doing that for the 11 integer d-values between -5 and 5, we obtained the points shown by the dots in Fig. 6.8a. The thin solid line in that figure is the best least-squares linear fit to those dots. The heavy solid line is the theoretical curve predicted by Eq. (5.36a). We see that while the simulation results do indeed fall on a straight line through the origin, the slope of that line is substantially smaller than the slope of the line predicted by formula (5.36a). But the line computed from the simulated data agrees *extremely well* with the least-squares straight-line fit to the experimental data in Fig. 5.10a, which we have extrapolated to negative values of $(n_i - n_{i+1})$ and inserted as the dashed line in Fig. 6.8a.

Equation (5.36b) predicts that the *variance* of $AJ_i(N_i(t_k), N_{i+1}(t_k); l, \Delta t)$ should be proportional to the *sum* $[N_i(t_k) + N_{i+1}(t_k)]$, and further that the proportionality constant should be $\kappa_l/\Delta t$. To test that prediction, we again scan the matrix

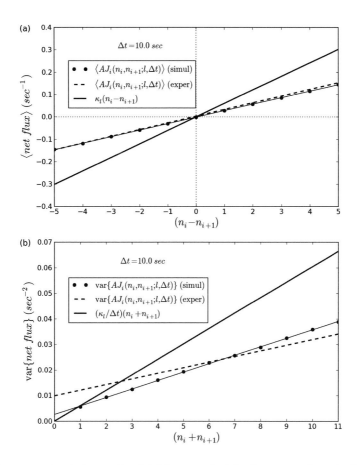

Fig. 6.8 *Computer simulation study of $AJ_i(n_i, n_{i+1}; l, \Delta t)$, the net number of solute molecules that will move from cell i to cell $i + 1$ during the next Δt, given n_i solute molecules in cell i, n_{i+1} solute molecules in cell $i + 1$, and cell length l. The system is the same model of the microfluidics diffusion experiment of Section 5.9 that was used in the preceeding three figures. The value of Δt used here was 10 seconds, the same as was used in the experimental study. (**a**) The mean of $AJ_i(n_i, n_{i+1}; l, \Delta t = 10s)$ as a function of $(n_i - n_{i+1})$. The dots were obtained from simulated data generated in one long run of the many-molecule stochastic simulation algorithm of Section 6.9. During that run, all cell populations were recorded at time intervals of Δt, and then the net numbers of $i \to i + 1$ transfers during each time interval were calculated using the recursion relations (5.39). The dots show averages of those results as a function of $(n_i - n_{i+1})$, and the light solid line has been fitted to the dots. The dashed line is adapted from the microfluidics diffusion experiment result in Fig. 5.10a. The heavy solid line is the theoretical prediction of formula (5.36a). We see that the results from simulation and experiment agree very well with each other, but not so well with the theoretical formula (5.36a). (**b**) The variance of $AJ_i(n_i, n_{i+1}; l, \Delta t = 10s)$ as a function of $(n_i + n_{i+1})$. The dots and their fitted line were obtained from the stochastic simulation run in the same way as in (a), but now as a function of $(n_i + n_{i+1})$ instead of $(n_i - n_{i+1})$. The dashed line is the microfluidics diffusion experiment result in Fig. 5.10b, and the heavy solid line is the result predicted by formula (5.36b). We see that for the variance, the predictions of simulation and experiment do not agree with each other as well as they did for the mean in (a); however, they are closer to each other than they are to the theory formula.*

of simulated values $N_i(t_k)$, but this time to find all pairs $(n_{i,k}, n_{i+1,k})$ satisfying $n_{i,k} + n_{i+1,k} = s$. For each such pair, we record the corresponding value of $AJ_i(N_i(t_k), N_{i+1}(t_k); l, \Delta t) \equiv f_{i,k}$. Then we compute the *variance* of all the $f_{i,k}$ values associated with the sum s, var$\{f_\mathrm{s}\}$, and we plot var$\{f_\mathrm{s}\}$ versus s. Doing that for all integer s-values from 1 to 11, we obtained the points shown by the dots in Fig. 6.8b. The thin solid line in that figure is the best least-squares linear fit to those dots, while the heavy solid line is the theoretical curve predicted by Eq. (5.36b). Again we see that the simulation results do indeed fall on a straight line; however, the slope of that line is substantially smaller than what is theoretically predicted, and the line does not pass through the origin as theory predicts. The dashed line in Fig. 6.8b is the least-squares straight-line fit to the experimental data in Fig. 5.10b. Although the agreement between the simulation results and the experimental results for the variance is not as good as it is for the mean in Fig. 6.8a, the experimental results for the variance are much closer to the simulation prediction than to the theory prediction.

What can we conclude from the fact that the simulation results and the experimental results agree fairly well with each other, but not with the theoretically predicted formulas (5.36)? In the absence of the simulation results, one possible conclusion would be that the discrete-stochastic hypothesis (5.3), from which formulas (5.36) were derived, is somehow defective. But our simulation results were obtained using a mathematical algorithm that was rigorously derived from that hypothesis. So the finding that the simulation results disagree with the predictions of formulas (5.36) in much the same way that the experimental results do tells us that it is not hypothesis (5.3) that is at fault, but rather formulas (5.36). The cause of the difficulty is not hard to identify: The derivation of formulas (5.36) from hypothesis (5.3) required the time interval Δt to be *small* in the sense that $\Delta t \ll 1/\kappa_l$. As we noted in the discussion of the experiment in Section 5.9, we have in this case $1/\kappa_l = 16.6\,\mathrm{s}$; thus, the chosen observation time interval $\Delta t = 10$ s poorly satisfies the condition $\Delta t \ll 1/\kappa_l$.

If a too-large value for Δt is indeed the cause of the problem, then a simulation study using a smaller value for Δt should show better agreement with the theoretically predicted curves. We therefore repeated the simulation described above *using the same simulated trajectory* but now sampling that trajectory at the smaller time interval of $\Delta t = 1$ s. The results obtained are shown in Fig. 6.9. The agreement with the theoretically predicted curves is dramatically improved. Notice that the theoretically predicted curve for the *mean* here is the same curve as in Fig. 6.8a, but the theoretically predicted curve for the *variance* has changed in that its slope has increased by a factor of 10 in accordance with Eq. (5.36b). A third windowing of the same simulated trajectory, this time with $\Delta t = 0.1$ s, produced the results shown in Fig. 6.10. At this observation interval, which satisfies the smallness criterion $\Delta t \ll 1/\kappa_l$ by a comfortable two orders of magnitude, differences between the simulation results and the theoretically predicted results are practically non-existent.

We conclude that the apparent shortcomings of the stochastic Fick's Law formulas (5.36) in Fig. 6.8 are indeed a consequence of not satisfying the condition $\Delta t \ll 1/\kappa_l$ which was assumed in deriving those formulas. We also conclude that, when the condition $\Delta t \ll 1/\kappa_l$ *is* satisfied, formulas (5.36) *are accurate*. Notice that having a smallness condition on Δt for formulas (5.36) is not at all unreasonable: The key

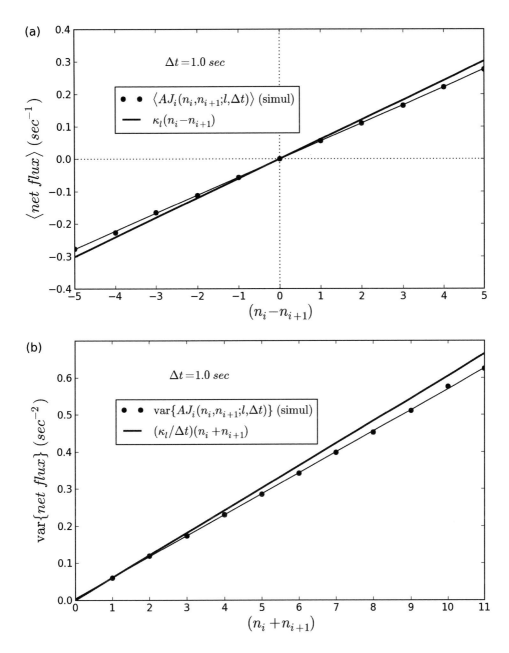

Fig. 6.9 *Results obtained from the same stochastic simulation run made in Fig. 6.8, but now with the cell populations sampled at intervals of* $\Delta t = 1$ s. *The aim here is to test the conjecture that the discrepancies in Fig. 6.8 between the simulated results (dots with fitted light line) and the predictions of formulas (5.36) (heavier line) are due to inaccuracies in the latter arising from* $\Delta t = 10$ s *being too large to satisfy the condition* $\Delta t \ll 1/\kappa_l = 16.6$ s, *a condition that was assumed in deriving formulas (5.36). The agreement between the simulated data and the theoretical formulas for this smaller* Δt *value is indeed substantially improved over Fig. 6.8.*

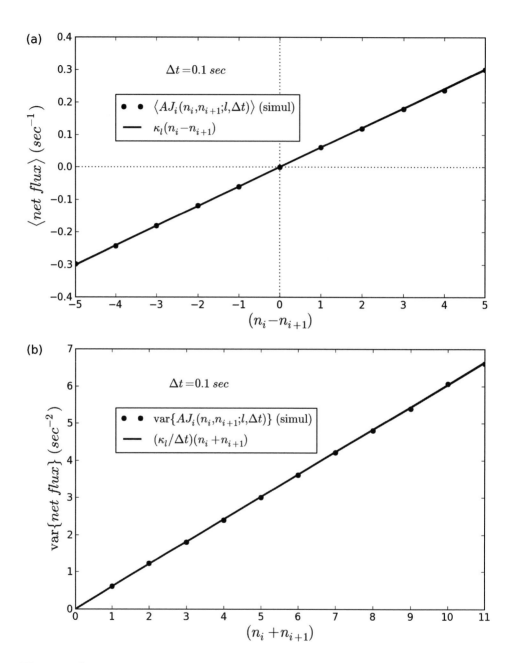

Fig. 6.10 *As in Fig. 6.9, but now with the cell populations sampled at $\Delta t = 0.1$ s intervals. Here Δt is a comfortable two orders of magnitude smaller than $1/\kappa_l = 16.6$ s. The excellent agreement between the simulated results and formulas (5.36) confirms the conjecture that Δt in Fig. 6.8 was simply too large to expect high accuracy from formulas (5.36). A too-large value for Δt will allow some solute molecules to transfer to non-adjacent cells during Δt, and such transfers are not accounted for in Eqs (5.36).*

physical quantity J_i in those formulas is a *flux*, which, like for instance a velocity, is an amount of something *per unit of time*; consequently, it is most usefully defined over a time interval Δt that is by some measure small. So the failure of formulas (5.36) for large Δt should not be seen as a defect of those formulas, but simply as proof that the conditions for the validity of the discrete-stochastic approach need to be taken seriously. For instance, if $\Delta t = 10\,\mathrm{s}$, so that $\kappa_l \Delta t = 0.6$, then interpreting the latter as "the probability that a solute molecule will exit its current cell through its right boundary in the next Δt" would imply that the probability that the molecule will leave its current cell through *either* boundary in the next Δt is $0.6 + 0.6 = 1.2$. It is of course considered extremely bad form for a probability to exceed 1!

It would be satisfying to derive from hypothesis (5.3) analytical formulas for the fitted simulation lines in Figs 6.8 and 6.9—formulas that would effectively generalize Eqs (5.27) and (5.28) to finite time increments. Unfortunately, attempts to do that have thus far been unsuccessful. We therefore have in those fitted simulation lines an example of an interesting consequence of hypothesis (5.3) that has been revealed by simulation but not successfully derived analytically. Of course, even if an analytical derivation is eventually found, much of the credit for that derivation must be given to the experiment that first exposed the result, and the stochastic simulation that validated it in the context of the discrete-stochastic model of diffusion. All of which illustrates the potent synergism that can arise by judiciously combining experiment, analysis, and simulation.

Appendix 6A: General solution to the single-molecule master equation

The single-molecule master equation (6.3) can be written in matrix form as

$$\frac{d}{dt}\begin{bmatrix} p(1,t|i_0) \\ p(2,t|i_0) \\ \vdots \\ p(M-1,t|i_0) \\ p(M,t|i_0) \end{bmatrix} = \kappa_l \begin{bmatrix} -1 & 1 & \cdots & 0 & 0 \\ 1 & -2 & 1 & \cdots & 0 \\ \vdots & \ddots & \ddots & \ddots & \vdots \\ 0 & \cdots & 1 & -2 & 1 \\ 0 & 0 & \cdots & 1 & -1 \end{bmatrix}\begin{bmatrix} p(1,t|i_0) \\ p(2,t|i_0) \\ \vdots \\ p(M-1,t|i_0) \\ p(M,t|i_0) \end{bmatrix}, \qquad (6A.1a)$$

or more compactly,

$$\frac{d\mathbf{p}(t)}{dt} = \kappa_l \mathbf{A}\,\mathbf{p}(t), \qquad (6A.1b)$$

with obvious definitions for the vector $\mathbf{p}(t)$ and the matrix $\mathbf{A}$. It can be shown that the equations $\mathbf{A}\,\mathbf{v}_i = \lambda_i \mathbf{v}_i$ $(i = 1,\ldots,M)$ which define the eigenvalues λ_i and eigenvectors $\mathbf{v}_i \equiv [v_{i1},\ldots,v_{iM}]^{\mathrm{T}}$ of $\mathbf{A}$ (here T denotes the matrix transpose) have the solutions

$$\lambda_i = 2 \left(\cos \frac{(i-1)\pi}{M} - 1 \right) \quad (i = 1, \ldots, M), \tag{6A.2a}$$

$$v_{ik} = M^{-1/2} \cos \left(\frac{(i-1)\pi}{2M} - \frac{(i-1)k\pi}{M} \right) \quad (i, k = 1, \ldots, M). \tag{6A.2b}$$

It can further be shown from Eqs (6A.2b) that the eigenvectors are orthonormal; i.e., each has unit norm, and any two are orthogonal:

$$\mathbf{v}_i^\mathrm{T} \mathbf{v}_j \equiv \sum_{k=1}^M v_{ik} v_{jk} = \delta_{ij}.$$

We introduce a matrix $\mathbf{V}$ which has for its columns the eigenvectors $\mathbf{v}_i$, and also a matrix $\mathbf{E}(t)$ that is diagonal with its ith diagonal element $e^{\kappa_l \lambda_i t}$:

$$(\mathbf{V})_{ij} \equiv v_{ji} \quad \text{and} \quad (\mathbf{E}(t))_{ij} \equiv \delta_{ij} e^{\kappa_l \lambda_i t} \quad (i, j = 1, \ldots, M). \tag{6A.3}$$

Because of the orthonormality of the eigenvectors, we have

$$\left(\mathbf{V}^\mathrm{T} \mathbf{V} \right)_{ij} = \sum_{k=1}^M \left(\mathbf{V}^\mathrm{T} \right)_{ik} (\mathbf{V})_{kj} = \sum_{k=1}^M v_{ik} v_{jk} = \delta_{ij}.$$

This means that $\mathbf{V}^\mathrm{T} \mathbf{V} = \mathbf{1}$ and hence that $\mathbf{V}^\mathrm{T} = \mathbf{V}^{-1}$. Given these results, it can be proved [see e.g. John C. Strikwerda, *Finite Difference Schemes and Partial Differential Equations* (Chapman & Hall, 1989)] that the solution to the differential equation (6A.1) for the initial condition that the solute molecule is in cell i_0 at time 0 is

$$\mathbf{p}(t) = \mathbf{V} \mathbf{E}(t) \mathbf{V}^\mathrm{T} \mathbf{e}_{i_0}. \tag{6A.4}$$

Here, $\mathbf{e}_i$ is the vector whose i^th component is 1 and all other components are 0. Formulas (6A.2), (6A.3), and (6A.4) evidently provide a way to obtain explicit expressions for the solution to the single-molecule master equation (6.3) for any $M \geq 2$.

Using the above formulas, the asymptotic result (6.11) can be proved for any $M \geq 2$ by reasoning as follows: Equation (6A.2a) implies that $\lambda_1 = 0$, and $\lambda_i < 0$ for $i = 2, \ldots, M$. The second of Eqs (6A.3) then implies that $\mathbf{E}(t \to \infty)$ is the diagonal matrix that has 1 for its (1,1) element (because $e^{\kappa_l \lambda_1 t} \equiv 1$ for all finite t, so it never deviates from that value in the limit $t \to \infty$), and 0 for all its other elements. Therefore,

$$\left(\mathbf{V} \mathbf{E}(\infty) \mathbf{V}^\mathrm{T} \right)_{ij} \equiv \sum_{l=1}^M \sum_{k=1}^M (\mathbf{V})_{ik} \left(\mathbf{E}(\infty) \right)_{kl} \left(\mathbf{V}^\mathrm{T} \right)_{lj},$$

$$= \sum_{l=1}^M \sum_{k=1}^M v_{ki} \left(\delta_{k1} \delta_{l1} \right) v_{lj},$$

$$= v_{1i} v_{1j} = M^{-1}.$$

The last step here follows from the implication of Eq. (6A.2b) that $v_{1k} = M^{-1/2}$ for all $k = 1, \ldots, M$. We therefore have from Eq. (6A.4) that the ith component of $\mathbf{p}(\infty)$ is

$$p(i, \infty | i_0) = \sum_{k=1}^{M} \left(\mathbf{V} \, \mathbf{E}(\infty) \, \mathbf{V}^{\mathrm{T}} \right)_{ik} \delta_{k \, i_0} = \sum_{k=1}^{M} M^{-1} \delta_{k \, i_0} = M^{-1}.$$

This is the asymptotic result (6.11).

The symmetry result (6.12) can be proved for any $M \geq 2$ by reasoning as follows: In Eq. (6A.4), let $\mathbf{B}(t) \equiv \mathbf{V} \, \mathbf{E}(t) \, \mathbf{V}^{\mathrm{T}}$, and take $i_0 = i$. Then we have for the jth component of $\mathbf{p}(t)$,

$$p(j, t | i) = \sum_{k=1}^{M} B_{jk}(t) \, \delta_{ki} = B_{ji}(t). \tag{6A.5}$$

But the transpose of $\mathbf{B}(t) \equiv \mathbf{B}$ is

$$\mathbf{B}^T \equiv (\mathbf{V} \, \mathbf{E} \, \mathbf{V}^T)^T = \mathbf{V}^{TT} \, \mathbf{E}^T \, \mathbf{V}^T = \mathbf{V} \, \mathbf{E} \, \mathbf{V}^T = \mathbf{B}.$$

Therefore, $\mathbf{B}(t)$ is a symmetric matrix. The consequent relation $B_{ji}(t) \equiv B_{ij}(t)$, together with Eq. (6A.5), then implies that $p(j, t | i) = p(i, t | j)$, as asserted by Eq. (6.12).

Appendix 6B: Confidence intervals in Monte Carlo averaging

Consider an experiment, either a real laboratory experiment or a Monte Carlo simulated experiment, in which K independent measurements of some quantity Y yield the values $y_1, \ldots, y_K$. Assuming these values can be regarded as "sample values" of a "random variable" Y which has a well-defined mean, then that mean will be given by

$$\langle Y \rangle = \lim_{K \to \infty} \frac{1}{K} \sum_{k=1}^{K} y_k. \tag{6B.1}$$

This suggests that for K large but *finite*, we can say that

$$\langle Y \rangle \approx \frac{1}{K} \sum_{k=1}^{K} y_k. \tag{6B.2}$$

But just how good is this finite-K approximation?

To answer this question, we will make the additional assumption that Y also has a well-defined variance, var $\{Y\}$. Then letting $Y_1, \ldots, Y_K$ be K statistically independent *copies* of the random variable Y, so that $\langle Y_k \rangle = \langle Y \rangle$ and var $\{Y_k\} =$ var $\{Y\}$, we can regard y_k to be a sample value of Y_k for each $k = 1, \ldots, K$. Next we define a new random variable Z_K by

$$Z_K \equiv \frac{1}{K} \sum_{k=1}^{K} Y_k. \tag{6B.3}$$

Then the "data average" on the right side of Eq. (6B.2), namely

$$\frac{1}{K}\sum_{k=1}^{K} y_k \equiv z_k, \tag{6B.4}$$

can be regarded as a sample value of Z_K. Now, according to the Central Limit Theorem (2.27),

$$\lim_{K\to\infty} Z_K = \mathcal{N}\left(\langle Y\rangle, \mathrm{var}\{Y\}/K\right). \tag{6B.5}$$

Therefore, if K is "sufficiently large", we can make the approximation

$$Z_K \approx \mathcal{N}\left(\langle Y\rangle, \mathrm{var}\{Y\}/K\right). \tag{6B.6}$$

This result tells us that sampling Z_K, which is essentially what we do when we compute the data average (6B.4), will be approximately like sampling a *normal* random variable with mean $\langle Y\rangle$ and variance $\mathrm{var}\{Y\}/K$.

In general, sampling the *normal* random variable $X = \mathcal{N}(\mu, \sigma^2)$ will yield a value x in the interval $[\mu - \alpha\sigma, \mu + \alpha\sigma]$ with probability 68% if $\alpha = 1$, 95% if $\alpha = 2$, and 99.7% if $\alpha = 3$. This fact is often expressed by writing $\mu \approx x \pm \alpha\sigma$, and saying that the (possibly unknown) mean μ of X will be within $\pm\alpha\sigma$ of any sample value x of X 68% of the time for $\alpha = 1$, 95% of the time for $\alpha = 2$, and 99.7% of the time for $\alpha = 3$. Applying this rule to the result (6B.6), we can say that *provided K is sufficiently large*,

$$\langle Y\rangle \approx \frac{1}{K}\sum_{k=1}^{K} y_k \pm \alpha\sqrt{\frac{\mathrm{var}\{Y\}}{K}}. \tag{6B.7}$$

Equation (6B.7) evidently *quantifies* the approximation (6B.2). But in any practical application of (6B.7) where $\langle Y\rangle$ is not known, $\mathrm{var}\{Y\}$ will not be known either. Therefore, we make *one more approximation*, one that also requires K to be "large":

$$\mathrm{var}\langle Y\rangle \equiv \langle Y^2\rangle - \langle Y\rangle^2 \approx \frac{1}{K}\sum_{k=1}^{K} y_k^2 - \left(\frac{1}{K}\sum_{k=1}^{K} y_k\right)^2. \tag{6B.8}$$

Substituting this into Eq. (6B.7) gives us our final formula for quantifying the accuracy of the finite-K estimate of $\langle Y\rangle$ in Eq. (6B.2):

$$\langle Y\rangle \approx \frac{1}{K}\sum_{k=1}^{K} y_k \pm \frac{\alpha}{\sqrt{K}}\sqrt{K^{-1}\sum_{k=1}^{K} y_k^2 - \left(K^{-1}\sum_{k=1}^{K} y_k\right)^2}. \tag{6B.9}$$

It is important to keep in mind that *the validity of formula (6B.9) hinges on K being "large enough"*. An unsophisticated but pragmatic way of deciding whether or not that condition is satisfied is to obtain more than one independent data set $\{y_1,\ldots,y_K\}$, and then check to see if the scatter in their separate estimates (6B.9) is roughly consistent with the predicted $\alpha = 1$ spread. If it is not, then K probably needs to be taken larger. Since the quantity under the large radical sign in (6B.9) becomes independent of K as $K \to \infty$, the size of the uncertainty interval given by (6B.9) is proportional to $1/\sqrt{K}$. Thus, in order to *halve* the size of the uncertainty interval, we would have to increase the number of samples K by a factor of 4. The usefulness of (6B.9) lies in the fact that it allows an *estimate of the accuracy* of the approximation (6B.2) at the relatively small additional cost of computing, in addition to the sum $\sum_{k=1}^{K} y_k$, also the sum $\sum_{k=1}^{K} y_k^2$.

An often used application of Eq. (6B.9) is when Y takes on only the two values 1 and 0, these with respective probabilities p and $(1 - p)$. In other words, Y is the binomial random variable

$$Y = \mathcal{B}(p, 1) \tag{6B.10}$$

as defined in Eq. (2.38). In that case, we will evidently have

$$\sum_{k=1}^{K} y_k = \sum_{k=1}^{K} y_k^2 = N_K, \tag{6B.11}$$

where N_K is the number of the K sample values $\{y_1, \dots, y_K\}$ that have the value 1. When (6B.11) is substituted into (6B.9), the result is found to be

$$\langle Y \rangle \approx \frac{N_K}{K} \pm \frac{\alpha}{\sqrt{K}} \sqrt{\frac{N_K}{K}\left(1 - \frac{N_K}{K}\right)}. \tag{6B.12}$$

Thus, the estimate of $\langle Y \rangle$ in this case is the fraction of the sample values that are 1, and the $\pm$uncertainty interval around that estimate is equal to $\alpha/\sqrt{K}$ times the square root of {the fraction of the sample values that are 1} times {the fraction of the sample values that are 0}.

Appendix 6C: Derivation of the first moment equation (6.31)

Differentiating definition (6.30a) with respect to t and then invoking the master equation (6.20) gives, suppressing for brevity the variables t and $\mathbf{n}_0$,

$$\frac{d \langle N_i \rangle}{dt} = \kappa_l \sum_{\mathbf{n}} \sum_{k=1}^{M} \varepsilon_{k1} n_i (n_k + 1) P(n_{k-1} - 1, n_k + 1, \tilde{\mathbf{n}})$$

$$+ \kappa_l \sum_{\mathbf{n}} \sum_{k=1}^{M} \varepsilon_{kM} n_i (n_k + 1) P(n_k + 1, n_{k+1} - 1, \tilde{\mathbf{n}})$$

$$- \kappa_l \sum_{\mathbf{n}} \sum_{k=1}^{M} (\varepsilon_{k1} + \varepsilon_{kM}) n_i n_k P(\mathbf{n}). \tag{6C.1}$$

Writing this

$$\frac{d\langle N_i\rangle}{dt} \equiv \alpha_i + \beta_i - \gamma_i, \tag{6C.2}$$

where α_i, β_i, and γ_i denote the respective double-sum terms in (6C.1), we will evaluate these three terms separately.

For α_i, we separate from the k-summation the two cases $k = i$ and $k = i+1$, recognizing that the latter case will not occur if $i = M$:

$$\alpha_i \equiv \kappa_l \sum_{\substack{k \neq i \\ k \neq i+1}} \sum_{\mathbf{n}} \varepsilon_{k1} n_i (n_k + 1) P(n_{k-1} - 1, n_k + 1, \tilde{\mathbf{n}})$$

$$+ \kappa_l \sum_{\mathbf{n}} \varepsilon_{i1} n_i (n_i + 1) P(n_{i-1} - 1, n_i + 1, \tilde{\mathbf{n}})$$

$$+ \kappa_l \varepsilon_{iM} \sum_{\mathbf{n}} \varepsilon_{i+1,1} n_i (n_{i+1} + 1) P(n_i - 1, n_{i+1} + 1, \tilde{\mathbf{n}}).$$

Next we shift the summation variables $\mathbf{n}$ in each of the above terms so that they coincide with the arguments of the function P in that term, and we use in particular the facts that

$$\sum_{\mathbf{n}} P(n_k \pm 1, \tilde{\mathbf{n}}) \equiv \sum_{\mathbf{n}} P(\mathbf{n})$$

and

$$\sum_{\mathbf{n}} n_k P(n_k \pm 1, \tilde{\mathbf{n}}) \equiv \sum_{\mathbf{n}} (n_k \mp 1) P(\mathbf{n}).$$

After noting from the definition (6.2) that $\varepsilon_{i+1,1} \equiv 1$, we obtain

$$\alpha_i = \kappa_l \sum_{\substack{k \neq i \\ k \neq i+1}} \varepsilon_{k1} \langle N_i N_k \rangle + \kappa_l \varepsilon_{i1} \langle (N_i - 1) N_i \rangle + \kappa_l \varepsilon_{iM} \langle (N_i + 1) N_{i+1} \rangle. \tag{6C.3}$$

For β_i, we separate from the k-summation the two cases $k = i$ and $k = i-1$, recognizing that the latter case will not occur if $i = 1$:

$$\beta_i \equiv \kappa_l \sum_{\substack{k \neq i \\ k \neq i-1}} \sum_{\mathbf{n}} \varepsilon_{kM} n_i (n_k + 1) P(n_k + 1, n_{k+1} - 1, \tilde{\mathbf{n}})$$

$$+ \kappa_l \sum_{\mathbf{n}} \varepsilon_{iM} n_i (n_i + 1) P(n_i + 1, n_{i+1} - 1, \tilde{\mathbf{n}})$$

$$+ \kappa_l \varepsilon_{i1} \sum_{\mathbf{n}} \varepsilon_{i-1,M} n_i (n_{i-1} + 1) P(n_{i-1} + 1, n_i - 1, \tilde{\mathbf{n}}).$$

Again shifting some summation indices, and noting that $\varepsilon_{i-1,M} \equiv 1$, we get

$$\beta_i = \kappa_l \sum_{\substack{k \neq i \\ k \neq i-1}} \varepsilon_{kM} \langle N_i N_k \rangle + \kappa_l \varepsilon_{iM} \langle (N_i - 1)N_i \rangle + \kappa_l \varepsilon_{i1} \langle (N_i + 1)N_{i-1} \rangle. \qquad (6C.4)$$

Finally, for γ_i, we have

$$\gamma_i \equiv \kappa_l \sum_k \varepsilon_{k1} \sum_{\mathbf{n}} n_i n_k P(\mathbf{n}) + \kappa_l \sum_k \varepsilon_{kM} \sum_{\mathbf{n}} n_i n_k P(\mathbf{n})$$

$$= \kappa_l \sum_k \varepsilon_{k1} \langle N_i N_k \rangle + \kappa_l \sum_k \varepsilon_{kM} \langle N_i N_k \rangle.$$

Breaking up each of the last two terms according to the values of the index k, and noting that $\varepsilon_{i+1,1} \equiv 1$ and $\varepsilon_{i-1,M} \equiv 1$, we have

$$\gamma_i = \kappa_l \sum_{\substack{k \neq i \\ k \neq i+1}} \varepsilon_{k1} \langle N_i N_k \rangle + \kappa_l \varepsilon_{i1} \langle N_i^2 \rangle + \kappa_l \langle N_i N_{i+1} \rangle \varepsilon_{iM}$$

$$+ \kappa_l \sum_{\substack{k \neq i \\ k \neq i-1}} \varepsilon_{kM} \langle N_i N_k \rangle + \kappa_l \varepsilon_{iM} \langle N_i^2 \rangle + \kappa_l \langle N_i N_{i-1} \rangle \varepsilon_{i1}. \qquad (6C.5)$$

The final ε-factors on the third and sixth terms here are to kill those terms when the value of i requires the moments in them to be absent.

Substituting Eqs (6C.3), (6C.4), and (6C.5) into Eq. (6C.2), we get

$$\frac{d \langle N_i \rangle}{dt} = \kappa_l \sum_{\substack{k \neq i \\ k \neq i+1}} \varepsilon_{k1} \langle N_i N_k \rangle + \kappa_l \varepsilon_{i1} \langle (N_i - 1)N_i \rangle + \kappa_l \varepsilon_{iM} \langle (N_i + 1)N_{i+1} \rangle$$

$$+ \kappa_l \sum_{\substack{k \neq i \\ k \neq i-1}} \varepsilon_{kM} \langle N_i N_k \rangle + \kappa_l \varepsilon_{iM} \langle (N_i - 1)N_i \rangle + \kappa_l \varepsilon_{i1} \langle (N_i + 1)N_{i-1} \rangle$$

$$- \kappa_l \sum_{\substack{k \neq i \\ k \neq i+1}} \varepsilon_{k1} \langle N_i N_k \rangle - \kappa_l \varepsilon_{i1} \langle N_i^2 \rangle - \kappa_l \varepsilon_{iM} \langle N_i N_{i+1} \rangle$$

$$- \kappa_l \sum_{\substack{k \neq i \\ k \neq i-1}} \varepsilon_{kM} \langle N_i N_k \rangle - \kappa_l \varepsilon_{iM} \langle N_i^2 \rangle - \kappa_l \varepsilon_{i1} \langle N_i N_{i-1} \rangle. \qquad (6C.6)$$

The four terms with summation signs in (6C.6) evidently cancel in pairs. The remaining eight terms can be collected as follows:

$$\frac{d\langle N_i \rangle}{dt} = \kappa_l \langle \varepsilon_{i1}(N_i + 1)N_{i-1} - \varepsilon_{i1}N_i N_{i-1} + \varepsilon_{iM}(N_i + 1)N_{i+1} - \varepsilon_{iM}N_i N_{i+1} \rangle$$
$$+ \kappa_l \langle \varepsilon_{i1}(N_i - 1)N_i - \varepsilon_{i1}N_i^2 + \varepsilon_{iM}(N_i - 1)N_i - \varepsilon_{iM}N_i^2 \rangle .$$

In each of the four pairs of terms here the quadratic moments cancel each other, and we are left with

$$\frac{d\langle N_i \rangle}{dt} = \kappa_l \langle \varepsilon_{i1}N_{i-1} + \varepsilon_{iM}N_{i+1} - \varepsilon_{i1}N_i - \varepsilon_{iM}N_i \rangle .$$

This is Eq. (6.31).

7
Continuous Markov process theory

Three years after Einstein published his groundbreaking 1905 paper on diffusion, a different analysis of diffusion was published by Paul Langevin. Although it was not clearly appreciated at the time, Langevin's approach avoided the shortcomings of Einstein's approach that were described in Chapter 4, yet included Einstein's theory as a special limiting approximation. A comprehensive rendering of Langevin's approach to diffusion requires the mathematical machinery of *continuous Markov process theory*, which is essentially a generalization to the stochastic realm of the deterministic theory of ordinary differential equations. In this chapter we will develop the concepts of that theory that are needed to adequately describe the approach to diffusion initiated by Langevin, and to see how that approach connects to Einstein's approach. It might be argued that the field of continuous Markov process theory actually began with Langevin's 1908 paper on Brownian motion, a paper that innovatively framed Newton's Second Law as what mathematicians today call a "stochastic differential equation". In fact, stochastic differential equations are often referred to by physical scientists today as "Langevin equations", a terminology that we will use here.

The purely mathematical review of continuous Markov process theory presented in this chapter might seem like a big diversion in our exposition of the physical theory of diffusion. But the value of this excursion will become clear in Chapters 8 through 11, where we will use these mathematical results to develop in a rigorous and efficient way a more far-reaching theoretical picture of molecular diffusion.

7.1   The Chapman–Kolmogorov and Kramers–Moyal equations

A *process* can be defined as any function $X(t)$ whose independent real variable represents *time*, and whose values can therefore be classified as "past", "present", or "future", with all the usual connotations of those terms. For example, both the past and the present can influence the future, but the future cannot influence either the past or the present—at least that is what we will assume here. As we shall see, these attributes of the time variable t affect the character of a process $X(t)$ in a major way.

A process is said to be *Markov* if and only if its future values depend on its past values *only* through its present value. A Markov process can therefore be said to be "past-forgetting" or "memoryless", since it does not need to look to its past in order to decide what it will do next. This is *not* to say that a Markov process is independent of its past, but only that its future depends upon *nothing more* about its past than what is implied by its present. A Markov process can be either *deterministic*, meaning that its future is precisely determined by its present, or *stochastic*, meaning that its future is only probabilistically determined by its present. In the former case $X(t)$ will be an ordinary sure variable, while in the latter case $X(t)$ will be a random variable. But since a sure variable is a special case of a random variable—namely the case in which the probabilities of all possible values but one vanish—it suffices to focus on the stochastic case. So in what follows, we will regard $X(t)$ as a *time-dependent random variable*.

When the possible values of a Markov process $X(t)$ are real numbers, as we are assuming here, then $X(t)$ will be fully described by a singly-conditioned probability density function (PDF) P, which is defined so that[1]

$$P(x, t|x_0, t_0)\, dx \equiv \text{the probability that } X(t) \text{ will lie in the infinitesimal}$$

$$\text{interval } [x, x + dx), \text{ given that } X(t_0) = x_0 \text{ for } t_0 \le t. \quad (7.1)$$

This function P is required to satisfy not only the three obvious conditions for a PDF,

$$P(x_2, t_2|x_1, t_1) \ge 0, \quad (7.2a)$$

$$\int_{-\infty}^{\infty} P(x_2, t_2|x_1, t_1)\, dx_2 = 1, \quad (7.2b)$$

$$P(x_2, t_1|x_1, t_1) = \delta(x_2 - x_1), \quad (7.2c)$$

but also the somewhat less obvious condition

$$P(x_3, t_3|x_1, t_1) = \int_{-\infty}^{\infty} P(x_3, t_3|x_2, t_2)\, P(x_2, t_2|x_1, t_1)\, dx_2,$$

$$\text{for all } t_1 \le t_2 \le t_3. \quad (7.3)$$

Equation (7.3) is called the *Chapman–Kolmogorov equation*. It is a consequence of applying the addition and multiplication laws of probability to the definition (7.1) thusly:

$$[P(x_3, t_3|x_1, t_1)dx_3] = \underset{\text{all } x_2}{\text{Sum}}\left([P(x_3, t_3|x_2, t_2)dx_3] \cdot [P(x_2, t_2|x_1, t_1)\, dx_2]\right).$$

Here, the probability that the process will go from x_1 at time t_1 to the interval $[x_3, x_3 + dx_3)$ at time t_3 is expressed as the sum of the probabilities of all the mutually exclusive and collectively exhaustive ways this can happen via possible states x_2 at a *fixed* intermediate time t_2. The multiplication law of probability actually requires the probability factor $P(x_3, t_3|x_2, t_2)dx_3$ on the right side, which gives the probability that the process will transition in time $t_3 - t_2$ from x_2 to the interval $[x_3, x_3 + dx_3)$, to be conditioned also on $X(t_1) = x_1$; however, because of the memoryless nature of a Markov process, the t_2-conditioning obviates the earlier t_1-conditioning.

The Chapman–Kolmogorov equation (7.3) is the cornerstone of Markov process theory. Its form for $(x_1, t_1) \to (x_0, t_0)$, $(x_2, t_2) \to (x - \xi, t)$, and $(x_3, t_3) \to (x, t + dt)$,

$$P(x, t + dt|x_0, t_0) = \int_{-\infty}^{\infty} P(x, t + dt|x - \xi, t)\, P(x - \xi, t|x_0, t_0)\, d\xi, \quad (7.4)$$

was anticipated by Einstein in his Eq. (3.1), and is the starting point for deriving several fundamental time-evolution equations for P. (Here and below, dt is understood to be a real variable, distinct from t, which is confined to the interval $[0, \varepsilon]$ where ε is an arbitrarily small positive number.) Thus, suppose we let

$$f(x) \equiv P(x + \xi, t + dt | x, t) \, P(x, t | x_0, t_0),$$

so that the integrand in Eq. (7.4) can be written $f(x - \xi)$. Then the Taylor series expansion

$$f(x - \xi) = f(x) + \sum_{n=1}^{\infty} \frac{(-\xi)^n}{n!} \frac{\partial^n f(x)}{\partial x^n}$$

allows us to write Eq. (7.4) as

$$P(x, t + dt | x_0, t_0) = \int_{-\infty}^{\infty} P(x + \xi, t + dt | x, t) \, P(x, t | x_0, t_0) d\xi$$

$$+ \sum_{n=1}^{\infty} \frac{(-1)^n}{n!} \frac{\partial^n}{\partial x^n} \left[\int_{-\infty}^{\infty} \xi^n P(x + \xi, t + dt | x, t) \, P(x, t | x_0, t_0) \, d\xi \right].$$

The first term on the right evidently integrates to $P(x, t | x_0, t_0)$. Transposing that term, dividing through by dt, and then taking the limit $dt \to 0^+$, we get

$$\frac{\partial}{\partial t} P(x, t | x_0, t_0) = \sum_{n=1}^{\infty} \frac{(-1)^n}{n!} \frac{\partial^n}{\partial x^n} \left[B_n(x, t) P(x, t | x_0, t_0) \right] \quad (t_0 \le t), \qquad (7.5)$$

where we have defined

$$B_n(x, t) \equiv \lim_{dt \to 0^+} \frac{1}{dt} \int_{-\infty}^{\infty} \xi^n P(x + \xi, t + dt | x, t) \, d\xi, \quad (n = 1, 2, \ldots). \qquad (7.6)$$

Equation (7.5) is called the *forward Kramers–Moyal equation*. It shows that the time evolution of $P(x, t | x_0, t_0)$ is fully determined by the set of functions $B_n(x, t)$ in Eq. (7.6). The adjective "forward" naturally prompts the question, is there a "backward" Kramers–Moyal equation? The answer is yes. In fact, we will make use of the backward Kramers–Moyal equation in a later chapter. But because that equation is somewhat aside from the main path of our present exposition, we will defer its derivation until the final section of this chapter.

7.2   The process increment and its PDF

The functions $B_n(x, t)$ defined in Eq. (7.6), which we have just seen collectively determine $P(x, t | x_0, t_0)$ through the forward Kramers–Moyal equation (7.5), are evidently themselves determined by the function

$$P(x + \xi, t + dt | x, t) \equiv \Pi(\xi | dt; x, t). \qquad (7.7)$$

This function $\Pi(\xi|dt; x, t)$ is evidently the PDF of the *increment* (ξ) in the process from time t to the *infinitesimally later* time $t + dt$, given that $X(t) = x$. This increment is itself a random variable, and we shall denote it formally by

$$\Xi(dt; x, t) \equiv X(t + dt) - X(t), \text{ given } X(t) = x. \tag{7.8}$$

Thus, $\Pi(\xi|dt; x, t)d\xi$ gives the probability that $\Xi(dt; x, t) \in [\xi, \xi + d\xi)$. Since the integral in Eq. (7.6) is, according to Eq. (2.5), the nth moment of the random variable $\Xi(dt; x, t)$, we can also write the functions $B_n(x, t)$ in Eq. (7.6) as

$$B_n(x, t) \equiv \lim_{dt \to 0^+} \frac{1}{dt} \int_{-\infty}^{\infty} \xi^n \Pi(\xi|dt; x, t) \, d\xi \equiv \lim_{dt \to 0^+} \frac{1}{dt} \langle \Xi^n(dt; x, t) \rangle. \tag{7.9}$$

This formula makes it clear that specifying the increment $\Xi(dt; x, t)$ in Eq. (7.8), or equivalently its PDF $\Pi(\xi|dt; x, t)$ in Eq. (7.7), will uniquely determine $P(x, t|x_0, t_0)$, and hence the Markov process $X(t)$, for all $t > t_0$.

7.3 The self-consistency requirement

Although we have considerable latitude in how we might specify the increment $\Xi(dt; x, t)$, we must do that in a way that satisfies the following *self-consistency condition*: If the infinitesimal interval $[t, t + dt)$ is divided into two subintervals $[t, t + \alpha dt)$ and $[t + \alpha dt, t + dt)$, where $\alpha \in (0, 1)$, then the increment incurred over the full interval $[t, t + dt)$ should be computable as the *sum* of the successive increments incurred over those two subintervals, at least to lowest order in dt:[2]

$$\Xi(dt; X(t), t) = \Xi(\alpha dt; X(t), t) + \Xi\left[(1 - \alpha)dt; X(t) + \Xi(\alpha dt; X(t), t), \ t + \alpha dt\right]. \tag{7.10a}$$

This requirement is in fact a direct consequence of the Chapman–Kolmogorov condition (7.3). To see this, note first that Eq. (7.3) requires

$$P(x + \xi, t + dt|x, t) = \int_{-\infty}^{\infty} P(x + \xi, t + dt|x + \xi_1, t + \alpha dt)P(x + \xi_1, t + \alpha dt|x, t)d\xi_1.$$

This equation can be written in terms of the PDF Π defined in Eq. (7.7) as

$$\Pi(\xi|dt; x, t) = \int_{-\infty}^{\infty} \Pi(\xi - \xi_1|(1 - \alpha)dt; x + \xi_1, t + \alpha dt)\Pi(\xi_1|\alpha dt; x, t) \, d\xi_1. \tag{7.10b}$$

And this equation has the general form (replacing $\Pi \to Q$ and $\xi \to z$)

$$Q(z) = \int Q_2(z - z_1)Q_1(z_1)dz_1$$

$$\equiv \int dz_1 \int dz_2 Q_2(z_2)Q_1(z_1) \, \delta\left(z - (z_1 + z_2)\right).$$

The last equation implies, by the RVT theorem (2.11), that Q is the PDF of the random variable $Z = Z_1 + Z_2$, where Z_1 and Z_2 are statistically independent random variables with respective PDFs Q_1 and Q_2. Thus we see that Eq. (7.10b) is mathematically equivalent to Eq. (7.10a).

In summary: $\Xi(dt; x, t)$ as defined in Eq. (7.8) completely specifies the Markov process $X(t)$, but $\Xi(dt; x, t)$ must satisfy the self-consistency condition (7.10a). It turns out that condition (7.10a) is a surprisingly strong constraint on the acceptable mathematical forms of $\Xi(dt; x, t)$. Two acceptable forms are known. They give rise to the two major classes of Markov process: *jump* and *continuous*.

The states of a *jump* Markov process can form either a discrete lattice or a continuum. Either way, the process jumps instantaneously from one state to another, but stays put between jumps. It is therefore possible to specify a history or trajectory of a jump Markov process in complete detail: we simply list the states visited in the order visited, along with the times that the jumps occur. Jump Markov processes are described by "master equations", which in turn can be derived from the Chapman–Kolomogorov equation (7.3) and the $\Xi(dt; x, t)$ formalism outlined above, although we will not do that here. We have already encountered jump Markov processes in Chapter 6, in connection with the master equations (6.3) and (6.20).

But we are interested now in *continuous* Markov processes. They are described not by master equations, but rather by equations bearing the names "Langevin" and "Fokker–Planck". The states of a continuous Markov process always form a continuum. The process moves over those states in a continuous but "fitful" way, never pausing to "rest" in any state as jump Markov processes do. It is thus not possible to specify the full trajectory of a continuous Markov process, but only where it is at any sequence of discrete times. Our interest in continuous Markov processes here is motivated mainly by the fact that in Langevin's theory of diffusion, the *velocity* of a solute molecule is modeled as a particular kind of continuous Markov process.

7.4   Derivation of the Langevin equation

A *continuous* Markov process $X(t)$ is by definition a Markov process that satisfies the *continuity condition*

$$X(t + dt) \to X(t) \text{ as } dt \to 0. \tag{7.11a}$$

In terms of the process increment (7.8), this condition can equivalently be written

$$\Xi(dt; x, t) \to 0 \text{ as } dt \to 0. \tag{7.11b}$$

We will now state and prove a theorem that specifies how a Markov process that satisfies this continuity condition—i.e., a continuous Markov process—*must* evolve in time. This theorem will immediately give us the Langevin equation, and with the help of Eqs (7.5) and (7.9) the Fokker–Planck equation as well. The Langevin and Fokker–Planck equations are the principal equations of continuous Markov process theory, and the theorem that gives rise to them is this:

Theorem: Let the increment $\Xi(dt; x, t)$ be required to: (i) satisfy the continuity condition (7.11b); (ii) be a smooth function of its three arguments dt, x, and t; (iii) have well-defined first and second moments; and (iv) satisfy the self-consistency condition (7.10a). Then $\Xi(dt; x, t)$ *must* have the mathematical form

$$\Xi(dt; x, t) = \mathcal{N}\left(A(x, t)dt, D(x, t)dt\right), \tag{7.12}$$

where A and $D \geq 0$ are smooth functions of x and t.

According to this theorem, the seemingly mild restrictions (i)–(iv) on the increment $\Xi(dt; x, t)$ collectively force $\Xi(dt; x, t)$ to be a *normal* random variable with a mean and variance that are both *directly proportional to dt*. The function A in Eq. (7.12) is called the *drift function*, and the function D is called the *diffusion function*—the latter not to be confused with the "coefficient of diffusion" in the classical diffusion equation. Our freedom to construct different continuous Markov processes stems solely from our freedom to specify different functional forms for $A(x, t)$ and $D(x, t)$.

Before we prove this theorem, let us take note of its immediate important corollary: If we combine the definition of $\Xi(dt; x, t)$ in Eq. (7.8) with the normal random variable property (2.16), we can write Eq. (7.12) in the mathematically equivalent form

$$X(t + dt) - X(t) = \mathcal{N}\left(A(X(t), t)dt, D(X(t), t)dt\right)$$

$$= A(X(t), t)dt + \sqrt{D(X(t), t)dt}\mathcal{N}(0, 1),$$

whence

$$X(t + dt) = X(t) + A(X(t), t)dt + D^{1/2}(X(t), t)N(t)\sqrt{dt}, \tag{7.13}$$

where $N(t) \equiv \mathcal{N}(0, 1)$. This is the *standard-form Langevin equation.*[3] Since $N(t')$ must be statistically independent of $N(t)$ for $t' > t$, in order to satisfy the Markov requirement that the current increment in the process be independent of past increments, then $N(t)$ in Eq. (7.13) is understood to satisfy the conditions

$$\langle N(t) \rangle = 0, \tag{7.14a}$$

$$\langle N(t)N(t') \rangle = \begin{cases} 1, & \text{if } t = t', \\ 0, & \text{if } t \neq t'. \end{cases} \tag{7.14b}$$

We will explore the consequences of the Langevin equation (7.13) after we prove the theorem from which it was deduced.

Proof of theorem: With $X(t) = x$, we divide the infinitesimal interval $[t, t + dt)$ into $n \geq 2$ subintervals of equal length dt/n by the points $t_i \equiv t + i(dt/n)$ for $i = 0, \ldots, n$. Then $t_0 \equiv t$ and $t_n \equiv t + dt$, so

$$\Xi(dt; x, t) \equiv X(t + dt) - X(t)$$

$$\equiv X(t_n) - X(t_0)$$

$$\equiv \sum_{i=1}^{n} [X(t_i) - X(t_{i-1})]$$

$$\equiv \sum_{i=1}^{n} [X(t_{i-1} + dt/n) - X(t_{i-1})]$$

$$\Xi(dt; x, t) \equiv \sum_{i=1}^{n} \Xi(dt/n; X(t_{i-1}), t_{i-1}). \tag{7.15}$$

Equation (7.15) can be viewed as a kind of "iterated" version of the self-consistency condition (7.10a), which $\Xi(dt; x, t)$ must satisfy because of requirement (iv) of the theorem. The mathematical rigor of the steps here that lead to Eq. (7.15) makes clear the logical necessity of that self-consistency condition.

Now, we know that the process X is continuous in t by requirement (i), and also that the increment Ξ is a continuous function of its arguments x and t by requirement (ii). Therefore, we can take dt so small that all the t_i's are so close to $t_0 \equiv t$ that, in Eq. (7.15), we can replace t_{i-1} with t and $X(t_{i-1})$ with $X(t) \equiv x$ without spoiling the lowest-order-in-dt accuracy of that equation. Doing that gives

$$\Xi(dt; x, t) = \sum_{i=1}^{n} \Xi_i(dt/n; x, t). \tag{7.16}$$

Here $\Xi_1(dt/n; x, t)$, ..., $\Xi_n(dt/n; x, t)$ are n *statistically independent copies* of the random variable $\Xi(dt/n; x, t)$; their statistical independence follows from our implicit assumption that the process X is Markov, so that any increment must be statistically independent of previous increments. Furthermore, each Ξ_i has a well-defined mean and variance, by assumption (iii). Therefore, by taking n *arbitrarily large*, we can invoke the Central Limit Theorem (2.26) to conclude that $\Xi(dt; x, t)$ in Eq. (7.16) must be a *normal* random variable, and moreover one whose mean and variance satisfy

$$\langle \Xi(dt; x, t) \rangle = n \langle \Xi(dt/n; x, t) \rangle, \quad \text{var}\{\Xi(dt; x, t)\} = n \,\text{var}\{\Xi(dt/n; x, t)\}. \tag{7.17}$$

Notice that both of Eqs (7.17) have the form $h(z) = nh(z/n)$, with z playing the role of the variable dt. We now prove the following lemma: If h is a smooth function of z that satisfies $h(z) = nh(z/n)$ for all positive integers n, then it must be so that $h(z) = Cz$ where C is independent of z. The proof of this lemma starts by differentiating the hypothesized relation $h(z) = nh(z/n)$ with respect to z:

$$h'(z) = nh'(z/n)(1/n) = h'(z/n).$$

In order for this relation to hold for *arbitrarily large* n, it must be so that $h'(z) = h'(z/\infty) = h'(0)$; in words, the slope of the function $h(z)$ must be the same for all z. This means that h must have the form $h(z) = Cz$, where C is independent of z, and that establishes the lemma. Now applying this lemma, with z replaced by dt, to the two smooth functions of dt in Eqs (7.17), we conclude that

$$\langle \Xi(dt; x, t) \rangle = A(x, t)dt, \quad \text{var}\{\Xi(dt; x, t)\} = D(x, t)dt. \tag{7.18}$$

Here, $A(x, t)$ and $D(x, t)$ play the role of the constant C in the lemma—they are constants with respect to the variable dt. By requirement (ii) A and D must be smooth functions of x and t. Also, since variances must be non-negative and dt is non-negative, then the function D must be non-negative. Combining Eqs (7.18) with the fact that $\Xi(dt; x, t)$ is normal, which we inferred earlier from Eq. (7.16) and the Central Limit Theorem, we obtain the result asserted in Eq. (7.12).

7.5 Implications of the Langevin equation

The Langevin equation (7.13) evidently has the character of an "updating formula"—a recipe for computing, from the state of the process at time t, its state at an infinitesimally later time $t + dt$. When Eq. (7.13) is used in that way, $N(t)$ in the last term on the right gets replaced at each step by a *sample value* of the random variable $\mathcal{N}(0, 1)$; a good idea of what those sample values typically are can be inferred from Fig. 2.4. The increment in the process from time t to time $t + dt$ is seen to be composed of two terms: a deterministic *drift term*, which is proportional to dt; and a stochastic *diffusion term*, which is proportional to $\sqrt{dt}$ and also to the random variable $\mathcal{N}(0, 1)$. The fact that the process X defined by the updating formula (7.13) satisfies the continuity condition (7.11a) is obvious. The "memoryless" nature of X is manifested in the fact that the right side of Eq. (7.13) makes no reference to past values of the process. The operation of taking the square root of dt in the diffusion term might seem bizarre, but it is mathematically quite proper: the square root of any non-negative number, no matter how small, is perfectly well defined.

This last observation does, however, bring up an important point: Since dt *is* infinitesimally small, then $dt \ll \sqrt{dt}$; e.g., $10^{-30} \ll 10^{-15}$. Therefore, the magnitude of the deterministic drift term in Eq. (7.13) will practically always be negligibly small compared to the magnitude of the stochastic diffusion term. So why do we not just discard the drift term in Eq. (7.13)? The reason we cannot do that is that the larger diffusion term is also proportional to the random variable $N(t) = \mathcal{N}(0, 1)$, and its sample values tend to be distributed symmetrically about the origin. Over a *succession* of very many dt-increments, the resultant indifferent sign switchings of the diffusion term reduce its *net* contribution to a level that is comparable to the *net* contribution of the drift term. In the imagery of the fable of the tortoise and the hare, the dt-term in Eq. (7.13) is the slow-but-steady tortoise, while the $\sqrt{dt}$-term is the fast-but-erratic hare. Over a long run, it's an even match.

Although the form of the Langevin equation (7.13) might appear to be highly arbitrary, the constructive nature of the proof of Eq. (7.12) in Section 7.4 shows that it is not: *Every* self-consistent continuous Markov process $X(t)$ *must* evolve according to an updating formula of the form (7.13). This means that if the two exponents on dt in Eq. (7.13) are changed in any way, or if the zero-mean *normal* random variable $N(t)$ is replaced by some other zero-mean random variable, the resulting formula will not describe a process that evolves in a self-consistent way. That lack of self-consistency might be subtle, but it would ultimately manifest itself in the finding that, if the value of dt in any *modified* updating formula were reduced, the simulated behavior of the process over any *fixed finite* interval of time will change in a statistically significant way; in other words, the process constructed using an updating formula *not* of the form (7.13) would depend on the size chosen for the infinitesimal dt.[4] By contrast, the process constructed using an updating formula of the form (7.13) will be statistically insensitive to the size of dt, provided of course that dt is "small enough".

For $dt \neq 0$, we can rearrange the standard-form Langevin equation (7.13) to read

$$\frac{X(t + dt) - X(t)}{dt} = A(X(t), t) + D^{1/2}(X(t), t)\frac{N(t)}{\sqrt{dt}} \quad (dt \neq 0). \tag{7.19}$$

But if we try to take the limit of this equation as $dt \to 0^+$ (which of course is not the same as setting $dt = 0$) in order to form the *derivative* $dX(t)/dt$, we see that *this limit will not exist unless* $D = 0$. This implies that a *truly stochastic* continuous Markov process—i.e., one for which the diffusion function D is not zero—is a function that is *everywhere continuous but nowhere differentiable*. But the paradigm of the derivative is so deeply ingrained in science and engineering that it is sometimes convenient to "pretend" that the limit of Eq. (7.19) as $dt \to 0^+$ exists. The ploy for accomplishing that feat is to define a new stochastic process $\Gamma(t)$ by

$$\Gamma(t) \equiv \lim_{dt \to 0^+} \frac{N(t)}{\sqrt{dt}} = \lim_{dt \to 0^+} (1/dt)^{1/2}\mathcal{N}(0,1) = \lim_{dt \to 0^+} \mathcal{N}(0, 1/dt). \tag{7.20}$$

It then "follows" from Eq. (7.19) that the derivative of $X(t)$ is given by

$$\frac{dX(t)}{dt} = A(X(t), t) + D^{1/2}(X(t), t)\Gamma(t). \tag{7.21}$$

This equation is called the *white-noise form Langevin equation*, and the strange process $\Gamma(t)$ defined in Eq. (7.20) is called *Gaussian white noise*. From the conditions on $N(t)$ in Eqs (7.14), it can be shown from the definition (7.20) that $\langle \Gamma(t) \rangle = 0$ and $\langle \Gamma(t)\Gamma(t') \rangle = \delta(t - t')$. The fact that $\langle \Gamma^2(t) \rangle = \delta(0)$ can be seen heuristically by first noting from the definition (7.20) that $\langle \Gamma^2(t) \rangle = (1/dt) \langle N^2(t) \rangle = 1/dt$, and then recalling that since $\delta(0)\, dt = 1$, then $\delta(0) = 1/dt$.

In spite of definition (7.20), the inconvenient fact remains that the white-noise form Langevin equation (7.21) is a formula for the derivative of a function whose derivative does not exist when $D \neq 0$. Some would therefore argue that Eq. (7.21) is nothing more than a mnemonic for the standard-form Langevin equation (7.13), and that Eq. (7.21) means nothing more, and nothing less, than Eq. (7.13). This is the view that we will take here. But at times, we will find the white-noise form (7.21) to be heuristically convenient. For example, Eq. (7.21) makes it obvious that the *deterministic subclass* of continuous Markov processes, which comprises all continuous Markov processes for which $D \equiv 0$, encompasses all functions $X(t)$ that satisfy a first order ordinary differential equation, and hence that *the Langevin equation is the stochastic generalization of the first-order ordinary differential equation.*

7.6   The forward Fokker–Planck equation

For a continuous Markov process, we have from Eqs (7.9) and (7.12) that

$$B_n(x, t) = \lim_{dt \to 0^+} \frac{1}{dt} \langle [\mathcal{N}(A(x, t)dt, D(x, t)dt)]^n \rangle \quad (n = 1, 2, \ldots). \tag{7.22}$$

Formulas for all the moments of $\mathcal{N}(\mu, \sigma^2)$ in terms of μ and σ are known,[5] and when one inserts them into Eq. (7.22) and then takes the limit $dt \to 0^+$ one finds that

$$B_1(x,t) = A(x,t), \quad B_2(x,t) = D(x,t), \quad B_{n \geq 3}(x,t) = 0. \tag{7.23}$$

Thus, for the process defined by the Langevin equation (7.13), the forward Kramers-Moyal equation (7.5) reduces to

$$\frac{\partial}{\partial t} P(x,t|x_0,t_0) = -\frac{\partial}{\partial x}[A(x,t)P(x,t|x_0,t_0)] + \frac{1}{2}\frac{\partial^2}{\partial x^2}[D(x,t)P(x,t|x_0,t_0)]. \tag{7.24}$$

This is called the *forward Fokker–Planck equation*. Notice that it is *exact* for continuous Markov processes; it is *not* an approximate truncation of Eq. (7.5), in the manner of Einstein's truncation of Eq. (3.3). As with the Kramers–Moyal equation, there is also a "backward" Fokker–Planck equation, but we will defer a discussion of it until the final section of this chapter.

The time evolution of a continuous Markov process $X(t)$ with drift function $A(x,t)$ and diffusion function $D(x,t)$ is *completely defined* by *either* the Langevin equation (7.13) *or* the forward Fokker–Planck equation (7.24). The former focuses on the process itself, while the latter focuses on the PDF of the process. As suggested earlier, the Langevin equation can be used to approximately simulate $X(t)$ by treating dt as a "suitably small" Δt. The Langevin equation can also be used to derive time-evolution equations for the moments of $X(t)$, as we shall see later. Another way to derive those moment evolution equations is to use the forward Fokker–Planck equation.

The forward Fokker–Planck equation is especially useful for describing the *equilibrium* or stationary behavior of *stable* processes, i.e., processes for which A and D have no explicit time dependence, and the limit

$$\lim_{t-t_0 \to \infty} P(x,t|x_0,t_0) \equiv P_{\text{eq}}(x) \tag{7.25}$$

exists. Notice that Eq. (7.25) does not imply that the process $X(t)$ eventually stops changing with time; it means only that *our estimate* of the state of the process, running unobserved, eventually stops changing with time. When Eq. (7.25) holds, the left side of the forward Fokker–Planck equation (7.24) vanishes in the limit $(t - t_0) \to \infty$, and the equation reduces to an ordinary differential equation for the *equilibrium* PDF, $P_{\text{eq}}(x)$:

$$-\frac{d}{dx}[A(x)P_{\text{eq}}(x)] + \frac{1}{2}\frac{d^2}{dx^2}[D(x)P_{\text{eq}}(x)] = 0.$$

As can be verified by taking derivatives, the quadrature solution of this equation is

$$P_{\text{eq}}(x) = \frac{K}{D(x)} \exp\left(\int^x \frac{2A(x')}{D(x')}dx'\right), \tag{7.26}$$

where K is a constant that must be chosen to make $\int_{-\infty}^{\infty} P_{\text{eq}}(x)\,dx = 1$. A process that is "in equilibrium" is of course more likely to be found at values of x where $P_{\text{eq}}(x)$ is relatively large. The local maximum(s) of $P_{\text{eq}}(x)$ are thus of special interest; they are called the *stable state(s)* of the process. An undisturbed process will typically spend most of its time in the vicinity of one or more of its stable states.

7.7   Multivariate continuous Markov processes

The theory outlined above for a *scalar* continuous Markov process $X(t)$ can be generalized to the case where the process is an M-component *vector*:

$$\mathbf{X}(t) \equiv [X_1(t), \ldots, X_M(t)]. \tag{7.27}$$

This multivariate extension of continuous Markov process theory is important for two reasons. First, it often happens that an individual component X_i of $\mathbf{X}$ is *not by itself* a Markov process. This will be so if the updating formula for X_i from time t to time $t + dt$ makes use of the time-t value of at least one other component $X_{j \neq i}$, since that value will inevitably contain information about X_i's past that is not contained in its current value $X_i(t)$. Indeed, this might be true of *every* component X_i of $\mathbf{X}$. But since the values of all the components at time t are *collectively* sufficient to determine the values of all the components at time $t + dt$, then $\mathbf{X}$ will be a Markov process. The multivariate generalization therefore brings under the tent of continuous Markov process theory many stochastic processes that, by themselves, are non-Markov. Second, just as any third-order ordinary differential equation for a function $z_1(t)$,

$$\frac{d^3 z_1}{dt^3} = f\left(z_1, \frac{dz_1}{dt}, \frac{d^2 z_1}{dt^2}\right)$$

can be viewed as the coupled set of three first-order ordinary differential equations

$$\frac{dz_1}{dt} = z_2, \quad \frac{dz_2}{dt} = z_3, \quad \frac{dz_3}{dt} = f(z_1, z_2, z_3)$$

for the three components of the vector function $(z_1(t), z_2(t), z_3(t))$, the multivariate extension of continuous Markov process theory allows that theory to be applied to dynamical systems that are described in a deterministic context by ordinary differential equations of *any* order.

In this section we will summarize, but not prove, the key results of multivariate continuous Markov process theory.[6] First, the standard-form *multivariate Langevin equation*, which generalizes Eq. (7.13), is the *set* of M coupled equations

$$X_i(t + dt) = X_i(t) + A_i(\mathbf{X}(t), t)dt + \sum_{j=1}^{M} b_{ij}(\mathbf{X}(t), t)N_j(t)\sqrt{dt} \quad (i = 1, \ldots, M). \tag{7.28}$$

Here the functions $A_i(\mathbf{x}, t)$ for $i = 1, \ldots, M$ and $b_{ij}(\mathbf{x}, t)$ for $i, j = 1, \ldots, M$ can be any smooth functions of their argument, and $N_1(t), \ldots, N_M(t)$ are M statistically independent, temporally uncorrelated unit normals $\mathcal{N}(0, 1)$. It is obvious from the form of Eq. (7.28) that the process $\mathbf{X}(t)$ as thus defined is *continuous*, in the sense that $\mathbf{X}(t + dt) \to \mathbf{X}(t)$ as $dt \to 0$, and also *memoryless*, in the sense that future values of $\mathbf{X}$ depend on its past values only through its present value. That Eq. (7.28) is the only form possible for a continuous, memoryless update formula that is *self-consistent* is not obvious, but this can be proved using a generalization of the theorem at Eq. (7.12).

The corresponding white-noise form of the multivariate Langevin equation, which generalizes Eq. (7.21), is the set of M coupled equations

$$\frac{dX_i(t)}{dt} = A_i(\mathbf{X}(t), t) + \sum_{j=1}^{M} b_{ij}(\mathbf{X}(t), t)\Gamma_j(t) \quad (i = 1, \ldots, M). \tag{7.29}$$

Here, $\Gamma_1(t), \ldots, \Gamma_M(t)$ are M statistically independent, temporally uncorrelated Gaussian white-noise processes, as defined in Eq. (7.20). They satisfy $\langle \Gamma_i(t)\Gamma_j(t') \rangle = \delta_{ij}\,\delta(t - t')$, where the first δ is Kronecker's (which equals 1 if $i = j$ and 0 otherwise), and the second is Dirac's.

It turns out that any two different sets of the M^2 functions $b_{ij}(\mathbf{x}, t)$ that yield the same $\frac{1}{2}M(M + 1)$ functions

$$D_i(\mathbf{x}, t) \equiv \sum_{j=1}^{M} b_{ij}^2(\mathbf{x}, t) \quad (i = 1, \ldots, M), \tag{7.30a}$$

$$C_{ij}(\mathbf{x}, t) \equiv \sum_{k=1}^{M} b_{ik}(\mathbf{x}, t)b_{jk}(\mathbf{x}, t) \quad (i < j = 1, \ldots, M), \tag{7.30b}$$

will describe the same continuous Markov process $\mathbf{X}(t)$. This is a consequence of the fact that the last term on the right side of Eq. (7.28) is a normal random variable whose representation as a linear combination of unit normals is not unique. An application of the normal linear combination theorem (2.25) to Eq. (7.28) will reveal that, with $X_i(t) = x_i$, the increment $X_i(t + dt) - x_i$ is a normal random variable with mean $A_i(\mathbf{x}, t)dt$ and variance $D_i(\mathbf{x}, t)dt$. A little more work will reveal that $C_{ij}(\mathbf{x}, t)dt$ is the *covariance* of the two increments $X_i(t + dt) - x_i$ and $X_j(t + dt) - x_j$.

The corresponding multivariate forward Fokker-Planck equation, which generalizes Eq. (7.24), reads

$$\frac{\partial P(\mathbf{x}, t | \mathbf{x}_0, t_0)}{\partial t} = -\sum_{i=1}^{M} \frac{\partial}{\partial x_i} [A_i(\mathbf{x}, t)P(\mathbf{x}, t | \mathbf{x}_0, t_0)] + \frac{1}{2}\sum_{i=1}^{M} \frac{\partial^2}{\partial x_i^2}[D_i(\mathbf{x}, t)P(\mathbf{x}, t | \mathbf{x}_0, t_0)]$$

$$+ \sum_{\substack{i, j = 1 \\ [i < j]}}^{M} \frac{\partial^2}{\partial x_i \partial x_j}[C_{ij}(\mathbf{x}, t)P(\mathbf{x}, t | \mathbf{x}_0, t_0)]. \tag{7.31}$$

There is also a multivariate *backward* Fokker-Planck equation; we will discuss it in Section 7.13.

7.8   The driftless Wiener process

Two specific continuous Markov processes will be of special interest to us here. The first, and the simplest of all non-deterministic continuous Markov processes, is the *driftless Wiener process*. It is defined by the drift and diffusion functions

$$A(x,t) \equiv 0, \quad D(x,t) = c, \tag{7.32}$$

where c is a positive constant. (A Wiener process "with drift" would have A equal to a non-zero constant.) The Langevin equation (7.13) for this process thus reads

$$X(t + dt) = X(t) + c^{1/2}N(t)\sqrt{dt}, \tag{7.33a}$$

or in the white-noise form of Eq. (7.21),

$$\frac{dX(t)}{dt} = c^{1/2}\Gamma(t). \tag{7.33b}$$

The latter equation shows that the driftless Wiener process with $c = 1$ has the honor of being the anti-derivative or integral of the Gaussian white-noise process $\Gamma(t)$. The forward Fokker–Planck equation (7.24) reads

$$\frac{\partial}{\partial t}P(x,t|x_0,t_0) = \frac{c}{2}\frac{\partial^2 P(x,t|x_0,t_0)}{\partial x^2}. \tag{7.34}$$

As can be verified by directly computing the derivatives, the solution to Eq. (7.34) for the required initial condition $P(x,t_0|x_0,t_0) = \delta(x - x_0)$ is

$$P(x,t|x_0,t_0) = \frac{1}{\sqrt{2\pi c(t - t_0)}}\exp\left(-\frac{(x - x_0)^2}{2c(t - t_0)}\right). \tag{7.35}$$

Thus, by Eq. (2.4), the driftless Wiener process is the normal random variable with mean x_0 and variance $c(t - t_0)$:

$$X(t) = \mathcal{N}\left(x_0, c(t - t_0)\right). \tag{7.36}$$

The Fokker–Planck equation (7.34) for the driftless Wiener process is evidently the Einstein diffusion equation (3.9) with $c = 2D$, where D is the solute molecule's coefficient of diffusion—not to be confused with the generic "diffusion function" $D(x,t)$ of a continuous Markov process. Comparing Eq. (7.36) with Eqs (3.11), we see that the key result of Einstein's analysis of the physical phenomenon of diffusion can be stated thusly: *Each rectilinear component of a diffusing molecule's position is an independent driftless Wiener process with $c = 2D$.* For this reason, the driftless Wiener process is sometimes called the "Brownian motion process"; however, as we will see in the following chapters, that terminology is a bit misleading, because there is another stochastic process that can also lay claim to that name.

7.9 The Ornstein–Uhlenbeck process

Another kind of continuous Markov process that will be important to us is the *Ornstein–Uhlenbeck (OU) process.*[7] It has drift and diffusion functions of the forms

$$A(x,t) \equiv -\tau^{-1}x, \quad D(x,t) = c, \tag{7.37}$$

where τ and c are both positive constants. The Langevin equation (7.13) for the OU process thus reads

$$X(t+dt) = X(t) - \tau^{-1}X(t)dt + c^{1/2}N(t)\sqrt{dt}, \tag{7.38a}$$

or in the white-noise form of Eq. (7.21),

$$\frac{dX(t)}{dt} = -\tau^{-1}X(t) + c^{1/2}\Gamma(t). \tag{7.38b}$$

The corresponding Fokker–Planck equation is obtained by substituting Eqs (7.37) into Eq. (7.24). From that partial differential equation, one can verify by computing derivatives that its solution for the initial condition $P(x,t_0|x_0,t_0) = \delta(x - x_0)$ is the PDF of the normal random variable

$$X(t) = \mathcal{N}\left(x_0 e^{-(t-t_0)/\tau}, \frac{c\tau}{2}\left(1 - e^{-2(t-t_0)/\tau}\right)\right). \tag{7.39}$$

But it is rather more instructive to derive this result from the Langevin equation (7.38a), as we will now see.

To prove from Eq. (7.38a) that $X(t)$ is *normal*, we write that equation in the form

$$X(t+dt) = \left(1 - \tau^{-1}dt\right)X(t) + (c\,dt)^{1/2}\mathcal{N}(0,1).$$

Since $X(t_0) = x_0 \equiv \mathcal{N}(x_0,0)$, then this equation for $t = t_0$ shows that $X(t_0 + dt)$ is a linear combination of two statistically independent normal random variables. So from Eq. (2.25), it follows that $X(t_0 + dt)$ is a normal random variable. The same will then be true of $X(t_0 + 2dt)$, and $X(t_0 + 3dt)$, etc., thus proving that $X(t)$ is normal for all $t > t_0$. Note, however, that this argument could *not* be made if either of the factors multiplying $X(t)$ and $\mathcal{N}(0,1)$ in the above equation involved $X(t)$, because the product of a normal random variable with any other random variable will not be normal.

To prove from Eq. (7.38a) that the mean and variance of $X(t)$ are as asserted in Eq. (7.39), we will make use of the averaging relations

$$\langle N(t)\rangle = 0, \quad \langle N^2(t)\rangle = 1, \quad \langle X(t)N(t)\rangle = \langle X(t)\rangle\langle N(t)\rangle = 0. \tag{7.40}$$

The first two of these relations follow from Eqs (7.14), while the last is a consequence of the statistical independence of $N(t)$ and $X(t)$. Averaging Eq. (7.38a) gives

$$\langle X(t+dt)\rangle = \langle X(t)\rangle - \tau^{-1}\langle X(t)\rangle\,dt + 0.$$

This is just the ordinary differential equation

$$\frac{d\langle X(t)\rangle}{dt} = -\frac{1}{\tau}\langle X(t)\rangle \quad (t_0 \le t). \tag{7.41}$$

The solution to this equation for the initial condition $\langle X(t_0)\rangle = x_0$ is easily obtained, and is the function asserted for the mean in Eq. (7.39).

To compute $\operatorname{var}\{X(t)\} \equiv \langle X^2(t) \rangle - \langle X(t) \rangle^2$, we begin by taking the time derivative of both sides of this equation:

$$\frac{d\operatorname{var}\{X(t)\}}{dt} \equiv \frac{d\langle X^2(t) \rangle}{dt} - 2\langle X(t) \rangle \frac{d\langle X(t) \rangle}{dt}$$
$$= \frac{d\langle X^2(t) \rangle}{dt} + \frac{2}{\tau}\langle X(t) \rangle^2,$$

where the last equality follows from Eq. (7.41). To evaluate the derivative term on the right, we first square Eq. (7.38a), retaining only terms up to first order in dt:

$$X^2(t+dt) = X^2(t) - 2\tau^{-1}X^2(t)dt + 2c^{1/2}X(t)N(t)(dt)^{1/2} + cN^2(t)dt.$$

Averaging this using Eqs (7.40) gives

$$\langle X^2(t+dt) \rangle = \langle X^2(t) \rangle - 2\tau^{-1}\langle X^2(t) \rangle\, dt + 0 + cdt,$$

and this in turn implies the ordinary differential equation

$$\frac{d\langle X^2(t) \rangle}{dt} = -\frac{2}{\tau}\langle X^2(t) \rangle + c.$$

Substituting this into the above equation for the derivative of $\operatorname{var}\{X(t)\}$ gives, again using $\operatorname{var}\{X(t)\} \equiv \langle X^2(t) \rangle - \langle X(t) \rangle^2$,

$$\frac{d\operatorname{var}\{X(t)\}}{dt} = -\frac{2}{\tau}\operatorname{var}\{X(t)\} + c \quad (t_0 \le t).$$

The solution to this differential equation for the initial condition $\operatorname{var}\{X(t_0)\} = 0$ is easily obtained, and is the function asserted for the variance in Eq. (7.39). This completes the derivation of Eq. (7.39) from the OU Langevin equation (7.38a).

Unlike the Wiener process, the OU process is *stable*. This means that, as can be seen from Eq. (7.39), its limit as $t \to \infty$ is a well-defined time-independent random variable:

$$\lim_{(t-t_0)\to\infty} X(t) \equiv X(\infty) = \mathcal{N}\left(0, \frac{c\tau}{2}\right). \tag{7.42}$$

We should note that substituting the OU drift and diffusion functions (7.37) into the quadrature solution (7.26) of the stationary forward Fokker–Planck equation will also yield the PDF of $\mathcal{N}(0, c\tau/2)$. But the result in Eq. (7.42), which we obtained from Eq. (7.39), is stronger: It says that the OU process *always* approaches that stationary normal form as $t \to \infty$. The formulas for the mean and variance of $X(t)$ in Eq. (7.39) also show that $X(t)$ effectively reaches its stationary or equilibrium form after a time of order τ. For that reason, τ is called the *relaxation time* of the OU process. Of course, Eq. (7.42) does *not* imply that the process $X(t)$ eventually stops changing with time; it only says that if we do not observe the process for a time much greater than τ, then our estimate of the value of the process stops changing with time.

7.10 The time-integral of the Ornstein–Uhlenbeck process

Also important for our work in later chapters will be the *time-integral* of the OU process. In general, the time-integral $Y(t)$ of any function $X(t')$ from $t' = t_0$ to $t' = t$ can be written

$$Y(t) \equiv y_0 + \int_{t_0}^t X(t')dt' \quad (t \geq t_0), \tag{7.43a}$$

which implicitly defines $Y(t_0) \equiv y_0$. Since differentiating this gives $dY(t)/dt = X(t)$, then to lowest order in dt we can also define $Y(t)$ through the differential updating relation

$$Y(t + dt) = Y(t) + X(t)dt \quad (t \geq t_0), \tag{7.43b}$$

which we "initialize" by stipulating that $Y(t_0) = y_0$. Equation (7.43b) may be considered the operational definition of the time-integral Y of *any* process X, although our present interest is the OU process. Equation (7.43b) is evidently an "updating formula" which defines Y in much the same way that the standard-form Langevin equation (7.38a) defines X. But notice that Eq. (7.43b) does not have the canonical Langevin form (7.13): its right side involves not only $Y(t)$ but also $X(t)$, which contains information about Y's past that is not contained in Y's present value $Y(t)$. Therefore, Y by itself, although obviously continuous, is *not* a Markov process. But since $X(t)$ and $Y(t)$ *together* suffice to determine $X(t + dt)$ and $Y(t + dt)$, then the *bivariate* continuous process $[X(t), Y(t)]$ *is* Markov; indeed, Eqs (7.38a) and (7.43b) *together* make up a multivariate Langevin equation of the form (7.28) for $M = 2$.

It turns out that the time-integral $Y(t)$ of the OU process, unlike the time-integral of the vast majority of continuous Markov processes, is a *normal* random variable. We can prove this quite easily by reasoning as follows: For $t = t_0$, Eq. (7.43b) gives

$$Y(t_0 + dt) = y_0 + x_0 dt = \mathcal{N}(y_0 + x_0 dt, 0);$$

thus, $Y(t_0 + dt)$ is normal. Then for $t = t_0 + dt$, Eq. (7.43b) gives

$$Y(t_0 + 2dt) = Y(t_0 + dt) + X(t_0 + dt)dt$$
$$= \mathcal{N}(y_0 + x_0 dt, 0) + X(t_0 + dt)dt.$$

Since the OU process $X(t)$ is normal, this last equation tells us that $Y(t_0 + 2dt)$ is the sum of two normal random variables. Those two normal random variables are *not* independent of each other, because both depend on x_0; nevertheless, as we proved in Chapter 2 in connection with Eqs (2.30), the sum of two normal random variables, whether they are independent or not, will be normal. Therefore, $Y(t_0 + 2dt)$ is normal. A straightforward iteration of this argument proves that $Y(t)$ must be normal for *all* $t > t_0$. But this normal random variable $Y(t)$ will *not* be statistically independent of the normal random variable $X(t)$.

As discussed in Chapter 2 in connection with Eqs (2.28) and (2.29), two statistically dependent normal random variables are completely determined by five parameters: their two means, their two variances, and their covariance. So, since the mean and variance of the OU process $X(t)$ are already determined, it remains only to determine the mean and variance of its integral $Y(t)$, and $\mathrm{cov}\left\{X(t), Y(t)\right\}$.

To compute the mean of $Y(t)$, we start by taking the average of Y's updating formula (7.43b):

$$\langle Y(t+dt)\rangle = \langle Y(t)\rangle + \langle X(t)\rangle\, dt.$$

This implies the ordinary differential equation

$$\frac{d\langle Y(t)\rangle}{dt} = \langle X(t)\rangle. \tag{7.44}$$

If we substitute on the right the explicit formula for $\langle X(t)\rangle$ given in Eq. (7.39), the result can easily be integrated, subject to the initial condition $\langle Y(t_0)\rangle = y_0$, yielding an explicit formula for the mean of $Y(t)$.

To compute $\mathrm{cov}\left\{X(t), Y(t)\right\} \equiv \langle X(t)Y(t)\rangle - \langle X(t)\rangle\langle Y(t)\rangle$, we start by taking the time derivative of both sides of this equation:

$$\frac{d\,\mathrm{cov}\left\{X(t), Y(t)\right\}}{dt} \equiv \frac{d\langle X(t)Y(t)\rangle}{dt} - \frac{d\langle X(t)\rangle}{dt}\langle Y(t)\rangle - \langle X(t)\rangle\frac{d\langle Y(t)\rangle}{dt}$$

$$= \frac{d\langle X(t)Y(t)\rangle}{dt} + \frac{1}{\tau}\langle X(t)\rangle\langle Y(t)\rangle - \langle X(t)\rangle^2,$$

where the last step has invoked Eqs (7.41) and (7.44). To evaluate the derivative term on the right side, we first take the product of Eqs (7.38a) and (7.43b), retaining only terms up to first order in dt:

$$X(t+dt)Y(t+dt) = X(t)Y(t) - \tau^{-1}X(t)Y(t)dt$$
$$+ c^{1/2}N(t)Y(t)(dt)^{1/2} + X^2(t)dt.$$

Averaging this equation with the help of Eqs (7.40), and noting that $\langle N(t)Y(t)\rangle = 0$ since $N(t)$ is statistically independent of $Y(t)$, we obtain

$$\frac{d\langle X(t)Y(t)\rangle}{dt} = -\frac{1}{\tau}\langle X(t)Y(t)\rangle + \langle X^2(t)\rangle.$$

Substituting this expression into the above formula for the derivative of the covariance yields

$$\frac{d\,\mathrm{cov}\left\{X(t), Y(t)\right\}}{dt} = -\frac{1}{\tau}\mathrm{cov}\left\{X(t), Y(t)\right\} + \mathrm{var}\left\{X(t)\right\}. \tag{7.45}$$

If we now substitute on the right the explicit formula for $\mathrm{var}\left\{X(t)\right\}$ given in Eq. (7.39), the result can straightforwardly be integrated subject to the initial condition $\mathrm{cov}\left\{X(t_0), Y(t_0)\right\} = 0$, yielding an explicit formula for $\mathrm{cov}\left\{X(t), Y(t)\right\}$.

Finally, to compute $\mathrm{var}\left\{Y(t)\right\} \equiv \langle Y^2(t)\rangle - \langle Y(t)\rangle^2$, we start by taking the time derivative of both sides of this equation:

$$\frac{d \operatorname{var}\{Y(t)\}}{dt} \equiv \frac{d\langle Y^2(t)\rangle}{dt} - 2\langle Y(t)\rangle \frac{d\langle Y(t)\rangle}{dt}$$

$$= \frac{d\langle Y^2(t)\rangle}{dt} - 2\langle Y(t)\rangle\langle X(t)\rangle,$$

where the last step invokes Eq. (7.44). To evaluate the derivative term on the right side, we take the square of Eq. (7.43b), retaining as usual only terms up to first order in dt:

$$Y^2(t+dt) = Y^2(t) + 2X(t)Y(t)dt.$$

Averaging this gives

$$\frac{d\langle Y^2(t)\rangle}{dt} = 2\langle X(t)Y(t)\rangle.$$

Substituting this into the above formula for the derivative of $\operatorname{var}\{Y(t)\}$ yields

$$\frac{d\operatorname{var}\{Y(t)\}}{dt} = 2\operatorname{cov}\{X(t), Y(t)\}. \tag{7.46}$$

If we now substitute on the right the explicit formula for $\operatorname{cov}\{X(t), Y(t)\}$ obtained by solving Eq. (7.45), the result can straightforwardly be integrated subject to the initial condition $\operatorname{var}\{Y(t_0)\} = 0$, yielding an explicit formula for the variance of $Y(t)$.

The result of the foregoing straightforward but algebraically tedious integrations are the following explicit formulas for the means, variances, and covariance of the OU process $X(t)$ and its integral $Y(t)$:

$$\mu_X(t) \equiv \langle X(t)\rangle = x_0 e^{-(t-t_0)/\tau}, \tag{7.47a}$$

$$\sigma_X^2(t) \equiv \operatorname{var}\{X(t)\} = \frac{c\tau}{2}\left(1 - e^{-2(t-t_0)/\tau}\right), \tag{7.47b}$$

$$\mu_Y(t) \equiv \langle Y(t)\rangle = y_0 + x_0\tau\left(1 - e^{-(t-t_0)/\tau}\right), \tag{7.47c}$$

$$\sigma_Y^2(t) \equiv \operatorname{var}\{Y(t)\} = c\tau^2\left[(t-t_0) - 2\tau\left(1 - e^{-(t-t_0)/\tau}\right)\right.$$
$$\left. + \frac{\tau}{2}\left(1 - e^{-2(t-t_0)/\tau}\right)\right], \tag{7.47d}$$

$$c_{XY}(t) \equiv \operatorname{cov}\{X(t), Y(t)\} = \frac{c\tau^2}{2}\left[1 - e^{-(t-t_0)/\tau}\right]^2. \tag{7.47e}$$

Some noteworthy features of these formulas: All depend on t only through the difference $t - t_0$; μ_X depends on x_0; μ_Y depends on y_0 *and* x_0; μ_X and μ_Y both depend on τ but not on c; and σ_X^2, σ_Y^2, and c_{XY} all depend on both τ and c but not on x_0 or y_0. Formulas (7.47), together with the *normality* of both $X(t)$ and $Y(t)$, constitute a complete and exact specification of the Ornstein–Uhlenbeck process and its integral.

As discussed in Section 2.7, there are two ways to analytically describe two correlated normal random variables. One way is to use the formula (2.28) for their joint PDF. That gives for the joint PDF of the OU process $X(t)$ and its integral $Y(t)$:

$$P_{X,Y}(x, y, t | x_0, y_0, t_0) = \frac{1}{2\pi \sqrt{\sigma_X^2 \sigma_Y^2 - c_{XY}^2}}$$

$$\times \exp\left(-\frac{\sigma_Y^2 (x - \mu_X)^2 + \sigma_X^2 (y - \mu_Y)^2 - 2c_{XY}(x - \mu_X)(y - \mu_Y)}{2\left(\sigma_X^2 \sigma_Y^2 - c_{XY}^2\right)}\right). \quad (7.48)$$

Here, μ_X, μ_Y, σ_X^2, σ_Y^2, and c_{XY} are the t-dependent functions in Eqs (7.47). But for us, a more useful way to analytically describe the OU process X and its integral Y is to make use of Eqs (2.29): With N_1 and N_2 two statistically independent normal random variables with means 0 and variances 1, we have

$$X(t) = \mu_X(t) + \sigma_X(t) N_1, \quad (7.49a)$$

$$Y(t) = \mu_Y(t) + \left(\frac{c_{XY}(t)}{\sigma_X(t)}\right) N_1 + \left(\sigma_Y^2(t) - \frac{c_{XY}^2(t)}{\sigma_X^2(t)}\right)^{1/2} N_2. \quad (7.49b)$$

As we shall see shortly, formulas (7.49) allow exact numerical simulation of $X(t)$ and $Y(t)$ for any $t \geq t_0$.

7.11 Numerically simulating the driftless Wiener process

The representation (7.36) of the driftless Wiener process can be written, with the help of Eq. (2.16), as

$$X(t) = \mathcal{N}\left(x_0, c(t - t_0)\right) = x_0 + \sqrt{c(t - t_0)} \mathcal{N}(0, 1).$$

It follows that if $X(t) = x_t$, then $X(t + \Delta t)$ for any $\Delta t > 0$ can be computed as

$$x_{t+\Delta t} = x_t + \sqrt{c \Delta t} \, n, \quad (7.50)$$

where n is a sample value of the normal random variable $\mathcal{N}(0, 1)$. Formula (7.50) can be applied iteratively to generate a *trajectory* or *realization* of a driftless Wiener process which is *exact* for any $\Delta t > 0$. We have of course already encountered Eq. (7.50) in the simulation formulas (4.11) for Einstein diffusion.

The trajectory constructed from Eq. (7.50) will consist of points separated in time by Δt. It is important to understand that *we may not interpolate between successive points with any smooth curve*. To see why, suppose we "zoom in" on the trajectory generated using Eq. (7.50) by reducing the *length*-scale by some factor $f < 1$ and the *time*-scale by a factor f^2. This is equivalent to making the variable transformations $x \to x' = fx$ and $t \to t' = f^2 t$. Multiplying Eq. (7.50) through by f shows that that equation becomes under this transformation $x'_{t'+\Delta t'} = x'_{t'} + \sqrt{c \Delta t'} \, n$; thus, the trajectory on this zoomed-in scale will be *statistically identical* to the original trajectory. It is an example of a "statistical fractal". This is a manifestation of the fact that continuous Markov processes are not smooth enough to be differentiable, in spite of the contrary suggestion of the white-noise form Langevin equation (7.21). *Smooth interpolations are never appropriate for the simulated trajectory of a continuous Markov process.* But the updating formula (7.50) is nevertheless mathematically exact for all $\Delta t > 0$.

7.12 Numerically simulating the Ornstein–Uhlenbeck process and its integral

The representation (7.39) of the OU process leads to a simulation updating formula analogous to Eq. (7.50): If $X(t) = x_t$, then $X(t + \Delta t)$ for any $\Delta t > 0$ can be computed as

$$x_{t+\Delta t} = x_t e^{-\Delta t/\tau} + \sqrt{\tfrac{1}{2}c\tau \left(1 - e^{-2\Delta t/\tau}\right)}\, n, \tag{7.51}$$

where n is a sample value of the normal random variable $\mathcal{N}(0, 1)$. Iteratively applying this formula produces an exact trajectory of the OU process X for any $\Delta t > 0$, again with the proviso that the successive points along the trajectory should not be interpolated with any smooth curve. It is easy to show that if $\Delta t = dt$, Eq. (7.51) reduces to the OU Langevin equation (7.38a), as we should expect.

Simulating the integral Y of the OU process X can only be done jointly with the simulation of X. The updating procedure is Eqs (7.49) with $(t_0, t) \to (t, t + \Delta t)$, and with substitutions from Eqs (7.47). After simplifying the algebra, we find that if $X(t) = x_t$ and $Y(t) = y_t$, then $X(t + \Delta t)$ and $Y(t + \Delta t)$ for any $\Delta t > 0$ can be computed as

$$x_{t+\Delta t} = x_t e^{-\Delta t/\tau} + \left[\frac{c\tau}{2}(1 - e^{-2\Delta t/\tau})\right]^{1/2} n_1, \tag{7.52a}$$

$$y_{t+\Delta t} = y_t + x_t \tau (1 - e^{-\Delta t/\tau}) + \left[c\tau^3 \frac{(1 - e^{-\Delta t/\tau})^3}{2(1 + e^{-\Delta t/\tau})}\right]^{1/2} n_1$$

$$+ \left\{ c\tau^3 \left[\frac{\Delta t}{\tau} + (1 - e^{-\Delta t/\tau})\left(-2 + \tfrac{1}{2}(1 + e^{-\Delta t/\tau}) - \frac{(1 - e^{-\Delta t/\tau})^2}{2(1 + e^{-\Delta t/\tau})}\right)\right]\right\}^{1/2} n_2. \tag{7.52b}$$

Here, n_1, which appears in both equations, and n_2, which appears in only the second equation, are statistically independent samples of the random variable $\mathcal{N}(0, 1)$. Notice that X's updating formula (7.52a) is the same as Eq. (7.51), as we should expect. The appearance of x_t on the right side of Y's updating formula (7.52b) is the reason why Y is not a Markov process: x_t contains information about Y's past that is not contained in Y's present value y_t. But since Y is differentiable, by virtue of its definition in Eq. (7.43a), a smooth interpolation between its successive points *is* allowed, provided of course that Δt is small enough.

We emphasize that the updating formulas (7.52) for the OU process X and its integral Y are *exact* for *any* value of $\Delta t > 0$. If the simulation is carried out with Δt constant, then the coefficients of x_t, n_1, and n_2 in formulas (7.52) can all be pre-calculated, and the simulation can be performed quite efficiently.

7.13 The backward Fokker–Planck equation[8]

We have seen that the forward Fokker-Planck equation (7.24) occupies a position of major importance in the theory of continuous Markov processes. As the time-evolution equation for the conditioned PDF $P(x, t|x_0, t_0)$ of $X(t)$, it determines, at least in principle, everything we can possibly know about $X(t)$. But it turns out that problems dealing with "first passages", which are concerned with when the process will first reach some specified state, are more conveniently handled using the *backward Fokker–Planck equation*, which reads:

$$-\frac{\partial P(x, t|x_0, t_0)}{\partial t_0} = A(x_0, t_0)\frac{\partial P(x, t|x_0, t_0)}{\partial x_0} + \frac{1}{2}D(x_0, t_0)\frac{\partial^2 P(x, t|x_0, t_0)}{\partial x_0^2}. \quad (7.53)$$

Whereas in the forward Fokker–Planck equation (7.24) all partial derivatives are taken with respect to variables in *front* of the conditioning bar in $P(x, t|x_0, t_0)$, in the backward Fokker–Planck equation (7.53) all partial derivatives are taken with respect to variables *behind* the conditioning bar. There are several other important differences between Eqs (7.24) and (7.53), but their overall similarity prompts one to wonder if the latter can be derived from the former by making some simple changes of variable. It cannot. Equation (7.53) must be derived from the Chapman–Kolmogorov equation separately from Eq. (7.24). The derivation proceeds via the "backward" version of the Kramers–Moyal equation (7.5), and it goes as follows.

We start by writing the Chapman–Kolmogorov equation (7.3) in the form

$$P(x, t|x_0, t_0) = \int_{-\infty}^{\infty} P(x, t|x_0 + \xi, t_0 + dt_0)\, P(x_0 + \xi, t_0 + dt_0|x_0, t_0)\, d\xi, \quad (7.54)$$

which may be contrasted with the form (7.4) that we used to derive the forward Kramers–Moyal equation. Next we define the function h by

$$h(x_0) \equiv P(x, t|x_0, t_0 + dt_0),$$

and we observe that the *first factor* in the integrand of Eq. (7.54) can be written $h(x_0 + \xi)$. That in turn can be expanded in a Taylor series about x_0:

$$h(x_0 + \xi) = h(x_0) + \sum_{n=1}^{\infty}\frac{\xi^n}{n!}\frac{\partial^n h(x_0)}{\partial x_0^n}.$$

When this expansion is substituted for the first factor in the integrand of Eq. (7.54), that equation becomes

$$P(x, t|x_0, t_0) = \int_{-\infty}^{\infty} P(x, t|x_0, t_0 + dt_0)\, P(x_0 + \xi, t_0 + dt_0|x_0, t_0)d\xi$$

$$+ \sum_{n=1}^{\infty}\frac{1}{n!}\int_{-\infty}^{\infty}\xi^n\frac{\partial^n P(x, t|x_0, t_0 + dt_0)}{\partial x_0^n}P(x_0 + \xi, t_0 + dt_0|x_0, t_0)d\xi\,.$$

The first term on the right evidently integrates to $P(x, t|x_0, t_0 + dt_0)$. Transposing that term, dividing through by dt_0, and then taking the limit $dt_0 \to 0^+$, we get

$$-\frac{\partial}{\partial t_0}P(x, t|x_0, t_0) = \sum_{n=1}^{\infty}\frac{1}{n!}B_n(x_0, t_0)\frac{\partial^n P(x, t|x_0, t_0)}{\partial x_0^n} \quad (t_0 \leq t), \quad (7.55)$$

where the functions B_n are as defined in Eq. (7.6). This equation is called the *backward Kramers–Moyal equation*.

For the special case of a *continuous* Markov process, we found in Eq. (7.23) that B_1 coincides with the drift function A, B_2 coincides with the diffusion function D, and B_n for all $n \geq 3$ vanishes identically. Therefore, for a *continuous* Markov process, the backward Kramers–Moyal equation (7.55) collapses to the backward Fokker–Planck equation (7.53).

In cases where neither the drift function A nor the diffusion function D depends explicitly on t—i.e., there is no dependence of the dynamics of the system on some external clock—the dependence of $P(x, t | x_0, t_0)$ on the two variables t and t_0 is only through their difference $(t - t_0)$. The system is then said to be *temporally homogeneous*. In that case, it is easy to show from the chain rule of differentiation that

$$\frac{\partial}{\partial t_0} P(x, t | x_0, t_0) = -\frac{\partial}{\partial t} P(x, t | x_0, t_0).$$

The backward Fokker–Planck equation (7.53) can then be written

$$\frac{\partial}{\partial t} P(x, t | x_0, t_0) = A(x_0) \frac{\partial P(x, t | x_0, t_0)}{\partial x_0} + \frac{1}{2} D(x_0) \frac{\partial^2 P(x, t | x_0, t_0)}{\partial x_0^2} \quad (t_0 \leq t). \quad (7.56)$$

This is the form of the backward Fokker–Planck equation that will be used in Chapter 11.

The foregoing results generalize straightforwardly, if tediously, for *multivariate* continuous Markov processes: The backward counterpart of the multivariate forward Fokker–Planck equation (7.31) turns out to be

$$-\frac{\partial P(\mathbf{x}, t | \mathbf{x}_0, t_0)}{\partial t_0} = \sum_{i=1}^{M} A_i(\mathbf{x}_0, t_0) \frac{\partial P(\mathbf{x}, t | \mathbf{x}_0, t_0)}{\partial x_{0i}} + \frac{1}{2} \sum_{i=1}^{M} D_i(\mathbf{x}_0, t_0) \frac{\partial^2 P(\mathbf{x}, t | \mathbf{x}_0, t_0)}{\partial x_{0i}^2}$$
$$+ \sum_{\substack{i, j = 1 \\ [i < j]}}^{M} C_{ij}(\mathbf{x}_0, t_0) \frac{\partial^2 P(\mathbf{x}, t | \mathbf{x}_0, t_0)}{\partial x_{0i} \partial x_{0j}}, \quad (7.57)$$

where the functions D_i and C_{ij} are as defined in Eqs (7.30). And if the process is temporally homogeneous, this simplifies to

$$\frac{\partial P(\mathbf{x}, t | \mathbf{x}_0, t_0)}{\partial t} = \sum_{i=1}^{M} A_i(\mathbf{x}_0) \frac{\partial P(\mathbf{x}, t | \mathbf{x}_0, t_0)}{\partial x_{0i}} + \frac{1}{2} \sum_{i=1}^{M} D_i(\mathbf{x}_0) \frac{\partial^2 P(\mathbf{x}, t | \mathbf{x}_0, t_0)}{\partial x_{0i}^2}$$
$$+ \sum_{\substack{i, j = 1 \\ [i < j]}}^{M} C_{ij}(\mathbf{x}_0) \frac{\partial^2 P(\mathbf{x}, t | \mathbf{x}_0, t_0)}{\partial x_{0i} \partial x_{0j}}. \quad (7.58)$$

In our review of continuous Markov process theory in this chapter, we have derived the Langevin equation in its standard and white-noise forms, as well as the forward and backward Fokker–Planck equations, and we have developed explicit expressions for the Wiener and Ornstein–Uhlenbeck processes. In the following chapters we will make use of these results, beginning in Chapter 8 with a presentation of Langevin's theory of diffusion.

Notes to Chapter 7

[1] The singly-conditioned PDF $P(x, t | x_0, t_0)$ is called the "transition probability" by some authors, who prefer to accord the title "the PDF of $X(t)$" to an *unconditioned* PDF, which they write $P(x, t)$. The problem with doing that is that no useful meaning can be attached to an unconditioned PDF of a stochastic process. That is because it is practically impossible to assign a value to $P(x, t)dx$, the probability that the process will be in the interval $[x, x + dx)$ at time t, without making *some* assumption or stipulation as to where the process was at some earlier time. For example, consider a particle moving in a zero-force field along the x-axis with a constant velocity v: What can we say about the location $x(t)$ of this particle at time t that does *not* involve making *some* assumption about the particle's position at some earlier time t_0? The answer: Nothing. As will be seen later in this chapter, none of the core equations of continuous Markov process theory make reference to an unconditioned PDF. And equations that do involve PDFs—the Chapman–Kolmogorov equation, the Kramers–Moyal equations, and the Fokker–Planck equations—all involve only the conditioned PDF defined in Eq. (7.1). So we will take the view that the function $P(x, t | x_0, t_0)$ is "the PDF" of $X(t)$; it specifies $X(t)$ as completely as is possible. Of course, if the value x_0 of the process at the initial time t_0 is itself declared to be a random variable, say with PDF Q_0, then the laws of probability imply that the "effective PDF" of the process at time t can be computed from the singly-conditioned PDF as $\int P(x, t | x_0, t_0)Q_0(x_0)dx_0$. But that PDF is *not* "unconditioned": it is conditioned on the explicit form of the function Q_0.

[2] The notational complexity of Eq. (7.10a) tends to obscure its logical simplicity. If we relabel the three times t, $t + \alpha dt$, and $t + dt$ respectively as t_0, t_1, and t_2, then all that Eq. (7.10a) is asserting is that

$$X(t_2) - X(t_0) = [X(t_2) - X(t_1)] + [X(t_1) - X(t_0)]. \tag{a}$$

This relation is obviously an *algebraic identity*; it *must* hold true. But this relation should not be confused with the "triangle inequality" relation

$$|X(t_2) - X(t_0)| \le |X(t_2) - X(t_1)| + |X(t_1) - X(t_0)|, \tag{b}$$

which arises from the fact that $t_{i+1} > t_i$ does not imply that $X(t_{i+1}) > X(t_i)$. The inequality (b) would be relevant if we were concerned with the total distance traveled by $X(t)$ in the time interval $[t_1, t_2]$, as for instance we were in Section 4.2. But our concern here is to find out what constraint the algebraic identity (a), in the form of

Eq. (7.10a), might impose on the mathematical form of the random variable $\Xi(dt; x, t)$. As we will see, that constraint turns out to be surprisingly restrictive.

[3]A commonly encountered alternative form of the Langevin equation (7.13) is obtained by using the random variable identity $c\mathcal{N}(0, 1) = \mathcal{N}(0, c^2)$ to write

$$N(t)\sqrt{dt} = \sqrt{dt}\mathcal{N}(0, 1) = \mathcal{N}(0, dt) \equiv dW(t).$$

As thus defined, $dW(t)$, which some authors denote by $dB(t)$, is a temporally uncorrelated normal random variable with mean 0 and variance dt. In place of Eqs (7.14), we would then have the auxiliary relations

$$\langle dW(t)\rangle = 0, \text{ and } \langle dW(t)dW(t')\rangle = \begin{cases} dt, \text{ if } t = t', \\ 0, \text{ if } t \neq t'. \end{cases}$$

This notation affords some measure of comfort to those who are uneasy about the propriety of taking the square root of an infinitesimal. But with our definition of dt as a real variable that is confined to the interval $[0, \varepsilon)$ where $\varepsilon > 0$ is arbitrarily small, the square root operation is really not a problem; e.g., the square root of the "very small" number 10^{-42} is 10^{-21}, and indeed, the square root of 0 is 0. Our preference for writing $N(t)\sqrt{dt}$ instead of $dW(t)$ is that the former makes more clear the actual dependence of the last term in Eq. (7.13) on dt.

[4]For a simple illustration of this point, suppose the form of the $D \equiv 0$ case of the updating formula (7.13) were changed from $X(t + dt) = X(t) + A(X(t), t)dt$ to

$$X(t + dt) = X(t) + A(X(t), t)\sqrt{dt}.$$

Although the new process X thus defined would not be differentiable, it would be both *continuous* and *Markov* (memoryless), which is all we care about here. But there is a problem with this infinitesimal updating rule. To see what that problem is, let's apply the rule successively to the two intervals $[t, t + \frac{1}{2}dt)$ and $[t + \frac{1}{2}dt, t + dt)$:

$$X(t + \tfrac{1}{2}dt) = X(t) + A(X(t), t)\sqrt{\tfrac{1}{2}dt},$$

$$X(t + dt) = X(t + \tfrac{1}{2}dt) + A\left(X(t + \tfrac{1}{2}dt), t + \tfrac{1}{2}dt\right)\sqrt{\tfrac{1}{2}dt}.$$

Since A is assumed to be a continuous function of its two arguments, and X is continuous in t, then we can replace the A factor in the second equation by $A(X(t), t)$ without spoiling the order-$\sqrt{dt}$ accuracy of that equation. Upon doing that and then substituting for $X(t + \frac{1}{2}dt)$ from the first equation, we find after simplifying the algebra that

$$X(t + dt) = X(t) + \sqrt{2}A(X(t), t)\sqrt{dt}.$$

The problem is now obvious: This last formula constitutes an updating rule for the full interval $[t, t + dt)$ which is *different* from the rule we started with. By simply halving the infinitesimal increment, we have changed the effective updating rule, and hence also the process that is generated by that rule. That won't happen with the updating formula $X(t + dt) = X(t) + A(X(t), t)dt$, nor with its more general version (7.13).

[5]For $X = \mathcal{N}(\mu, \sigma^2)$, it can be proved that the nth moment of X is given by

$$\langle X^n \rangle = n! \sum_{k(\text{even})=0}^{n} \frac{\mu^{n-k}(\sigma^2)^{k/2}}{(n-k)!\,(k/2)!\,2^{k/2}}.$$

Using this relation in Eq. (7.22), the results (7.23) follow immediately, without any approximations.

[6]Proofs of the results summarized in Section 7.7 can be found in the article "The multivariate Langevin and Fokker–Planck equations", *American Journal of Physics* **64**:1246–1257 (1996), by D. Gillespie.

[7]In 1930, Dutch physicist Leonard Ornstein (1880–1941) and Dutch-born American physicist George Uhlenbeck (1900–1988) coauthored a seminal paper on the type of Markov process that would later bear their names. Their paper was entitled "On the theory of the Brownian motion" [*Physical Review* **36**:823–841 (1930)], and its first author was the younger Uhlenbeck. Ornstein was then a well-established professor at the University of Utrecht, while Uhlenbeck was a junior faculty member at the University of Michigan. Uhlenbeck's student days had all been in Holland, mainly at the University of Leiden where he received his Ph.D. in 1927. Enroute to that degree, Uhlenbeck and fellow graduate student Samuel Goudsmit discovered the intrinsic spin of the electron, a major discovery in physics. After getting his Ph.D., Uhlenbeck made his home in the United States, save only for the period 1935–38 when he was a professor at the University of Utrecht. The Nazi invasion of Holland in May of 1940 was devastating for Ornstein, who was a Jew. He turned down an invitation to escape to the United States, out of a sense of loyalty to his laboratory at Utrecht. But in September of 1940, the occupying powers summarily dismissed him from the University and barred him from his laboratory. Ornstein withdrew in seclusion to his home, where he died six months later.

In 1945, Uhlenbeck and post-doctoral student Ming Chen Wang (1906–2010) at the University of Michigan published a sequel to the 1930 paper: "On the theory of the Brownian motion II", *Reviews of Modern Physics* **17**:323–342 (1945). That paper summarized further developments and clarifications regarding the OU process that had been made in the intervening 15 years. Both papers have been reprinted in the book *Selected Papers on Noise and Stochastic Processes* (edited by N. Wax, Dover, 1954). The prominence of the Ornstein–Uhlenbeck process in physics arises from the fact that the drift function of this process has the "spring-force" form $-kx$, where $k > 0$; that, coupled with the fact that its diffusion function has the simplest possible non-trivial form, casts the OU process in the role of the "simple harmonic oscillator" of stochastic process theory. The OU process is also used in physics to describe thermal-electric (Johnson) noise in conductors.

[8]The material in this section will be needed only for Chapter 11, and can therefore be deferred until then.

8
Langevin's theory of diffusion

Although Einstein's theory of diffusion is adequate for many purposes, we saw in Sections 4.2 and 4.4 that it is physically incorrect on "small" time-scales. That failing makes it impossible for Einstein's theory to accurately describe a solute molecule's velocity and finely resolved trajectory; in addition, it imposes some lower limit on the cell length l in the discrete-stochastic model's conventional formula for κ_l. We reasoned in Section 4.6 that this shortcoming of Einstein's theory stems from the fact that, on sufficiently small spatial-temporal scales, the motion of a solute molecule is ballistic instead of diffusional, and is therefore not described by the diffusion equation. In Section 4.7 we characterized that small-scale ballistic behavior well enough to derive a formula for the diffusional bimolecular chemical reaction probability rate (Section 4.8), and also a refinement of the formula $\kappa_l = D/l^2$ for smaller values of l (Section 5.6). But those ad hoc extensions of Einstein's theory highlight the need for a theory of diffusion that has a firmer foundation in the physics of molecular motion. What subsequently turned out to be such a theory was proposed in 1908 by Paul Langevin.[1]

The point of departure of Langevin's analysis was Newton's Second Law for the solute molecule, innovatively framed as what today would be called a "stochastic" differential equation. Our presentation of Langevin's theory will be updated and refined in the sense that it will exploit some mathematical insights (described in Chapter 7) that were not fully appreciated in Langevin's time. In this chapter, we will begin by describing the key assumption underlying Langevin's theory. We will corroborate that assumption by demonstrating its agreement with a rudimentary but plausible physical model of how solute molecules move in a sea of many much smaller solvent molecules. Then we will show how Langevin's theory of diffusion unfolds via the mathematics of the Ornstein–Uhlenbeck process of Section 7.9. Finally, we will show how a basic requirement of statistical thermodynamics fixes the only free parameter in Langevin's theory, thus making the theory complete. In Chapters 9 and 10 we will derive further implications of Langevin's theory, and show how it connects to and goes beyond Einstein's theory.

8.1 Langevin's key assumption

We consider a single solute molecule of mass m that is subject only to forces arising from collisions of that solute molecule with the many surrounding much smaller solvent molecules. Following Langevin, we invoke Newton's Second Law to assert that the x-component of the solute molecule's velocity $V_x(t)$ obeys the equation

$$m\frac{dV_x(t)}{dt} = -\gamma V_x(t) + f\Gamma_x(t). \tag{8.1}$$

The key assumption being made in this "$F = ma$" equation is that the x-component of the instantaneous force exerted on the solute molecule by the solvent molecules can be written as the sum of two separate forces. The first of those forces is directly proportional to, but oppositely directed from, the solute molecule's velocity $V_x(t)$. The positive constant of proportionality γ is called the *drag coefficient*. This force is just the conventional *drag force* that arises in classical fluid dynamics for macroscopic objects moving not too rapidly in a homogeneous fluid. The second force was originally envisioned by Langevin to be a rather nondescript, zero-mean, randomly fluctuating force whose fluctuations are independent of $V_x(t')$ for all $t' \leq t$. However, in Eq. (8.1) we have made the nature of this force more specific: We have taken it to be some as yet unspecified constant f times *Gaussian white noise* $\Gamma_x(t)$, which we defined in Chapter 7 in Eq. (7.20). One might well wonder what the physical justification is for this Gaussian white-noise assumption, and indeed for the assumption that the overall effect on the solute molecule of its collisions with surrounding solvent molecules can be accurately portrayed by the sum of these two forces. In the next section we will attempt to supply a physical rationale for all these assumptions. But first we want to take note of an important immediate implication of Eq. (8.1).

If we divide Eq. (8.1) through by m and then compare with Eq. (7.21), we will see that Eq. (8.1) is a "white-noise form Langevin equation" for $V_x(t)$. To get the corresponding "standard-form" Langevin equation (7.13), we write the derivative on the left side of Eq. (8.1) as a ratio of the infinitesimals $[V_x(t + dt) - V_x(t)]$ and dt, then multiply through by dt, and finally invoke the definition (7.20) of Gaussian white noise. Doing that gives the standard-form Langevin equation version,

$$V_x(t + dt) = V_x(t) - \left(\frac{\gamma}{m}\right) V_x(t) dt + \left(\frac{f}{m}\right) N_x(t)\sqrt{dt}, \tag{8.2}$$

where $N_x(t) = \mathcal{N}(0, 1)$. This "infinitesimal updating formula" is completely equivalent to Eq. (8.1). Or as intimated in Section 7.5, Eq. (8.2) is what Eq. (8.1) *really means*. Upon comparing Eq. (8.2) with Eq. (7.38a), we see that Langevin's assumption amounts to asserting that $V_x(t)$ is an *Ornstein–Uhlenbeck process* with parameters

$$\tau = \frac{m}{\gamma}, \quad c = \left(\frac{f}{m}\right)^2. \tag{8.3}$$

Therefore, all the results derived for the Ornstein–Uhlenbeck process and its integral in Chapter 7 can be immediately applied to $V_x(t)$ and its integral $X(t)$ simply by making the substitutions (8.3). Later in this chapter, we will show that those results, combined with some thermodynamic reasoning, allow us to determine Langevin's provisionally unspecified parameter f. Once that is done, Langevin's theory will be mathematically complete, and we can proceed to catalogue its physical implications. But first we want to address the problem of finding a rational physical basis for the assumed form of Langevin's force in Eq. (8.1).

8.2   A physical rationale for Langevin's assumption

As we just noted, the form of the force on the right side of Eq. (8.1) implies that $V_x(t)$, the x-component of the velocity of a solute molecule, is a continuous Markov process. In this section, we will construct another mathematical model of $V_x(t)$, this one a *jump* Markov process, by reasoning more carefully about the way in which the velocity of a solute molecule should behave in response to collisions of that molecule with the many smaller, less massive solvent molecules. After developing our jump Markov process model, we will show that its master equation becomes, in a plausible continuum limit, the Fokker–Planck equation for an Ornstein–Uhlenbeck process. That finding will validate the mathematical form of Eqs (8.1) and (8.2). In the course of this analysis, we will discover that the two force components on the right side of Eq. (8.1), and more specifically the two parameters γ and f, must be mathematically connected in a very specific way. The quantitative confirmation of that connection that will be demonstrated in Section 8.3 will provide a further validation of Langevin's theory of diffusion. We emphasize that the model of simple diffusion which we are about to describe is intended only to *corroborate* Langevin's Eq. (8.1), not replace it.

In our new model, $V_x(t)$ will be allowed to assume only the $2N + 1$ discrete values that divide the interval $[-v_{\max}, v_{\max}]$ into subintervals of equal length

$$\Delta = v_{\max}/N. \tag{8.4}$$

Here N is some large positive integer, and $v_{\max}$ is some large positive real number. The possible states (values) of $V_x(t)$ in our model are therefore

$$v_n = n\Delta, \quad (n = 0, \pm 1, \ldots, \pm N). \tag{8.5}$$

We will later let both N and $v_{\max}$ go to ∞ in such a way that $\Delta \to 0$. In that limit, $V_x(t)$ will have access to a *continuum* of values, and thus will be converted from a *jump* Markov process to a *continuous* Markov process.

We will model the movement of $V_x(t)$ over the discrete states (8.5) as a simple "birth–death" type jump Markov process—the same kind of process that we encountered earlier in Chapter 5—in which only jumps to adjacent states are allowed. For such a process, the dynamics will be fully specified by two functions, $W_+(v)$ and $W_-(v)$, which are *defined* by the following statement:

$$W_\pm(v_n)dt \equiv \text{Prob}\,\{V_x(t + dt) = v_{n\pm1},\ \text{given that}\ V_x(t) = v_n\}, \tag{8.6}$$

where dt is a positive infinitesimal; i.e., $W_+(v_n)dt$ is the probability that the velocity of the molecule will change from its present value v_n to v_{n+1} in the next dt, and similarly for $W_-(v_n)dt$. Our first task will be to devise formulas for these two functions that plausibly quantify how collisions of the solute molecule with the solvent molecules affect the solute molecule's velocity. Since symmetry considerations imply that

$$W_-(-v) \equiv W_+(v), \tag{8.7}$$

it suffices to focus on just one of these functions, say $W_+(v)$.

Suppose first that the solute molecule is at rest, i.e., is in the state $v_0 = 0$. Then there will be some probability that, during the next infinitesimal time dt, it will be struck on its left (negative-x) side by a solvent molecule sufficiently hard that its velocity will jump from $v_0 = 0$ to $v_1 = \Delta$. Since the definition (8.6) implies that this probability will be proportional to dt, we will take that probability to be of the form $a\,dt$, where a is some positive constant; thus, we take

$$W_+(0) = a. \tag{8.8}$$

By Eq. (8.7), with the same probability $a\,dt$ the solute molecule at rest will be struck during the next dt on its *right* (positive-x) side by a solvent molecule sufficiently hard that its velocity will jump from $v_0 = 0$ to $v_{-1} = -\Delta$. Given this physical interpretation of $a\,dt$, we should expect a to be a *decreasing* function of the solute molecule's mass m, since the solvent molecule will have a harder time getting a heavier solute molecule moving; however, for now we need not inquire into the properties of a.

Suppose next that the solute molecule is moving with velocity $v_N = v_{\max}$. Then because $v_{\max}$ is the maximum allowed value of V_x, the probability that the molecule will be struck on its left side sufficiently hard to further increase its velocity in the next dt must be zero. Thus we must demand that

$$W_+(v_{\max}) = 0. \tag{8.9}$$

The implication of Eqs (8.9) and (8.8), that $W_+(v)$ *decreases* from a to 0 as v *increases* from 0 to $v_{\max}$, is consistent with the fact that *left-side* collisions of the solute molecule with solvent molecules should, on average, become not only less frequent but also less hard as the solute molecule's *right-going* speed increases. Indeed, that should be true not just for v in the interval $[0, v_{\max}]$, but for v everywhere in $[-v_{\max}, v_{\max}]$. The simplest way to secure that property in our model here would be to assume that $W_+(v)$ decreases *linearly* with v on the entire interval $[-v_{\max}, v_{\max}]$. So that is what we will do. This assumption, together with the two specifications (8.8) and (8.9), suffice to completely determine the function W_+, and the function W_- then follows from Eq. (8.7):

$$\left.\begin{array}{l} W_+(v_n) = a(1 - v_n/v_{\max}) \\ W_-(v_n) = a(1 + v_n/v_{\max}) \end{array}\right\} \quad (-v_{\max} \le v_n \le v_{\max}). \tag{8.10}$$

Plots of these two functions are shown in Fig. 8.1. We will next work out the mathematical implications of this simple but physically plausible model of diffusion. Let

$$P(v_n, t \,|\, v^{(0)}, 0) \equiv \mathrm{Prob}\left\{ V_x(t) = v_n, \text{ given that } V_x(0) = v^{(0)} \right\}. \tag{8.11}$$

Then the laws of probability imply the following relation:

$$\begin{aligned} P(v_n, t + dt \,|\, v^{(0)}, 0) = {} & P(v_n, t \,|\, v^{(0)}, 0) \cdot [1 - W_-(v_n)dt - W_+(v_n)dt] \\ & + P(v_{n+1}, t \,|\, v^{(0)}, 0) \cdot W_-(v_{n+1})dt \\ & + P(v_{n-1}, t \,|\, v^{(0)}, 0) \cdot W_+(v_{n-1})dt. \end{aligned}$$

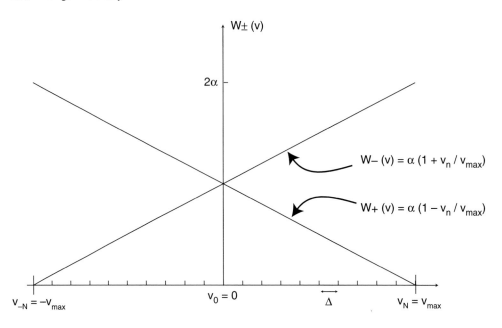

Fig. 8.1 *Plots of the functions $W_+(v)$ and $W_-(v)$ in Eqs (8.10). $W_+(v_n)\,dt$ is defined to be the probability that the solute molecule, moving along the x-axis with velocity v_n, will in the next dt be moving with the slightly greater velocity v_{n+1} owing to collisions with solvent molecules during dt. On physical grounds, we expect this probability to decrease monotonically to zero as v_n increases from its minimum possible value to its maximum possible value. Our model makes the simplifying assumption that the decrease is linear in v. The form of the function $W_-(v)$ then follows by symmetry. The additional modeling assumptions that the possible values of the velocity are discrete and bounded will later be removed by a limiting procedure.*

Here, the first term on the right is the probability that the velocity is v_n at time t, and then does not change in the next dt; the second term is the probability that the velocity is v_{n+1} at time t, and then jumps down to v_n in the next dt; and the last term is the probability that the velocity is v_{n-1} at time t and then jumps up to v_n in the next dt. All other routes from time t to the state v_n at time $t + dt$ will require more than one jump in $[t, t + dt)$, and so will have probabilities that are of order > 1 in dt, and so can be ignored. Transposing the first term on the right, dividing through by dt, and then taking the limit $dt \to 0^+$, we get the following *master equation* for our jump Markov process:

$$\frac{\partial P(v_n, t \mid v^{(0)}, 0)}{\partial t} = P(v_{n+1}, t \mid v^{(0)}, 0)\, W_-(v_{n+1}) - P(v_n, t \mid v^{(0)}, 0)\, W_+(v_n)$$

$$+ P(v_{n-1}, t \mid v^{(0)}, 0)\, W_+(v_{n-1}) - P(v_n, t \mid v^{(0)}, 0)\, W_-(v_n). \quad (8.12)$$

Next we substitute into this equation the $W_\pm$ functions in Eqs (8.10). Then, using $\Delta = v_{\max}/N$, we algebraically rearrange the right side to get the following equation:

$$\frac{\partial P(v_n, t \mid v^{(0)}, 0)}{\partial t} = \frac{2a}{N} \left[\frac{v_{n+1} P(v_{n+1}, t \mid v^{(0)}, 0) - v_{n-1} P(v_{n-1}, t \mid v^{(0)}, 0)}{2\Delta} \right]$$

$$+ \frac{av_{\max}^2}{N^2} \left[\frac{P(v_{n-1}, t \mid v^{(0)}, 0) - 2P(v_n, t \mid v^{(0)}, 0) + P(v_{n+1}, t \mid v^{(0)}, 0)}{\Delta^2} \right].$$

$$(8.13)$$

Now we want to pass to the *triple limit* $N \to \infty$, $v_{\max} \to \infty$, $\Delta \to 0$, in such a way that the two terms on the right side of Eq. (8.13) neither blow up nor disappear. To that end, we first note that as $\Delta \to 0$, the two factors in brackets in Eq. (8.13) become, respectively, first and second derivatives with respect to v. Therefore, we need only arrange things so that the coefficients of those two factors become finite, non-zero constants in that limit. The coefficient of the first term will evidently stay constant and non-zero as $N \to \infty$ if and only if we make $a \propto N$. And if we take $a \propto N$, the coefficient of the second term will stay constant and non-zero as $N \to \infty$ if and only if we make $v_{\max}^2 \propto N$. Therefore, we take

$$a = \alpha N, \ v_{\max} = u\sqrt{N}, \tag{8.14}$$

where α is some positive constant with dimensions of inverse time, and u is some positive constant with dimensions of velocity. Notice that the second of Eqs (8.14) implies that $\Delta \equiv v_{\max}/N = u/\sqrt{N}$; therefore, with this new parameterization the *single* limit $N \to \infty$ automatically accomplishes the two limits $\Delta \to 0$ and $v_{\max} \to \infty$.

Substituting Eqs (8.14) into Eq. (8.13) and then taking the limit $N \to \infty$, we obtain the partial differential equation

$$\frac{\partial P(v, t \mid v^{(0)}, 0)}{\partial t} = -\frac{\partial}{\partial v} \left[(-2\alpha v) P(v, t \mid v^{(0)}, 0) \right] + \frac{1}{2} \frac{\partial^2}{\partial v^2} \left[(2\alpha u^2) P(v, t \mid v^{(0)}, 0) \right].$$

$$(8.15)$$

A comparison of this equation with Eq. (7.24) reveals that it is a *forward Fokker–Planck equation* for a *continuous* Markov process $V_x(t)$ whose drift and diffusion functions are

$$A(v, t) = -2\alpha v, \ D(v, t) = 2\alpha u^2. \tag{8.16}$$

This implies, by Eq. (7.21), that $V_x(t)$ in this continuum limit satisfies the *Langevin equation*

$$\frac{dV_x(t)}{dt} = -2\alpha V_x(t) + (2\alpha u^2)^{1/2} \Gamma_x(t). \tag{8.17}$$

Furthermore, if we choose the two constants α and u so that

$$2\alpha = \frac{\gamma}{m} \text{ and } \sqrt{2\alpha} u = \frac{f}{m}, \tag{8.18}$$

then Eq. (8.17) becomes identical to Eq. (8.1), and hence also to Eq. (8.2).

Thus we see that a simple but physically plausible model of how the velocity of a solute molecule responds to collisions of that molecule with the many smaller solvent molecules implies the two-term form in Eq. (8.1) for the effective force exerted in the x-direction on the solute molecule by the surrounding solvent molecules. This analysis confirms in particular the rather dramatic assumption that the fluctuating force term is proportional to Gaussian white noise. Notice also that the first member of Eq. (8.14) and the first member of Eq. (8.18) together imply that $a \propto m^{-1}$; this corroborates our earlier observation that, for physical reasons, the variable a introduced in Eq. (8.8) ought to be a decreasing function of the mass of the solute molecule.

An important ancillary result of the foregoing analysis is the *quantitative connection* between the dissipative drag force component and the randomly fluctuating force component that is implied by Eq. (8.17): The appearance of the factor α in the first term on the right, coupled with the appearance of the factor $\sqrt{\alpha}$ in the second term, implies that *the factor f that multiplies $\Gamma_x(t)$ in Eq. (8.1) should be proportional to the square root of the factor γ that multiplies $V_x(t)$.* We will see in the next section that independent reasoning, based on statistical thermodynamics, leads precisely to that result.

8.3   Fixing the factor *f*: the fluctuation–dissipation theorem

We return now to Langevin's assumption (8.2), which to reiterate amounts to the assumption that $V_x(t)$ is an Ornstein–Uhlenbeck process whose relaxation time τ and diffusion constant c are as given in Eqs (8.3). From the generic formula (7.39) for the Ornstein–Uhlenbeck process, it follows upon substituting the parameter values (8.3) that

$$V_x(t) = \mathcal{N}\left(v_{x0}e^{-(t-t_0)\gamma/m}, \frac{f^2}{2m\gamma}\left(1 - e^{-2(t-t_0)\gamma/m}\right)\right), \tag{8.19}$$

where $V_x(t_0) = v_{x0}$. And from this it immediately follows that

$$V_x(t - t_0 \to \infty) = \mathcal{N}\left(0, \frac{f^2}{2m\gamma}\right). \tag{8.20}$$

Now, according to statistical thermodynamics, the solute molecule should asymptotically approach "thermal equilibrium" at the system temperature T. That means, mathematically, that each rectilinear component of the solute molecule's velocity should have a "Maxwell–Boltzmann distribution"; i.e., it should be *normally* distributed with mean zero and variance $k_B T/m$. So *physically* we should have

$$V_x(t - t_0 \to \infty) = \mathcal{N}\left(0, \frac{k_B T}{m}\right). \tag{8.21}$$

Equations (8.20) and (8.21) will evidently agree if and only if $f^2/(2m\gamma) = k_B T/m$. That equality can be secured simply by taking the free parameter f in Langevin's key assumption to be

$$f = \sqrt{2\gamma k_{\mathrm{B}} T}. \tag{8.22}$$

This renders Langevin's theory essentially complete. We can now write Langevin's equations (8.1) and (8.2) in the more specific forms,

$$m\frac{dV_x(t)}{dt} = -\gamma V_x(t) + \sqrt{2\gamma k_{\mathrm{B}} T}\,\Gamma_x(t), \tag{8.23}$$

$$V_x(t + dt) = V_x(t) - \left(\frac{\gamma}{m}\right) V_x(t)dt + \left(\frac{\sqrt{2\gamma k_{\mathrm{B}} T}}{m}\right) N_x(t)\sqrt{dt}. \tag{8.24}$$

Notice that the fluctuating force component in the "$F{=}ma$" equation (8.23), namely

$$F_x^{\mathrm{fluc}}(t) = \sqrt{2\gamma k_{\mathrm{B}} T}\,\Gamma_x(t), \tag{8.25}$$

is proportional to the square root of the drag coefficient γ. This agrees with the prediction in the preceding section of our simple model about the α-dependence of the two terms on the right side of Eq. (8.17). The existence of a connection between the drag force $-\gamma V_x(t)$ and the fluctuating force $F_x^{\mathrm{fluc}}(t)$ should not be surprising in view of the fact that those two forces have a common physical origin: both arise from the collective effects of collisional impacts of solvent molecules on the solute molecule. But it is far from obvious that the connection should have the specific quantitative form asserted by Eq. (8.25). Equation (8.25) is often referred to as the *fluctuation–dissipation theorem*. Its qualitative message is that the randomly fluctuating force component and the dissipative drag force component are not independent of each other: each increases monotonically with the other, and neither can exist without the other.

The fluctuation–dissipation theorem is sometimes expressed in a form different from Eq. (8.25). Although we shall not require that alternate form in our work here, we will derive it for the sake of completeness: Invoking the mathematical property of Gaussian white noise mentioned just after Eq. (7.21), we have from Eq. (8.25),

$$\left\langle F_x^{\mathrm{fluc}}(t) F_x^{\mathrm{fluc}}(t + t')\right\rangle = 2\gamma k_{\mathrm{B}} T\left\langle \Gamma_x(t)\Gamma_x(t + t')\right\rangle = 2\gamma k_{\mathrm{B}} T\delta(t'),$$

where δ is the Dirac delta function. Integrating this equation over all t' and then solving for γ, we get

$$\gamma = \frac{1}{2k_{\mathrm{B}} T} \int_{-\infty}^{\infty} \left\langle F_x^{\mathrm{fluc}}(t) F_x^{\mathrm{fluc}}(t + t')\right\rangle dt'. \tag{8.26}$$

This relation, which evidently relates the dissipative drag force coefficient to the "auto-covariance" of the fluctuating force, is often referred to as the *fluctuation–dissipation relation*. It is entirely equivalent to Eq. (8.25).

8.4   The Langevin diffusion formulas

With the parameter f now fixed by Eq. (8.22), we can write the OU parameters (8.3) for $V_x(t)$ in their final form,

$$\tau = \frac{m}{\gamma}, \quad c = \frac{2\gamma k_B T}{m^2}. \tag{8.27}$$

The x-coordinate of the solute molecule's position, $X(t)$, will of course be the *time-integral* of $V_x(t)$, and hence can be defined by the infinitesimal updating formula

$$X(t + dt) = X(t) + V_x(t)dt. \tag{8.28}$$

This updating formula for $X(t)$ together with the updating formula (8.24) for $V_x(t)$ comprise a *bivariate* Langevin equation for the bivariate continuous Markov process $(V_x(t), X(t))$; see Section 7.7. So we can invoke the results developed in Sections 7.9 and 7.10 for a generic OU process and its integral: $V_x(t)$ and $X(t)$ are both *normal* random variables:

$$V_x(t) = \mathcal{N}\left(\langle V_x(t)\rangle, \text{var}\{V_x(t)\}\right), \tag{8.29}$$

$$X(t) = \mathcal{N}\left(\langle X(t)\rangle, \text{var}\{X(t)\}\right). \tag{8.30}$$

And their five defining moments (means, variances, and covariance) are given by Eqs (7.47) with τ and c replaced by Eqs (8.27). Thus, where $v_{x0} = V_x(t_0)$ and $x_0 = X(t_0)$, we have:

$$\langle V_x(t)\rangle = v_{x0}\, e^{-(t-t_0)/\tau}, \tag{8.31a}$$

$$\text{var}\{V_x(t)\} = \frac{k_B T}{m}\left(1 - e^{-2(t-t_0)/\tau}\right), \tag{8.31b}$$

$$\langle X(t)\rangle = x_0 + v_{x0}\tau\left(1 - e^{-(t-t_0)/\tau}\right), \tag{8.31c}$$

$$\text{var}\{X(t)\} = 2\left(\frac{k_B T}{\gamma}\right)\left[(t - t_0) - 2\tau\left(1 - e^{-(t-t_0)/\tau}\right) + \tfrac{1}{2}\tau\left(1 - e^{-2(t-t_0)/\tau}\right)\right], \tag{8.31d}$$

$$\text{cov}\{V_x(t), X(t)\} = \left(\frac{k_B T}{\gamma}\right)\left(1 - e^{-(t-t_0)/\tau}\right)^2. \tag{8.31e}$$

Equations (8.29), (8.30), and (8.31) completely characterize $V_x(t)$ and $X(t)$ in Langevin's theory of diffusion. Analogous formulas hold for the variable pairs $(V_y(t), Y(t))$ and $(V_z(t), Z(t))$, and those three bivariate normal random variables are statistically independent of each other. But pair members, such as $V_x(t)$ and $X(t)$, are *not* statistically independent of each other, since the right side of Eq. (8.31e) is not identically zero.

8.5 The correlation between position and velocity

Equation (8.31e) implies that as t increases from t_0 to infinity, the covariance of $V_x(t)$ and $X(t)$ *increases* from 0 to $k_B T/\gamma$. That seems completely contrary to our expectation that the statistical dependency between $V_x(t)$ and $X(t)$ should diminish with time. However, as discussed in connection with Eqs (2.23) and (2.24), a better measure of the statistical dependency between $V_x(t)$ and $X(t)$ than their covariance is their *correlation*,

$$\mathrm{corr}\left\{V_x(t), X(t)\right\} \equiv \frac{\mathrm{cov}\left\{V_x(t), X(t)\right\}}{\sqrt{\mathrm{var}\{V_x(t)\}\mathrm{var}\{X(t)\}}}.$$

If we substitute into the right side of this formula the expressions given in Eqs (8.31) and then algebraically simplify, we get

$$\mathrm{corr}\left\{V_x(t), X(t)\right\} = \sqrt{\frac{(1 - e^{-s})^3}{(1 + e^{-s})\left[2s - 4(1 - e^{-s}) + (1 - e^{-2s})\right]}}, \quad s \equiv \frac{t - t_0}{\tau}. \quad (8.32)$$

Notice that when we measure the elapsed time $t - t_0$ in units of τ, corr $\left\{V_x(t), X(t)\right\}$ becomes independent of all physical parameters of the problem.

The behavior of corr $\left\{V_x(t), X(t)\right\}$ in the long-time limit $s \to \infty$ is easy to infer from Eq. (8.32): Replacing all factors e^{-s} on the right side by 0 shows that in the limit $s \to \infty$, the right side approximates to $\sqrt{1/(2s)}$. The short-time limit $s \to 0^+$ is harder to discern, because the right-hand side of Eq. (8.32) is severely indeterminate at $s = 0$. But if we Taylor-expand the exponentials in Eq. (8.32) to sufficiently high order in s (specifically, the numerator under the radical to second order in s, the first factor in the denominator to first order in s, and the second factor in the denominator to fourth order in s), we find after some algebra that to *first* order in s the right side of Eq. (8.32) approximates to $\sqrt{\frac{3}{4}}\left(1 - \frac{1}{8}s\right)$. We thus conclude that Eq. (8.32) implies the limiting behaviors

$$\mathrm{corr}\left\{V_x(t), X(t)\right\} \approx \begin{cases} \sqrt{\dfrac{3}{4}}\left(1 - \dfrac{(t - t_0)}{8\tau}\right), & \text{if } (t - t_0) \ll \tau, \\[3mm] \sqrt{\dfrac{\tau}{2(t - t_0)}}, & \text{if } (t - t_0) \gg \tau. \end{cases} \quad (8.33)$$

So we see that corr $\left\{V_x(t), X(t)\right\}$ starts at $t = t_0$ at the fairly highly correlated value $\sqrt{\frac{3}{4}} = 0.866$ (recall that the correlation always lies in the interval $[-1, 1]$), and then decreases monotonically to zero as $t \to \infty$. Initially that decrease is linear in $(t - t_0)$, but eventually it becomes proportional to $(t - t_0)^{-1/2}$. This monotonic decay of the correlation from some positive value to zero as t goes from t_0 to ∞ thus agrees with our physical intuition. But this decay to zero is slow—much slower than any exponential decay would be. The reason for this slow rate of decay is that relation (8.28), which defines $X(t)$ in terms of $V_x(t)$, establishes a strong connection between those two variables that is not easily "forgotten".

As discussed in connection with Eq. (2.31), the *square* of the correlation of two *normal* random variables can be interpreted as the fraction of the variance of either variable that "is associated with" the fluctuations in the other. Since $V_x(t)$ and $X(t)$ are both normal, we have plotted in Fig. 8.2 $\left[\mathrm{corr}\left\{V_x(t), X(t)\right\}\right]^2$ versus $(t - t_0)/\tau$; the solid curve is the *square* of the exact function in Eq. (8.32), and the two dashed curves show the *squares* of the small-t and large-t approximations in Eq. (8.33).

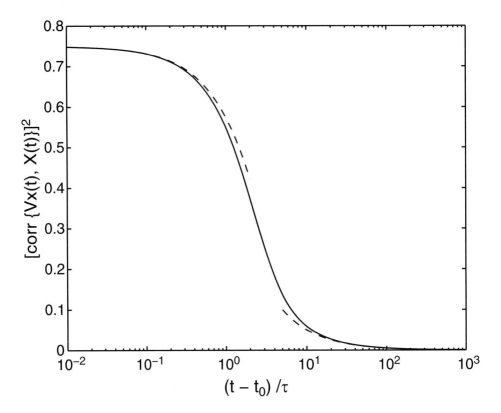

Fig. 8.2 *A plot of the square of* corr $\{V_x(t), X(t)\}$ *as a function of* $(t - t_0)/\tau$. *The solid curve is a plot of the square of the exact formula (8.32). The two dashed curves are plots of the squares of the small-t and large-t approximation formulas in Eqs (8.33). As discussed in the text, the quantity plotted here can be interpreted as the fraction of the variance of $X(t)$ that is caused by fluctuations in $V_x(t)$. The relatively slow fall-off of this quantity with time—note the logarithmic scale on the time axis—reflects the fact that the fundamental dynamical relation $X(t + dt) = X(t) + V_x(t)\,dt$ establishes a strong connection between $X(t)$ and $V_x(t)$ that is not easily forgotten.*

An examination of Eqs (8.24) and (8.28), which dynamically define $V_x(t)$ and $X(t)$, reveals that $V_x(t)$ influences the evolution of $X(t)$ but not vice versa. It therefore seems reasonable to conclude that the plot in Fig. 8.2 shows the fraction of var $\{X(t)\}$ that *is caused by* fluctuations in $V_x(t)$. Evidently, when t is very near t_0, three-quarters of var $\{X(t)\}$ is due to fluctuations in $V_x(t)$. As t increases beyond t_0, that fraction dies off to zero, eventually like $\frac{1}{2}\tau/(t - t_0)$.

8.6 Two-time auto-correlations

Also computable from the basic Langevin time-evolution equations (8.24) and (8.28) for $V_x(t)$ and $X(t)$ are explicit expressions for the *two-time auto-correlations*, corr $\{V_x(t_1), V_x(t_2)\}$ and corr $\{X(t_1), X(t_2)\}$, for $t_0 \leq t_1 \leq t_2$. We will conclude this chapter by showing how that is done. The derivations are a bit lengthy; the final formula for the velocity auto-correlation is exhibited in Eq. (8.37), and for the position auto-correlation in Eq. (8.42).

To compute the *velocity* auto-correlation, we start by writing Eq. (8.24) with $t = t_2$, and then multiplying the result through by $V_x(t_1)$. Using the parameter abbreviations in Eqs (8.27), we get

$$V_x(t_1)V_x(t_2 + dt_2) = V_x(t_1)V_x(t_2) - \left(\frac{1}{\tau}\right) V_x(t_1)V_x(t_2)dt_2 + c^{1/2}V_x(t_1)N_x(t_2)\sqrt{dt_2}.$$
$$(8.34)$$

Next we take the average of Eq. (8.34), using the fact that the zero-mean normal $N_x(t_2)$ is statistically independent of $V_x(t_1)$ so that $\langle V_x(t_1)N_x(t_2)\rangle = \langle V_x(t_1)\rangle \langle N_x(t_2)\rangle = 0$. In the resulting equation, we transpose the first term on the right, divide through by dt_2, and then take the limit $dt_2 \to 0^+$ to get

$$\frac{d}{dt_2} \langle V_x(t_1)V_x(t_2)\rangle = -\frac{1}{\tau} \langle V_x(t_1)V_x(t_2)\rangle \quad (t_2 \geq t_1). \qquad (8.35)$$

Now, since cov $\{V_x(t_1), V_x(t_2)\} \equiv \langle V_x(t_1)V_x(t_2)\rangle - \langle V_x(t_1)\rangle \langle V_x(t_2)\rangle$, then

$$\frac{d}{dt_2}\mathrm{cov} \{V_x(t_1), V_x(t_2)\} = \frac{d}{dt_2} \langle V_x(t_1)V_x(t_2)\rangle - \langle V_x(t_1)\rangle \frac{d}{dt_2} \langle V_x(t_2)\rangle.$$

For the first term on the right, we substitute from Eq. (8.35); for the second term, we invoke the implication of Eq. (8.31a) that $d \langle V_x(t)\rangle/dt = - \langle V_x(t)\rangle/\tau$:

$$\frac{d}{dt_2}\mathrm{cov} \{V_x(t_1), V_x(t_2)\} = -\frac{1}{\tau} \langle V_x(t_1)V_x(t_2)\rangle - \langle V_x(t_1)\rangle \left(-\frac{1}{\tau} \langle V_x(t_2)\rangle\right).$$

Thus,

$$\frac{d}{dt_2}\mathrm{cov} \{V_x(t_1), V_x(t_2)\} = -\frac{1}{\tau}\mathrm{cov} \{V_x(t_1), V_x(t_2)\} \quad (t_2 \geq t_1).$$

The solution to this ordinary differential equation for the required initial condition cov $\{V_x(t_1), V_x(t_2 = t_1)\} = $ cov $\{V_x(t_1), V_x(t_1)\} \equiv$ var $\{V_x(t_1)\}$ is

$$\mathrm{cov} \{V_x(t_1), V_x(t_2)\} = \mathrm{var} \{V_x(t_1)\} \, e^{-(t_2-t_1)/\tau} \quad (t_2 \geq t_1). \qquad (8.36)$$

Dividing this result by $\sqrt{\mathrm{var}\{V_x(t_1)\} \cdot \mathrm{var}\{V_x(t_2)\}}$, and then invoking the explicit formula (8.31b) for var $\{V_x(t)\}$, we conclude that

$$\mathrm{corr}\{V_x(t_1), V_x(t_2)\} = e^{-(t_2-t_1)/\tau}\sqrt{\frac{1 - e^{-2(t_1-t_0)/\tau}}{1 - e^{-2(t_2-t_0)/\tau}}} \quad (t_0 \leq t_1 \leq t_2). \qquad (8.37)$$

Two limiting cases of Eq. (8.37) are worth noting. First is the "equilibrium limit" $t_0 \to -\infty$, or more specifically, when $(t_1 - t_0) \gg \tau$ and hence also $(t_2 - t_0) \gg \tau$. In that case, the two exponentials under the radical in Eq. (8.37) can be neglected compared to 1, and the equation approximates to

$$\text{corr}\{V_x(t_1), V_x(t_2)\} \doteq e^{-(t_2-t_1)/\tau} \quad (t_1 \leq t_2;\ t_1 - t_0 \gg \tau). \tag{8.38a}$$

This formula gives $\text{corr}\{V_x(t_1), V_x(t_2)\}$ when the system was initialized, or last observed, "a long time ago". It implies that in this circumstance the correlation dies out in a time of order τ. So the relaxation time τ is also the "decorrelation time" for the velocity of the solute molecule.

The other limit of interest is the "small-time limit" $(t_2 - t_0) \to 0$, or more specifically, when $(t_2 - t_0) \ll \tau$ and hence also $(t_1 - t_0) \ll \tau$. In that circumstance, an expansion of all three exponentials in Eq. (8.37) to first order yields the approximation

$$\text{corr}\{V_x(t_1), V_x(t_2)\} \doteq \left(1 - \frac{t_2 - t_1}{\tau}\right) \sqrt{\frac{t_1 - t_0}{t_2 - t_0}} \quad (t_0 \leq t_1 \leq t_2;\ t_2 - t_0 \ll \tau). \tag{8.38b}$$

It can be seen from this result, and also from Eq. (8.37), that for any $t_2 = t_1 > t_0$, $\text{corr}\{V_x(t_1), V_x(t_2)\} = 1$. This is as expected. But it is also clear from those two equations that, for any $t_2 > t_1 = t_0$, $\text{corr}\{V_x(t_1), V_x(t_2)\} = 0$. That possibly puzzling result is explained by the fact that $V_x(t_0)$ is a *sure* variable, and hence is uncorrelated with any random variable.

Since $V_x(t_1)$ and $V_x(t_2)$ are both *normal* random variables, and since the earlier velocity $V_x(t_1)$ can influence the later velocity $V_x(t_2)$ but not vice versa, it seems reasonable to conclude from Eq. (2.31) that $[\text{corr}\{V_x(t_1), V_x(t_2)\}]^2$, the *square* of Eq. (8.37), gives the fraction of $\text{var}\{V_x(t_2)\}$ that is due to fluctuations in $V_x(t_1)$. Figure 8.3 shows a plot of this fraction as a function of t_1 and t_2, in units of τ, with $t_0 = 0$.

To compute the *position* auto-correlation, we start by writing Eq. (8.28) with $t = t_2$, and then multiplying the result through by $X(t_1)$:

$$X(t_1)X(t_2 + dt_2) = X(t_1)X(t_2) + X(t_1)V_x(t_2)dt_2.$$

Averaging this equation leads to the ordinary differential equation

$$\frac{d}{dt_2}\langle X(t_1)X(t_2)\rangle = \langle X(t_1)V_x(t_2)\rangle.$$

Since $\text{cov}\{X(t_1), X(t_2)\} \equiv \langle X(t_1)X(t_2)\rangle - \langle X(t_1)\rangle\langle X(t_2)\rangle$, then

$$\frac{d}{dt_2}\text{cov}\{X(t_1), X(t_2)\} = \frac{d}{dt_2}\langle X(t_1)X(t_2)\rangle - \langle X(t_1)\rangle\frac{d}{dt_2}\langle X(t_2)\rangle.$$

Substituting for the first term on the right from the preceding equation, and noting from Eq. (8.28) that $d\langle X(t)\rangle/dt = \langle V_x(t)\rangle$, we get

$$\frac{d}{dt_2}\text{cov}\{X(t_1), X(t_2)\} = \langle X(t_1)V_x(t_2)\rangle - \langle X(t_1)\rangle\langle V_x(t_2)\rangle,$$

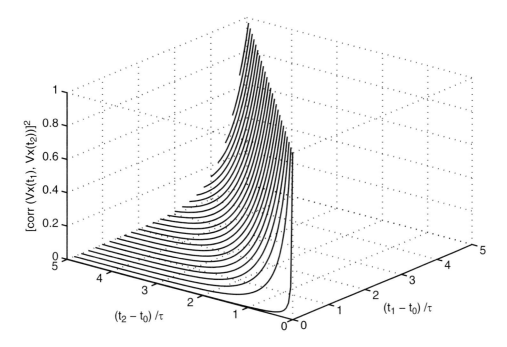

Fig. 8.3 *A plot of the square of* corr $\{V_x(t_1), V_x(t_2)\}$ *in Eq. (8.37) as a function of* $t_1 - t_0$ *$(t_1 \geq t_0)$ and $t_2 - t_0$ $(t_2 \geq t_1)$ in units of τ with $t_0 = 0$. As discussed in the text, the quantity plotted can be interpreted as the fraction of the variance of $V_x(t_2)$ that is caused by fluctuations in $V_x(t_1)$.*

whence

$$\frac{d}{dt_2} \mathrm{cov}\left\{X(t_1), X(t_2)\right\} = \mathrm{cov}\left\{X(t_1), V_x(t_2)\right\}. \tag{8.39}$$

Equation (8.39) tells us that in order to compute $\mathrm{cov}\left\{X(t_1), X(t_2)\right\}$, we must first compute $\mathrm{cov}\left\{X(t_1), V_x(t_2)\right\}$. To that end, we put $t = t_2$ in Eq. (8.24), and then multiply the result through by $X(t_1)$:

$$X(t_1)V_x(t_2 + dt_2) = X(t_1)V_x(t_2) - \left(\frac{1}{\tau}\right)X(t_1)V_x(t_2)dt_2 + c^{1/2}X(t_1)N_x(t_2)\sqrt{dt_2}.$$

Averaging this using $\langle X(t_1)N_x(t_2)\rangle = 0$ leads to the differential equation

$$\frac{d}{dt_2}\langle X(t_1)V_x(t_2)\rangle = -\frac{1}{\tau}\langle X(t_1)V_x(t_2)\rangle \quad (t_2 \geq t_1).$$

Since $\mathrm{cov}\left\{X(t_1), V_x(t_2)\right\} = \langle X(t_1)V_x(t_2)\rangle - \langle X(t_1)\rangle \langle V_x(t_2)\rangle$, then

$$\frac{d}{dt_2}\mathrm{cov}\left\{X(t_1), V_x(t_2)\right\} = \frac{d}{dt_2}\langle X(t_1)V_x(t_2)\rangle - \langle X(t_1)\rangle\frac{d}{dt_2}\langle V_x(t_2)\rangle.$$

On the right side of this equation, we substitute for the first term from the preceding equation, and we also note from Eq. (8.31a) that $d\langle V_x(t)\rangle/dt = -\langle V_x(t)\rangle/\tau$:

$$\frac{d}{dt_2}\text{cov}\left\{X(t_1), V_x(t_2)\right\} = -\frac{1}{\tau}\langle X(t_1)V_x(t_2)\rangle - \langle X(t_1)\rangle \left(-\frac{1}{\tau}\langle V_x(t_2)\rangle\right).$$

Thus,

$$\frac{d}{dt_2}\text{cov}\left\{X(t_1), V_x(t_2)\right\} = -\frac{1}{\tau}\text{cov}\left\{X(t_1), V_x(t_2)\right\} \quad (t_2 \geq t_1).$$

The solution to this differential equation for $\text{cov}\left\{X(t_1), V_x(t_2)\right\}$ for the obvious initial condition at $t_2 = t_1$ is

$$\text{cov}\left\{X(t_1), V_x(t_2)\right\} = \text{cov}\left\{X(t_1), V_x(t_1)\right\}e^{-(t_2-t_1)/\tau} \quad (t_2 \geq t_1). \tag{8.40}$$

Now substituting Eq. (8.40) into Eq. (8.39), we get

$$\frac{d}{dt_2}\text{cov}\left\{X(t_1), X(t_2)\right\} = \text{cov}\left\{V_x(t_1), X(t_1)\right\}e^{-(t_2-t_1)/\tau} \quad (t_2 \geq t_1).$$

Integrating this equation over t_2 is straightforward, and gives

$$\text{cov}\left\{X(t_1), X(t_2)\right\} - \text{cov}\left\{X(t_1), X(t_1)\right\}$$
$$= \text{cov}\left\{V_x(t_1), X(t_1)\right\}\tau\left[1 - e^{-(t_2-t_1)/\tau}\right].$$

Therefore, since $\text{cov}\left\{X(t_1), X(t_1)\right\} \equiv \text{var}\left\{X(t_1)\right\}$,

$$\text{cov}\left\{X(t_1), X(t_2)\right\} = \text{var}\left\{X(t_1)\right\}$$
$$+\text{cov}\left\{V_x(t_1), X(t_1)\right\}\tau\left[1 - e^{-(t_2-t_1)/\tau}\right] \quad (t_0 \leq t_1 \leq t_2). \tag{8.41}$$

Finally, dividing this covariance by the square root of the product of the variances in $X(t_1)$ and $X(t_2)$, we conclude that the correlation of $X(t_1)$ and $X(t_2)$ is

$$\text{corr}\left\{X(t_1), X(t_2)\right\} = \sqrt{\frac{\text{var}\{X(t_1)\}}{\text{var}\{X(t_2)\}}}$$
$$+ \frac{\text{cov}\left\{V_x(t_1), X(t_1)\right\}\tau\left[1 - e^{-(t_2-t_1)/\tau}\right]}{\sqrt{\text{var}\left\{X(t_1)\right\} \cdot \text{var}\left\{X(t_2)\right\}}} \quad (t_0 \leq t_1 \leq t_2).$$
$$\tag{8.42}$$

Equation (8.42) can be made explicit by substituting for the variances and the covariance on the right the appropriate formulas from Eqs (8.31). Since the resulting expression is more cumbersome than informative, we will not display it here. But two limiting cases are worth noting. First is the "equilibrium limit" $t_0 \to -\infty$, or more specifically, when $(t_1 - t_0) \gg \tau$ and hence also $(t_2 - t_0) \gg \tau$. In that case, Eqs (8.31d) and (8.31e) can be approximated for both $i = 1$ and 2 as

$$\text{var}\left\{X(t_i)\right\} \doteq 2\left(\frac{k_B T}{\gamma}\right)(t_i - t_0), \ \ \text{cov}\left\{V_x(t_i), X(t_i)\right\} \doteq \left(\frac{k_B T}{\gamma}\right).$$

With these approximations, Eq. (8.42) becomes

$$\text{corr}\left\{X(t_1), X(t_2)\right\} \doteq \sqrt{\frac{t_1 - t_0}{t_2 - t_0}} + \frac{\tau\left[1 - e^{-(t_2 - t_1)/\tau}\right]}{2\sqrt{(t_1 - t_0)\cdot(t_2 - t_0)}}.$$

But by assumption, $(t_1 - t_0)(t_2 - t_0) \gg \tau^2$, so the second term on the right can be neglected, and we get

$$\text{corr}\left\{X(t_1), X(t_2)\right\} \doteq \sqrt{\frac{t_1 - t_0}{t_2 - t_0}} \quad (t_1 \le t_2; \ t_1 - t_0 \gg \tau). \tag{8.43a}$$

We note that this approximate "long-time" limit result agrees with the prediction in Eq. (3.17) of Einstein's theory of diffusion. We will address the connection between the Einstein and Langevin theories in detail in Chapter 9.

The other interesting limiting case of Eq. (8.42) is the "small-time limit" $(t_2 - t_0) \to 0$, or more specifically, when $(t_2 - t_0) \ll \tau$ and hence also $(t_1 - t_0) \ll \tau$. It turns out that in this case, the Einstein formula (3.17) for $\text{corr}\left\{X(t_1), X(t_2)\right\}$ is not correct. Letting $s_i \equiv (t_i - t_0)/\tau$, and noting that in this case both s_1 and s_2 are $\ll 1$, we find that Taylor expansions of Eqs (8.31d) and Eq. (8.31e) in s_i (to at least third order in the first equation and at least first order in the second equation) give to lowest order,

$$\text{var}\left\{X(t_i)\right\} \doteq \frac{2k_B T\tau}{3\gamma}s_i^3, \ \ \text{cov}\left\{V_x(t_1), X(t_1)\right\} \doteq \frac{k_B T}{\gamma}s_1^2.$$

When these approximations are substituted into Eq. (8.42) and the resulting expression is simplified for the conditions $s_i \ll 1$, we get

$$\text{corr}\left\{X(t_1), X(t_2)\right\} \doteq \left(\frac{t_1 - t_0}{t_2 - t_0}\right)^{3/2}\left[1 + \frac{3}{2}\left(\frac{t_2 - t_1}{t_1 - t_0}\right)\right] \quad (t_0 \le t_1 \le t_2; \ t_2 - t_0 \ll \tau). \tag{8.43b}$$

This tells us that the *initial* fall-off with t_1 and t_2 in the two-time position auto-correlation is somewhat *slower* in the Langevin theory than what is predicted by the Einstein theory in Eq. (3.17).

Since $X(t_1)$ and $X(t_2)$ are both *normal* random variables, and since the earlier position $X(t_1)$ can influence the later position $X(t_2)$ but not vice versa, it follows from Eq. (2.31) that $[\text{corr}\{X(t_1), X(t_2)\}]^2$, the *square* of Eq. (8.42), can be viewed as the fraction of var $\left\{X(t_2)\right\}$ that is caused by fluctuations in $X(t_1)$. Figure 8.4 shows a plot of this fraction against t_1 and t_2, in units of τ, with $t_0 = 0$.

In the next chapter, we will continue exploring the implications of Langevin's theory of diffusion. Our first order of business will be to see how the classical diffusion coefficient D emerges in this theory, and how Langevin's theory connects to Einstein's theory.

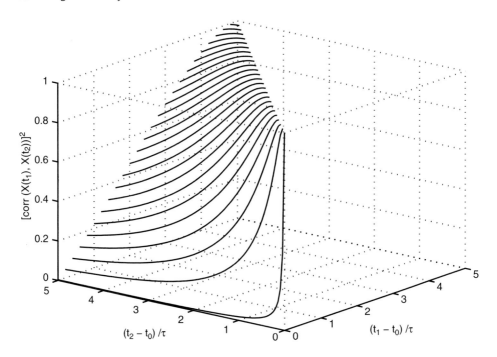

Fig. 8.4 *A plot of the square of* corr $\{X(t_1), X(t_2)\}$ *in Eq. (8.42) as a function of* $t_1 - t_0$ $(t_1 \geq t_0)$ *and* $t_2 - t_0$ $(t_2 \geq t_1)$ *in units of* τ *with* $t_0 = 0$. *As discussed in the text, the quantity plotted can be interpreted as the fraction of the variance of* $X(t_2)$ *that is caused by fluctuations in* $X(t_1)$.

Note to Chapter 8

[1]Paul Langevin (1872–1946) was a French physicist. He received his undergraduate education at the prestigious ESPCI and ENS schools in Paris, the city of his birth. After a year of postgraduate work at Cambridge University with J. J. Thompson (the discoverer of the electron), Langevin returned to Paris to begin work as a graduate student at the Sorbonne under Pierre Curie (the co-discoverer with his wife Marie of radioactivity). Langevin received his Ph.D. in 1902, developing a theory of paramagnetism and diamagnetism that is still in vogue today. In 1908, he wrote the paper that concerns us here: "Sur la théorie du movement brownien", *Comptes Rendus de l'Académie des Sciences (Paris)* **146**:530–533. An English translation of that paper, along with some insightful comments on its contents, has been given by D. Lemons and A. Gythiel in the *American Journal of Physics* **65**:1079–1081 (1997). Langevin's analysis in his paper is self-contained, clear, and remarkably brief: applying some clever but simple rules of calculus to the law $F = ma$, and making some insightful physical assumptions, he deduces Einstein's formula for the mean-square displacement in less than a page. Our analysis in this and the following two chapters will take a different route which is longer, but which leads to a much more comprehensive set of results.

Another reason to read the Lemons and Gythiel paper is the brief but fascinating sketch it gives of Langevin's life and work: his early active support of Einstein's new theory of relativity; his "scandalous" love affair in 1911 with Marie Curie, who had been widowed five years earlier when her husband Pierre was killed in a Paris street accident; his work on the piezoelectric effect (which had been discovered in 1880 by his mentor Pierre Curie and Pierre's brother Jacques), work which resulted in him and Russian scientist Constantin Chilowski being awarded two US patents in 1916 and 1917 for ultrasonic submarine detection; his outspoken opposition to fascism in the 1930s; his arrest by the Nazis after their invasion of France in 1940; and his subsequent escape to Switzerland to witness the end of the terrible war.

9

Implications of Langevin's theory

In this chapter we will examine the major implications of the Langevin theory of diffusion that was introduced in Chapter 8. We will begin by deriving the Langevin formula for the mean-square displacement of a solute molecule. That formula will enable us not only to identify the diffusion coefficient D, but also to see how Langevin's theory of diffusion contains Einstein's theory as a special limiting approximation. We will see how these considerations lead naturally to: a fundamental relation between the diffusion coefficient and the drag coefficient called the Einstein formula; a quantity called the characteristic diffusion length; a confirmation of the reasoning we used to derive the stochastic bimolecular chemical reaction rate in Chapters 3 and 4; and new information on the validity conditions for the discrete-stochastic model of diffusion discussed in Chapters 5 and 6. We will derive exact formulas for numerically simulating the trajectory of a solute molecule according to Langevin's theory. Using those formulas, we will compare and contrast Langevin trajectories with Einstein trajectories for the bead-in-water system considered in Chapters 5 and 6. We will reexamine the relative motion of two solute molecules in the context of the Langevin theory. We will show how the Langevin theory's more realistic rendering of the velocity of a diffusing molecule not only makes possible an alternative definition of the coefficient of diffusion, but also provides insight into the energetics of diffusion. Finally, we will give a careful critique of what it means to say that a diffusing molecule is "overdamped".

9.1 The Langevin mean-square displacement formulas

From an experimental standpoint, the principal result of Einstein's theory of diffusion is that the mean-square x-displacement of a solute molecule in time $t - t_0$ is equal to $2D(t - t_0)$. That result was also crucial for connecting the discrete-stochastic model of diffusion described in Chapters 5 and 6 to Einstein's model of diffusion via the relation $\kappa_l = D/l^2$.

To see what Langevin's theory of diffusion has to say about the mean-square x-displacement, we begin by substituting Eq. (8.31c) into Eq. (8.30), and then invoking the normal random variable property (2.15a). That gives the following expression for the x-displacement of a diffusing molecule in time $[t_0, t]$:

$$X(t) - x_0 = \mathcal{N}\left(v_{x0}\tau\left(1 - e^{-(t-t_0)/\tau}\right), \text{var}\left\{X(t)\right\}\right). \tag{9.1}$$

The identity $\langle Z^2 \rangle \equiv \text{var}\{Z\} + \langle Z \rangle^2$ then allows us to conclude that the mean-square x-displacement in time $[t_0, t]$ is

$$\left\langle (X(t) - x_0)^2 \right\rangle = \text{var}\{X(t)\} + \left[v_{x0}\tau \left(1 - e^{-(t-t_0)/\tau} \right) \right]^2 .$$

Upon substituting Eq. (8.31d) for the first term on the right, this becomes

$$\left\langle (X(t) - x_0)^2 \right\rangle = 2 \left(\frac{k_\mathrm{B}T}{\gamma} \right) \left[(t - t_0) - 2\tau \left(1 - e^{-(t-t_0)/\tau} \right) + \frac{\tau}{2} \left(1 - e^{-2(t-t_0)/\tau} \right) \right]$$
$$+ v_{x0}^2 \tau^2 \left(1 - e^{-(t-t_0)/\tau} \right)^2 . \tag{9.2}$$

This is Langevin's formula for the mean-square x-displacement of a solute molecule in time $[t_0, t]$.

9.2   The coefficient of diffusion: the connection to Einstein's theory

The Langevin mean-square displacement formula (9.2) is evidently more complicated than Einstein's formula (3.12). Among other things, the Langevin formula involves the x-velocity v_{x0} of the molecule at time t_0. But on reflection, that seems quite reasonable; indeed, it prompts the question: Why does the Einstein result (3.12) *not* involve v_{x0}? The plain answer is that Einstein's analysis totally ignores the velocity of the diffusing molecule. We cannot do that here. But we can, in effect, eliminate v_{x0} by *averaging* over it. In the common circumstance that we do not know v_{x0}, it will be reasonable to assume that the solute molecule at time t_0 is "in thermal equilibrium at temperature T". As discussed in connection with Eq. (8.21), that would mean that $V_x(t_0)$ can be regarded as a normal random variable with mean 0 and variance $k_\mathrm{B}T/m$. That in turn would imply that the average of the square of $V_x(t_0)$ is given by

$$\left\langle V_x^2(t_0) \right\rangle_{\text{eq}} \equiv \text{var}\{V_x(t_0)\}_{\text{eq}} + \langle V_x(t_0) \rangle_{\text{eq}}^2 = \frac{k_\mathrm{B}T}{m} . \tag{9.3}$$

It would then make sense to *average* Eq. (9.2) over this Maxwell–Boltzmann distribution of values for v_{x0}. An inspection of Eq. (9.2) reveals that the effect of doing that would be simply to make the following replacement in the last term of that equation:

$$v_{x0}^2 \leftarrow \frac{k_\mathrm{B}T}{m} . \tag{9.4}$$

Upon making this replacement, and then noting that

$$\frac{k_\mathrm{B}T}{m}\tau^2 = \frac{k_\mathrm{B}T}{m}\frac{m}{\gamma}\tau = \frac{k_\mathrm{B}T}{\gamma}\tau ,$$

we obtain for the "equilibrium average" of Eq. (9.2),

$$\left\langle \left(X(t) - x_0\right)^2 \right\rangle_{\text{eq}} = 2\left(\frac{k_{\mathrm{B}}T}{\gamma}\right)\left[(t - t_0) - 2\tau\left(1 - e^{-(t-t_0)/\tau}\right) + \frac{\tau}{2}\left(1 - e^{-2(t-t_0)/\tau}\right)\right]$$
$$+ \left(\frac{k_{\mathrm{B}}T}{\gamma}\right)\tau\left(1 - e^{-(t-t_0)/\tau}\right)^2.$$

And this complicated looking expression can be algebraically simplified to

$$\left\langle \left(X(t) - x_0\right)^2 \right\rangle_{\text{eq}} = 2\left(\frac{k_{\mathrm{B}}T}{\gamma}\right)\left[(t - t_0) - \tau\left(1 - e^{-(t-t_0)/\tau}\right)\right]. \tag{9.5}$$

It is this *equilibrium* mean-square x-displacement formula, in which v_{x0} in Eq. (9.2) has been averaged over its thermodynamically expected values, that should be compared to Einstein's v_{x0}-free formula (3.12).

Now, we found in Chapter 4 that the Einstein model of diffusion is valid only if $(t - t_0)$ is "sufficiently large". Since the only parameter in Eq. (9.5) that has dimensions of time is τ, then it must be the $(t - t_0) \gg \tau$ *limit* of Eq. (9.5) that *ought* to agree with the Einstein formula (3.12). That limit is evidently

$$\left\langle \left(X(t) - x_0\right)^2 \right\rangle_{\text{eq}} \approx 2\left(\frac{k_{\mathrm{B}}T}{\gamma}\right)(t - t_0) \quad (t - t_0) \gg \tau. \tag{9.6}$$

Comparing this result with Einstein's classical result $2D(t - t_0)$ reveals two things: First, Langevin's mean-square displacement formula replicates Einstein's linear-in-time behavior *only* in the limit $(t - t_0) \gg \tau$. And second, the classical diffusion coefficient D must be given by

$$D = \frac{k_{\mathrm{B}}T}{\gamma}. \tag{9.7}$$

Ironically, at least from our point of view here, Eq. (9.7) is known as the *Einstein formula*. It was in fact discovered by Einstein in the first of his several papers on Brownian motion (see Note 1 of Chapter 3). But Einstein's derivation of Eq. (9.7) used reasoning that was quite different from our reasoning here, as we will explain later in Chapter 10 in our discussion of Eqs (10.8). In any case, since Eq (9.7) relates the three experimentally measurable parameters D, T, and γ, it is an important result.

Complementing the "long-time" approximation (9.6) to Eq. (9.5) is the "short-time" approximation. It can be obtained by Taylor-expanding the exponential in the exact formula (9.5) to second order in $(t - t_0)/\tau$. The first-order term in that expansion exactly cancels the first term in brackets, and using $\tau = m/\gamma$ we find that Eq. (9.5) approximates to

$$\left\langle \left(X(t) - x_0\right)^2 \right\rangle_{\text{eq}} \approx \left(\frac{k_{\mathrm{B}}T}{m}\right)(t - t_0)^2 \quad (t - t_0) \ll \tau. \tag{9.8}$$

Thus, whereas for *long* times the equilibrium mean-square x-displacement depends *linearly* on $(t - t_0)$, for *short* times it depends *quadratically* on $(t - t_0)$. The physical import of Eq. (9.8) can be seen more clearly by taking its square root:

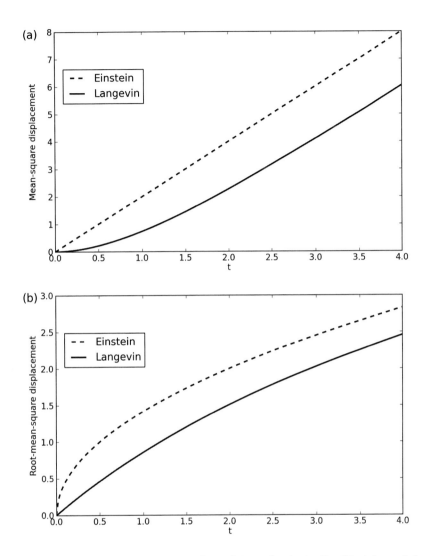

Fig. 9.1 *Comparing the average motion of a solute molecule in the Einstein and Langevin theories.* **(a)** *The Einstein and Langevin predictions for the average of the square of the x-displacement of a solute molecule in time t, in units such that $D = 1$ and $\tau = 1$, and with $t_0 = 0$. The Einstein curve (dashed) is a plot of Eq. (3.12). The Langevin curve (solid) is a plot of the "equilibrium" mean-square x-displacement in Eq. (9.5), which assumes that the initial mean-square velocity of the solute molecule has the Maxwell–Boltzmann equilibrium value $k_B T/m = D/\tau$. The two curves are evidently quite different for $t \leq \tau$. But for $t \gg \tau$, the Einstein curve is simply shifted upward from the Langevin curve by the relatively insignificant amount (for $t \gg \tau$) of $2D\tau$.* **(b)** *A comparison of the square roots of the two curves in (a), i.e., the root-mean-square x-displacements. The slope of each curve here provides an estimate of the average speed of the solute molecule. The initial average speed predicted by the Langevin curve is the "thermal" value $\sqrt{D/\tau}$, whereas the initial average speed predicted by the Einstein curve is infinite.*

$$\sqrt{\left\langle (X(t) - x_0)^2 \right\rangle_{\text{eq}}} \approx \sqrt{\frac{k_B T}{m}} (t - t_0) \quad (t - t_0) \ll \tau. \tag{9.9}$$

This says that, for sufficiently short times, the root-mean-square displacement of the solute molecule along the x-axis is given by the product of its root-mean-square speed $\sqrt{k_B T/m}$ along that axis times the elapsed time. In other words, *on sufficiently small time-scales the solute molecule moves ballistically*, just as we reasoned it should in Section 4.6. This prediction of the Langevin theory of diffusion independently confirms the reasoning we used in Sections 4.7 and 4.8 to derive a formula for the stochastic bimolecular chemical reaction rate.

In Fig. 9.1a, we compare the graph of the Langevin formula (9.5) for the equilibrium mean-square x-displacement (solid curve) with the graph of Einstein's formula (3.12) (dashed curve) for the parameter values $D = 1$ and $\tau = 1$, and with $t_0 = 0$. The overestimation of the Einstein formula is evidently severe for $t \leq \tau$, and still significant even at $t = 4\tau$. But for $t \gg \tau$, the exponential in Eq. (9.5) reduces to 0, and the second term inside the brackets becomes $-\tau$. The Langevin curve then becomes parallel to the Einstein curve, but down-shifted by $2D\tau$, which for $t \gg \tau$ will be a relatively insignificant amount. In Fig. 9.1b we compare the square roots of the two curves in (a)—i.e., the *root-mean-square x-displacements*. The slope of each curve here estimates the average speed of the solute molecule. The initial slope of the Langevin curve is the assumed "thermal" value, $\sqrt{k_B T/m} = \sqrt{D/\tau}$. But the initial slope of the Einstein curve is infinite, providing another illustration of the failure of the Einstein theory to describe the short-time motion of a solute molecule in a physically credible way.

Our conclusion here, that the Einstein theory agrees with the Langevin theory in the long-time limit but not in the short-time limit, is entirely in accord with our earlier findings in Section 8.6. There we found that $\text{corr}\{X(t_1), X(t_2)\}$ in Langevin's theory, which is displayed in Eq. (8.42), reduces to the Einstein formula (3.17) when $t_1 - t_0 \gg \tau$ [see Eq. (8.43a)], but when $t_2 - t_0 \ll \tau$ the Langevin and Einstein predictions differ [see Eq. (8.43b)].

9.3   The relaxation time and the characteristic diffusion length

Our finding that Einstein's model of diffusion is valid only over time intervals much longer than τ suggests that we finally have in τ a quantitative estimate of the time it takes for a solute molecule to have "many collisions" with solvent molecules—inasmuch as that was the crucial *physical* limitation on Einstein's analysis. Since Eq. (9.7) implies that $\gamma = k_B T/D$, then $\tau \equiv m/\gamma$ can also be written

$$\tau = \frac{mD}{k_B T}. \tag{9.10}$$

Thus, Einstein's theory of diffusion will be valid only for time increments Δt that satisfy

$$\Delta t \gg \frac{mD}{k_B T}. \tag{9.11}$$

We had an earlier hint of this result in Eq. (4.20), which we deduced by a very different line of reasoning.

Another manifestation of the physical significance of τ can be seen by comparing the solute molecule's root-mean-square x-*displacement* in time τ to its root-mean-square x-*distance traveled* in time τ, with both of those averages being computed on the assumption that the initial velocity is "thermal" in the sense of Eq. (9.4). For the rms x-displacement in time τ we have, putting $t = t_0 + \tau$ in Eq. (9.5) and invoking the Einstein formula (9.7),

$$\sqrt{\left\langle \left(X(t_0 + \tau) - X(t_0) \right)^2 \right\rangle_{\text{eq}}} = \sqrt{2D\left(\tau - \tau(1 - e^{-1})\right)} = \sqrt{\frac{2}{e}}\sqrt{D\tau}. \qquad (9.12a)$$

The rms x-distance actually traveled by the solute molecule in time τ is equal to the product of τ times the molecule's "effective speed" along the x-axis, as estimated by the square root of the mean of the square of $V_x(t)$. Since the Maxwell–Boltzmann distribution tells us that $\sqrt{\langle V_x^2(t) \rangle} = \sqrt{k_{\mathrm{B}}T/m}$, then the rms x-distance actually traveled by the solute molecule in time τ is

$$\sqrt{\frac{k_{\mathrm{B}}T}{m}} \cdot \tau = \sqrt{\frac{D\gamma}{m}}\,\tau = \sqrt{D\tau}, \qquad (9.12b)$$

where we have again used Eq. (9.7). The salient point here is that the two quantities (9.12a) and (9.12b) are *roughly equal*: The rms x-displacement of the solute molecule in time τ is only slightly smaller, by a factor of $\sqrt{2/e} \approx 0.86$, than the rms x-distance traveled in time τ. If the solute molecule were viewed over times much *larger* than τ, its rms displacement would be much *less* than the distance traveled,[1] as we would expect in *diffusional* motion. But if the molecule is viewed over times much *smaller* than τ, its rms displacement will be proportional to the observation time [see Eq. (9.9)], as we would expect in *ballistic* motion, and therefore approximately equal to the distance traveled.

Another way to see this result is to take a spatial perspective: Over distances that are significantly *larger* than $\sqrt{D\tau}$, the solute molecule appears to move *diffusionally*. Over distances that are significantly *smaller* than $\sqrt{D\tau}$, the solute molecule appears to move *ballistically*. The critical distance $\sqrt{D\tau}$ that separates, in this approximate way, the spatial scales of diffusional and ballistic motion is called the *characteristic diffusion length*:

$$\text{characteristic diffusion length} \equiv \sqrt{D\tau}. \qquad (9.13)$$

9.4   Implications for the discrete-stochastic model of diffusion

The restriction $\Delta t \gg \tau$ on Einstein's model of diffusion does not, of course, apply to Langevin's model. But it does apply to the discrete-stochastic model described in Chapters 5 and 6. To see why, recall that the discrete-stochastic model is based on the κ_l-hypothesis (5.3). That hypothesis asserts that, in a linear array of "cells" of length l, there will exist a constant κ_l which is such that $\kappa_l \delta t$ gives the probability that a randomly chosen molecule will jump to a specified adjacent cell in the next suitably small time δt. In Eq. (5.9), we found that this hypothesis implies that the mean-square x-displacement of a solute molecule in time δt will be $2 \left(\kappa_l l^2 \right) \delta t$. The fact that this last result can be made consistent with Einstein's result $2D\delta t$ by taking $\kappa_l = D/l^2$ is what validated the κ_l-hypothesis from the point of view of Einstein's theory. But with Langevin's more accurate result (9.5), we now know that the mean-square x-displacement of a solute molecule in time δt will be *linear* in δt only if $\delta t \gg \tau$ [see Eq. (9.6)]. At the other extreme $\delta t \ll \tau$, the mean-square x-displacement in time δt will be *quadratic* in δt [see Eq. (9.8)], and there is no way to reconcile that behavior with the linear δt behavior that is implicit in hypothesis (5.3). So we see that the discrete-stochastic hypothesis (5.3) can be valid only when $\delta t \gg \tau$.

In fact, we concluded in Section 5.5 that δt in the discrete-stochastic model must be "neither too large nor too small". But the only quantitative condition that we could discern in that regard was condition (5.15a), $\kappa_l \delta t \ll 1$, which gives an extreme *upper* bound on δt of $1/\kappa_l$. Now, with the requirement $\delta t \gg \tau$ clear from Langevin's theory, we also have an extreme *lower* bound on δt. So in the discrete-stochastic model δt must satisfy

$$\tau \ll \delta t \ll \frac{1}{\kappa_l}. \tag{9.14}$$

But this condition also implies an extreme *lower* bound on the cell length l, a parameter that we had likewise concluded in Section 5.5 should be "neither too large nor too small". This extreme lower bound on l comes about because condition (9.14) can be satisfied only if $\tau \ll 1/\kappa_l$. And if we insert into that strong inequality the relation $\kappa_l = D/l^2$, we obtain $\tau \ll 1/(D/l^2)$, or equivalently $l^2 \gg D\tau$, whence

$$l \gg \sqrt{D\tau}. \tag{9.15}$$

In words, *the cell length l must be comfortably larger than the characteristic diffusion length*, as defined in Eq. (9.13). Unless that is so, the cells will be so small that in any time δt which satisfies (9.14) we will have to worry about jumps to *non-adjacent* cells—an eventuality that is not allowed for in the derivations of the discrete-stochastic master equations (6.3) and (6.20).

In Section 5.6, we found in Eqs (5.22) and (5.23) that the formula $\kappa_l = D/l^2$ will be accurate only if $l \gg 4D/\bar{v}$. How does that lower bound on l compare with the lower bound (9.15)? Using first the formula for $\bar{v}$ in Eq. (5.16), and then the Einstein relation (9.7), and finally the definition $\tau \equiv m/\gamma$, we find that condition (5.23) can be written

$$l \gg 4D/\bar{v} = 4D\sqrt{\frac{\pi m}{8k_{\mathrm{B}}T}} = \sqrt{2\pi}D\sqrt{\frac{m}{D\gamma}} = \sqrt{2\pi}D\sqrt{\frac{\tau}{D}} = \sqrt{2\pi}\sqrt{D\tau}.$$

Since $\sqrt{2\pi} \approx 2.5$ is of order 1, we conclude that *condition (5.23) for the validity of the formula $\kappa_l = D/l^2$ is in fact the same as condition (9.15)*. This agreement provides reassuring confirmation of the reasoning that we used in both analyses.

So, does the cell length l chosen for the bead diffusion experiments in Chapters 5 and 6 satisfy conditions (9.15) or (5.23)? In those experiments, the 400 μm length of the system volume Ω was divided into 80 cells, giving $l = 5$ μm. For the microscopic polystyrene bead in water, $D = 1.51$ μm^2 s^{-1}, and as we will see later in this chapter (Section 9.7), $\tau = 4.9 \times 10^{-9}$ s; hence, the characteristic diffusion length is $\sqrt{D\tau} = 8.6 \times 10^{-5}$ μm. Since that is nearly five orders of magnitude smaller than l, we conclude that condition (9.15) was indeed satisfied in those experiments.

9.5   The Langevin picture of $V_x(t)$

In Section 9.2 we compared the Langevin and Einstein predictions for $X(t)$. In this section we will do the same for $V_x(t)$. In particular, we want to verify that $V_x(t)$ behaves the same in the two theories over time intervals that are much larger than τ.

It is not obvious that that is so: In Langevin's theory, we have from Eqs (8.29), (8.31a), and (8.31b) that $V_x(t)$ is the normal random variable

$$V_x^L(t) = \mathcal{N}\left(v_{x0}\ e^{-(t-t_0)/\tau}, \frac{k_B T}{m}\left(1 - e^{-2(t-t_0)/\tau}\right)\right). \qquad (9.16)$$

But in Einstein's theory, where $X(t)$ is the driftless Wiener process (7.36) with $c = 2D$, it follows from the corresponding Langevin equation (7.33b) that $V_x(t)$ is given by

$$V_x^E(t) \equiv \frac{dX(t)}{dt} = \sqrt{2D}\Gamma_x(t) \approx \mathcal{N}(0, \infty), \qquad (9.17)$$

where the last equality follows from the definition of Gaussian white noise in Eq. (7.20). The difference between the above two descriptions of $V_x(t)$ is striking: $V_x^L(t)$ in Eq. (9.16) is a *continuous* function of t, and for reasonable values of v_{x0} it rarely gets larger in absolute value than a few multiples of $\sqrt{k_B T/m}$. But $V_x^E(t)$ in Eq. (9.17) bounces *discontinuously* back and forth between $\pm\infty$. Also, $V_x^L(t)$ and $X(t)$ are not statistically independent, according to Eq. (8.31e), whereas $V_x^E(t)$, being proportional to Gaussian white noise, is statistically independent of $X(t)$.

In the domain $(t - t_0) \gg \tau$ where we expect the two descriptions of the velocity to agree, $V_x^L(t)$ in Eq. (9.16) evidently becomes

$$V_x^L(t) = \mathcal{N}\left(0, \frac{k_B T}{m}\right) \qquad (t - t_0) \gg \tau. \qquad (9.18)$$

This implies that for time increments $t \gg \tau$, $V_x^L(t)$ no longer looks continuous, but instead appears to bounce back and forth between values of order $\pm\sqrt{k_B T/m}$. But by Eq. (9.10), we have

$$\frac{k_B T}{m} = \frac{D}{\tau}. \qquad (9.19)$$

This formula implies that if D is held *fixed*, then on time-scales where τ can be considered to be "very small", $\sqrt{k_B T/m}$ will appear to be "very large". So under these conditions, $V_x^L(t)$ will indeed resemble $V_x^E(t)$ in Eq. (9.17).

As for the statistical dependency of $V_x^L(t)$ and $X^L(t)$ which is implied by Eq. (8.32), it follows from Eq. (8.33) that corr $\{V_x^L(t), X^L(t)\}$ becomes practically zero when $(t - t_0) \gg \tau$. Therefore, in that Einstein regime, $V_x^L(t)$ and $X^L(t)$ will be essentially statistically independent, just as $V_x^E(t) \propto \Gamma(t)$ and $X^E(t)$ are statistically independent. As for the statistical dependency of $V_x^L(t_1)$ and $V_x^L(t_2)$ for $t_2 \geq t_1 \geq t_0$ which is implied by Eq. (8.37), Eq. (8.38a) shows that if $t_1 - t_0 \gg \tau$, then corr $\{V_x^L(t_1), V_x^L(t_2)\} = e^{-(t_2-t_1)/\tau}$. Therefore, when $t_1 - t_0 \gg \tau$ and $t_2 - t_1 \gg \tau$, corr $\{V_x^L(t_1), V_x^L(t_2)\}$ will be essentially zero, and $V_x^L(t_1)$ and $V_x^L(t_2)$ will be essentially statistically independent, just as $V_x^E(t_1) \propto \Gamma(t_1)$ and $V_x^E(t_2) \propto \Gamma(t_2)$ are statistically independent.

Thus we see that the Einstein and Langevin predictions for $V_x(t)$ do agree in the long-time limit. But it turns out that only the Langevin description of $V_x(t)$ is useful in practical contexts. We will illustrate this point later in this chapter in two ways: First, we will show in Section 9.9 how the diffusion coefficient D can be defined in terms of Langevin's $V_x(t)$ in a way that does not involve the mean-square displacement formula. And second, we will show in Section 9.10 how the Langevin picture of the velocity illuminates the energetics of diffusion. But before addressing those points, let us see what the Langevin theory has to say about numerically simulating the motion of a diffusing solute molecule.

9.6   The Langevin simulation formulas

The standard-form Langevin equation for the continuous Markov process $V_x(t)$ in Langevin's model of diffusion is given in Eq. (8.24):

$$V_x(t + dt) = V_x(t) - \left(\frac{\gamma}{m}\right) V_x(t) dt + \left(\frac{\sqrt{2\gamma k_B T}}{m}\right) N_x(t) \sqrt{dt}.$$

Here, $N_x(t)$ is a temporally uncorrelated normal random variable with mean 0 and variance 1. Using the definition $\tau = m/\gamma$ and the Einstein relation $D = k_B T/\gamma$, the coefficients in this equation can be expressed solely in terms of τ and D. Upon doing that, and then recalling the updating formula (8.28) for the process $X(t)$, we obtain the following *bivariate Langevin equation* for the *bivariate Markov process* $(V_x(t), X(t))$:

$$V_x(t + dt) = V_x(t) - \frac{1}{\tau} V_x(t) dt + \frac{\sqrt{2D}}{\tau} N_x(t) \sqrt{dt}, \tag{9.20a}$$

$$X(t + dt) = X(t) + V_x(t) dt. \tag{9.20b}$$

An obvious *approximate* way to numerically simulate the process defined by this bivariate Langevin equation would be to simply replace dt in these two equations with a "suitably small" Δt. That would give the approximate updating formulas

$$v_x(t + \Delta t) \approx v_x(t) - \frac{1}{\tau} v_x(t) \Delta t + \frac{\sqrt{2D}}{\tau} n_x \sqrt{\Delta t}, \tag{9.21a}$$

$$x(t + \Delta t) \approx x(t) + v_x(t) \Delta t, \tag{9.21b}$$

where n_x denotes a sample value of the unit normal random variable $\mathcal{N}(0, 1)$. We may expect these updating formulas to be accurate if $\Delta t \ll \tau$.

But we found in Chapter 7 that an *exact* pair of updating formulas for any Ornstein–Uhlenbeck process and its integral is provided by Eqs (7.52). Those equations evidently involve the parameters τ and c in only the two combinations $c\tau/2$ and $c\tau^3$. It is easy to show from Eqs (8.27) and (9.7) that in our case, $c\tau^2 = 2D$. And from that it follows that $c\tau/2 = D/\tau$ and $c\tau^3 = 2D\tau$. Upon substituting these results into Eqs (7.52), after replacing $x \to v_x$ and $y \to x$ in those equations, we obtain the following *exact* updating formulas for simulating the trajectory of a diffusing molecule:

$$v_x(t + \Delta t) = v_x(t)e^{-\Delta t/\tau} + \left[\frac{D}{\tau}(1 - e^{-2\Delta t/\tau})\right]^{1/2} n_{x1}, \tag{9.22a}$$

$$x(t + \Delta t) = x(t) + v_x(t)\tau(1 - e^{-\Delta t/\tau}) + \left[D\tau\frac{(1 - e^{-\Delta t/\tau})^3}{(1 + e^{-\Delta t/\tau})}\right]^{1/2} n_{x1}$$

$$+ \left\{2D\left[\Delta t + \tau(1 - e^{-\Delta t/\tau})\left(-2 + \tfrac{1}{2}(1 + e^{-\Delta t/\tau}) - \frac{(1 - e^{-\Delta t/\tau})^2}{2(1 + e^{-\Delta t/\tau})}\right)\right]\right\}^{1/2} n_{x2}. \tag{9.22b}$$

In these two equations, n_{x1} and n_{x2} denote two statistically independent samples of the unit normal random variable $\mathcal{N}(0, 1)$. Analogous formulas of course hold for the y- and z-components of the velocity and position.

In the limit $\Delta t \to 0$, *both* pairs of updating formulas (9.21) and (9.22) imply that $v_x(t + \Delta t) \to v_x(t)$ and $x(t + \Delta t) \to x(t)$, in agreement with the continuous nature of V_x and X. And since X is not only continuous but also (by definition) differentiable, there is no "small Δt" problem of the kind we encountered with the Einstein updating formulas (4.11) with *either* X-updating formula (9.21b) or (9.22b). The advantage of formulas (9.22) over formulas (9.21) is that formulas (9.22) are exact for all $\Delta t > 0$, whereas formulas (9.21) require $\Delta t \ll \tau$. And although formulas (9.22) are more complicated than formulas (9.21), in a computer simulation where Δt is kept constant the factors multiplying the changing quantities $v_x(t)$, n_{x1}, and n_{x2} in formulas (9.22) will need to be calculated only at the first time step. So for simulation purposes, formulas (9.22) will practically always be preferable to formulas (9.21).

The structure of the exact updating formulas (9.22) is revealing. The coupling of X to V_x evidently occurs through the *second* and *third* terms on the right side of Eq. (9.22b): the second term couples $X(t + \Delta t)$ to the value $V_x(t)$, while the third term couples $X(t + \Delta t)$ to the update $V_x(t) \to V_x(t + \Delta t)$ via the shared random number n_{x1}. The *fourth* term on the right side of Eq. (9.22b), being proportional to a random number n_{x2} that is not used by V_x, forms a component of the $X(t) \to X(t + \Delta t)$ update that is statistically independent of V_x.

With these insights, we can now see how the "Einstein behavior" of the Langevin updating formulas emerges in the long-time limit $\Delta t \gg \tau$: In the v_x updating formula (9.22a) in that limit, the first term approaches zero, and the second term approaches $\sqrt{D/\tau}\,n_1 = \mathcal{N}(0, D/\tau)$. Since in this $\Delta t/\tau \gg 1$ limit the parameter τ will appear to be "very small", then D/τ will appear to be "very large"; thus, $v_x(t + \Delta t) \approx \mathcal{N}(0, D/\tau)$ will have the attributes of Gaussian white noise—a normal random variable with zero mean and a "very large" variance. But the fact that $D/\tau = k_{\mathrm{B}}T/m$ assures that the Maxwell–Boltzmann character of the velocity is strictly preserved. As for the $\Delta t \gg \tau$ limit of the x updating formula (9.22b), the second term on the right maxes out at $v_x(t)\tau$, and the third term maxes out at $\sqrt{D\tau}n_{x1}$. Both of those terms will be negligibly small because of the smallness of τ. And since those two terms are solely responsible for the coupling between x and v_x, then that coupling will be practically non-existent when $\Delta t \gg \tau$. Meanwhile, the fourth term on the right side of Eq. (9.22b) becomes $\sqrt{2D\Delta t}n_{x2}$, and when that is combined with the first term the result is the classical Einstein updating formula (4.11a).

A comparison of Eqs (9.22) and (4.11a) shows that another way to describe the difference between the Einstein simulation formula and the Langevin simulation formulas is to say that the former gives a *one*-variable (x) *one*-parameter (D) simulation, while the latter gives a *two*-variable (x and v_x) *two*-parameter (D and τ) simulation. The Langevin approach also brings onto the stage the three physical parameters m, γ, and T, which are related to each other, and to D and τ, through the Einstein relation (9.7) and the definition $\tau = m/\gamma$.

9.7 Examples of trajectory simulations in the Langevin and Einstein theories

In order to compare the predictions of the Einstein and Langevin simulation formulas (4.11a) and (9.22), we will use for T, m, and γ values from the microfluidics diffusion experiment described in Section 5.9. In that experiment, the solute molecules are actually polystyrene beads of radius $R = 0.145$ μm, which are in an aqueous solution at room temperature $T = 300$ K. From the density 1.05 g/cm^3 of polystyrene and the computed volume of the bead, we calculate the bead mass to be $m = 1.34 \times 10^{-17}$ kg.

As for the drag coefficient γ, there are at least two ways of determining its value. A direct experimental way would be to measure the average terminal speed $\bar{v}_{term}$ of the bead as it falls downward in a vertical column of water (downward because the specific gravity of polystyrene is greater than 1). Taking the "up" direction to be positive, there would then be three macroscopic forces acting on the bead: the gravitational force $-mg$, the buoyancy force $+\rho_{\mathrm{H_2O}}(\frac{4}{3}\pi R^3)g$, and the drag force $+\gamma\bar{v}_{term}$. Since the bead is moving with a constant velocity, these three forces must sum to zero: $-mg + \rho_{\mathrm{H_2O}}(\frac{4}{3}\pi R^3)g + \gamma\bar{v}_{term} = 0$. Solving for γ yields

$$\gamma = \frac{g}{\bar{v}_{term}}\left(m - \rho_{\mathrm{H_2O}}(\tfrac{4}{3}\pi R^3)\right). \tag{9.23}$$

A measurement of $\bar{v}_{term}$ thus suffices to determine γ. A less direct but for us more convenient way to calculate γ is to invoke the Stokes formula from fluid mechanics. That formula asserts that the drag forces exerted on a spherical object of radius R which is moving with a not too large velocity v through a fluid of viscosity η has magnitude $6\pi\eta Rv$. Since the drag coefficient γ is by definition the drag force magnitude divided by v, then the Stokes formula for γ is

$$\gamma = 6\pi\eta R. \tag{9.24}$$

Inserting into the Stokes formula the viscosity of water $\eta = 10^{-3} \ \text{kg}/(\text{m} \cdot \text{s})$ and the aforementioned radius of the bead, we obtain the value $\gamma = 2.73 \times 10^{-9} \ \text{kg/s}$.

With the above values for m, T, and γ, we find from the Einstein relation $D = k_{\text{B}}T/\gamma$ and the definition $\tau = m/\gamma$ that

$$D = 1.51 \times 10^{-12} \ \text{m}^2/\text{s} \quad \text{and} \quad \tau = 4.9 \times 10^{-9} \ \text{s}. \tag{9.25}$$

These are the values for D and τ that we will use in the Einstein and Langevin updating formulas (4.11a) and (9.22) in the simulations described below.

Notice that the Einstein updating formula (4.11a) requires a *single* sample n_x of the unit normal random variable $\mathcal{N}(0, 1)$, while the Langevin updating formulas (9.22) require *two* such samples, n_{x1} and n_{x2}. As discussed in the preceding section, when $\Delta t \gg \tau$ the n_{x2} term in Eq. (9.22b) becomes approximately $\sqrt{2D\Delta t}\,n_{x2}$, and that term completely dominates the other two Δt-dependent terms on the right side of Eq. (9.22b). Since $\sqrt{2D\Delta t}\,n_{x2}$ corresponds to the term $\sqrt{2D\Delta t}\,n_x$ in the Einstein formula (4.11a), *we will take $n_x = n_{x2}$ in our comparative simulations of the two updating formulas.* This will allow us to verify more easily that the Einstein and Langevin formulas agree for $\Delta t \gg \tau$, and also to see more clearly for all values of Δt, the extent of the "Langevin correction" to the Einstein formula. For the two-dimensional trajectories generated below, the y-components of the positions and velocities were constructed using these same procedures.

Figure 9.2 shows two 100-point simulations of the bead's trajectory in the xy-plane as computed from the *Einstein* updating formulas (4.11) in (a) and the *Langevin* updating formulas (9.22) in (b), using in both cases a time step of $\Delta t = 10^4\tau$. The bead was initially located at the origin (the open-circled dot in the figure)

$$x_0 = y_0 = 0. \tag{9.26a}$$

In the case of the Langevin trajectory the bead was given an initial velocity that would be typical at room temperature $T = 300\,\text{K}$:

$$v_{x0} = v_{y0} = \sqrt{k_{\text{B}}T/m} = 0.0176\,\text{m/s}. \tag{9.26b}$$

In each trajectory in Fig. 9.2, the small dots show the positions of the bead at successive time steps, and the large solid dot shows the bead's position after the final (100th) time step. Successive dots have been connected by thin straight-line segments to clarify the sequence of the points; however, these connecting lines are *not* meant to represent the actual trajectory of the bead between successive points. An indication of the bead's size on the scale of this figure is given by the double arrow. The message of the two plots in this figure is that for a time step Δt that is 10000 times larger than τ the Einstein trajectory is practically indistinguishable from the Langevin trajectory.

A "zoom-in" on the bead's first step away from the origin in Fig. 9.2a is shown in Fig. 9.3a, which fills in that first step with 100 points at successive time intervals of $\Delta t = 10^2\tau$, obtained once again using the Einstein updating formulas (4.11). The large

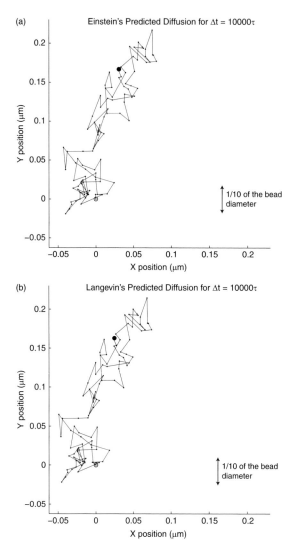

Fig. 9.2 *Comparative simulations of the trajectories of a single diffusing solute molecule in the xy-plane as predicted by* **(a)** *the Einstein theory and* **(b)** *the Langevin theory. We have taken D and τ to be the values in Eqs (9.25) for a polystyrene bead in the microfluidics diffusion experiment of Section 5.9. At time 0 the center of the bead is at the origin of the xy-plane (the open-circled dot). The plotted points in each figure show the locations of the center of the bead at 100 equally spaced time intervals of $\Delta t = 10^4 \tau$. The final position of the bead, at time $10^6 \tau$, is indicated by the large solid black dot. The line segments connecting the points are to make clear the sequential order of the points; they do* not *represent the actual path of the bead between observations. The Einstein trajectory in (a) was generated using formulas (4.11a) and (4.11b). The Langevin trajectory in (b) was generated using formulas (9.22a,b) for the x-motion, and a like pair of formulas for the y-motion. The Langevin formulas require that we specify an initial velocity for the bead, and for that we chose a typical "thermal" velocity $v_{x0} = v_{y0} = \sqrt{k_B T/m} = \sqrt{D/\tau}$ as given by Eq. (9.26b). For reasons explained in the text, the sample value n_x of $\mathcal{N}(0,1)$ in the Einstein simulation formula (4.11a) was taken to be the same as the sample value n_{x2} in the Langevin simulation formula (9.22b); likewise, $n_y = n_{y2}$. These two plots show that for time steps Δt that are very much larger than τ, the Einstein and Langevin trajectories are practically indistinguishable.*

solid dot in Fig. 9.3a is therefore the same as the small dot that ends the first step away from the origin in Fig. 9.2a. Figure 9.3a thus shows a more finely resolved trajectory for the bead along the first step in Fig. 9.2a. In like manner, the Langevin trajectory in Fig. 9.3b is a 100-point zoom-in on the first step of the Langevin trajectory in Fig. 9.2b, obtained from the Langevin updating formulas. At this smaller time step and more magnified spatial scale, the Einstein and Langevin trajectories are still quite similar to each other; however, differences between them are more obvious than in Fig. 9.2.

Figure 9.4 is related to Fig. 9.3 in exactly the same way that Fig. 9.3 is related to Fig. 9.2: Each trajectory in Fig. 9.4 was generated using $\Delta t = \tau$, and is a 100-step zoom-in on the first step of the corresponding trajectory in Fig. 9.3. We see that now the two trajectories are quite different, both in overall shape and in the final location of the bead. Whereas the average step length in the Einstein and Langevin trajectories in the preceding two figures were nearly the same, in this figure the average length of an Einstein step is about 50% larger than the average length of a Langevin step.

In Fig 9.5, the two trajectories were generated with $\Delta t = 10^{-2}\tau$, and are 100-step zoom-ins on the first steps of the corresponding trajectories in Fig. 9.4. Now the two trajectories are very different. The Langevin trajectory has evidently resolved itself into a reasonably smooth albeit meandering curve, whereas the Einstein trajectory looks like a stochastically scaled-down version of itself in the previous figures. Since the total time span for these trajectories is τ, we have indicated on each figure the size of the *characteristic diffusion length* $\sqrt{D\tau} = 8.6 \times 10^{-5}\mu\text{m}$ [see Eq. (9.13)]. As we saw in Section 9.3, a *Langevin* simulation of duration τ should produce trajectories whose *average total path length* is equal to the characteristic diffusion length, and whose *average net displacement* is about 86% of the characteristic diffusion length. The Langevin trajectory in Fig. 9.5b has path length $9.25 \times 10^{-5}\mu\text{m}$ and net displacement $7.05 \times 10^{-5}\mu\text{m}$, values that are quite consistent with the theoretical predictions.

Figure 9.6 shows a final 100-step zoom-in on the first step of each trajectory in Fig. 9.5. The Langevin trajectory in Fig. 9.6b in this $\Delta t = 10^{-4}\tau$ simulation now appears to be essentially ballistic. In sharp contrast, the corresponding Einstein trajectory in Fig. 9.6a continues to exemplify its statistically self-similar (fractal) nature, as was discussed in Section 7.11.

Notice that the five figures 9.2 through 9.6 actually show *only one* Einstein trajectory and *only one* Langevin trajectory: The single Langevin trajectory in all these figures was constructed by first generating points 1 through 100 using formulas (9.22) with $\Delta t = 10^{-4}\tau$; and then generating points 101 through 200 with $\Delta t = 10^{-2}\tau$ *without* resetting the position or the velocity or the time variable t; and then generating points 201 through 300 with $\Delta t = \tau$, again without resetting the position or the velocity or t; and so on until points 401 to 500 were generated with $\Delta t = 10^4\tau$ The single Einstein trajectory for Figs 9.2a through 9.6a was generated in the same way, except using formulas (4.11) instead of formulas (9.22).

Figure 9.7 shows a magnification of the Langevin plot in Fig. 9.6b during the short time interval $[0, 0.01\tau]$. The nearly 45-degree direction of that trajectory shows the strong influence of the bead's initial velocity (9.26b), and the even spacing between the dots signifies that the bead here is moving with a nearly constant speed. The

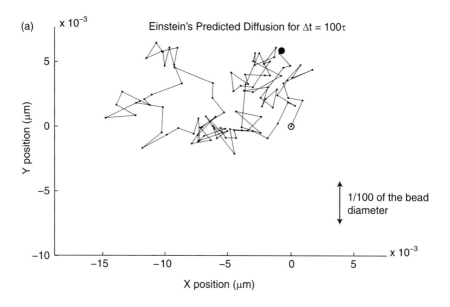

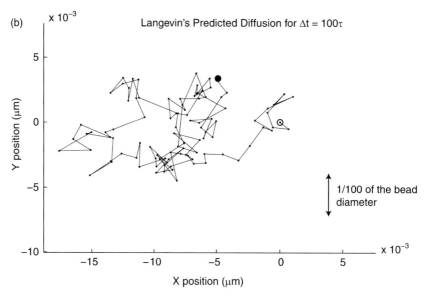

Fig. 9.3 *Comparative simulations of the trajectories of a single diffusing solute molecule in the xy-plane in the (a) Einstein and (b) Langevin theories, as in Fig. 9.2, but now with the time step reduced by two orders of magnitude to $\Delta t = 10^2\tau$. The trajectory in (a) is a "zoom-in" on the first step of the trajectory in Fig. 9.2a, so the end-point of the first step away from the origin in Fig. 9.2a is the same as the final (large) point of the trajectory in (a) here. Similarly, the trajectory in (b) is a zoom-in on the first step of the trajectory in Fig. 9.2b, effectively filling in that first step with 99 new observations of the bead's position. Although the Einstein and Langevin trajectories here are very similar, the differences between them are more noticeable than in Fig. 9.2. Both trajectories make clear the folly of assuming that the line segments connecting successive points in Fig. 9.2 are in any way representative of the true path of the bead between those points.*

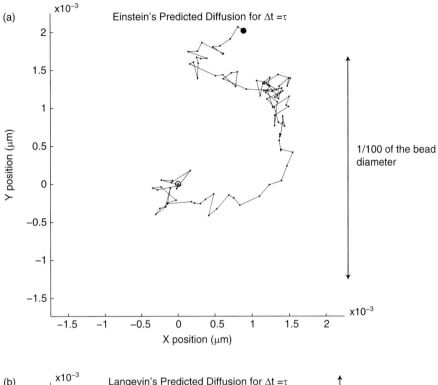

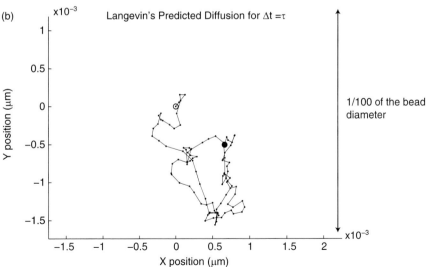

Fig. 9.4 *Comparative simulations of the trajectories of a single diffusing solute molecule in the xy-plane in the **(a)** Einstein and **(b)** Langevin theories, as in Fig. 9.3, but with the time step reduced by two orders of magnitude to $\Delta t = \tau$. The trajectory in (a) is a zoom-in on the first step of the trajectory in Fig. 9.3a, and the trajectory in (b) is a zoom-in on the first step of the trajectory in Fig. 9.3b. The Einstein and Langevin trajectories are now noticeably different.*

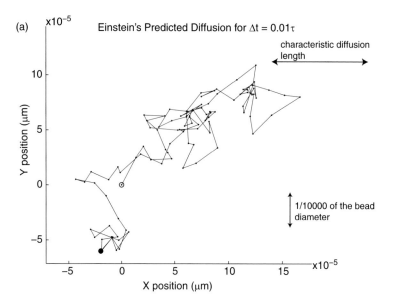

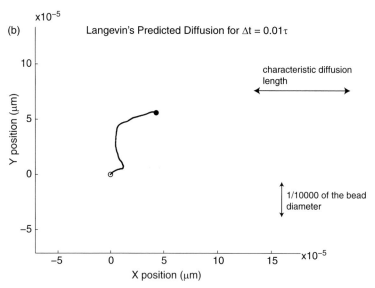

Fig. 9.5 *Comparative simulations of the trajectories of a single diffusing solute molecule in the xy-plane in the (a) Einstein and (b) Langevin theories, as in Fig. 9.4, but with the time step reduced by two orders of magnitude to $\Delta t = 10^{-2}\tau$. The trajectory in (a) is a zoom-in on the first step of the trajectory in Fig. 9.4a, and the trajectory in (b) is a zoom-in on the first step of the trajectory in Fig. 9.4b. At the scale of this small time step, the two trajectories, each spanning a total time τ, bear no resemblance at all to each other. The Einstein trajectory is a stochastically scaled down version of itself in the earlier figures, whereas the Langevin trajectory now has the character of a meandering but almost smooth path. Also shown on each plot is the size of the characteristic diffusion length $\sqrt{D\tau}$, discussed in Section 9.3; the Langevin theory predicts that that length should estimate the* average distance actually traveled *by the molecule in time τ, and also that the* rms net displacement *in time τ should be smaller by a factor of 0.86. The trajectory in (b) is consistent with those predictions, but the trajectory in (a) is not.*

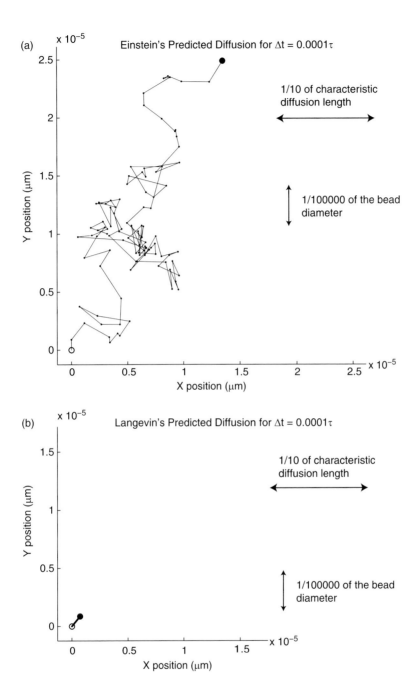

Fig. 9.6 *Comparative simulations of the trajectories of a single diffusing solute molecule in the xy-plane in the **(a)** Einstein and **(b)** Langevin theories, as in Fig. 9.5, but with the time step reduced by two orders of magnitude to $\Delta t = 10^{-4}\tau$. The trajectory in (a) is a zoom-in on the first step of the trajectory in Fig. 9.5a, and the trajectory in (b) is a zoom-in on the first step of the trajectory in Fig. 9.5b. The Langevin trajectory in (b) shows an essentially ballistically moving bead. The Einstein trajectory in (a) is, by comparison, wildly hyperactive.*

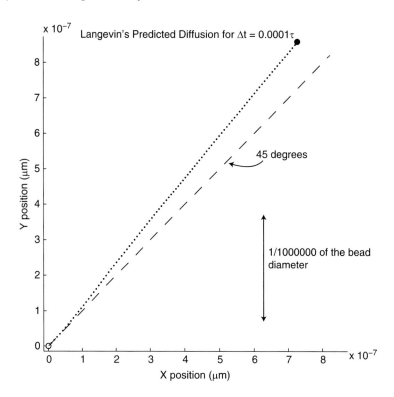

Fig. 9.7 *A magnification of the Langevin plot in Fig. 9.6b showing more clearly the position of the center of the bead at successive time intervals of $\Delta t = 10^{-4}\tau$. The regularity of the spacing of those positions shows that the bead is moving at a nearly constant speed, and in a nearly straight line. The direction of this trajectory shows the strong influence of the bead's assumed initial velocity, $v_{x0} = v_{y0} = \sqrt{D/\tau}$. The gradual deviation of the trajectory away from the exact 45-degree direction (dashed) is the result of relatively small fluctuations in the velocity components at each time step.*

slight deviation of the trajectory away from the exact 45-degree path (shown dashed) is caused by small velocity fluctuations at each time step, as will be seen more clearly shortly.

It is interesting to examine the effect of *reversing* the bead's initial velocity in the Langevin simulation: In Fig. 9.8a, the up-going dots from the initial position (open circle) show once again the trajectory in Fig. 9.7, while the down-going dots show the trajectory that resulted when both components of the initial velocity (9.26b) were multiplied by -1. The same chain of random numbers was used in both simulations. As expected on this very short time-scale, the two trajectories are approximately oppositely directed. Figure 9.8b shows these same two Langevin trajectories extended from total time $10^{-2}\tau$ to time $10^6\tau$ and now with $\Delta t = 10^4\tau$ On this much larger spatial scale, which is the same as the scale in Fig. 9.2, the two trajectories are practically

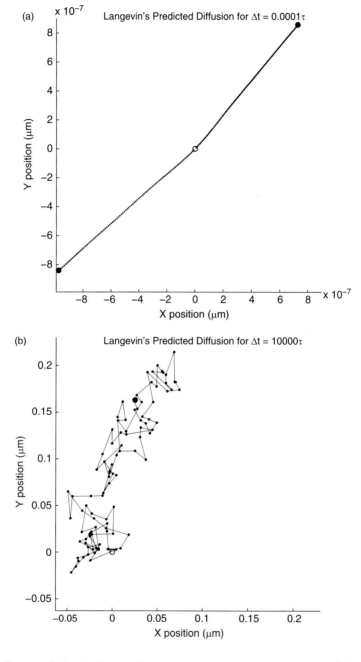

Fig. 9.8 *Influence of the bead's initial velocity on the Langevin trajectory.* (**a**) *The up-going trajectory is the Langevin trajectory in Fig. 9.7 for* $\Delta t = 10^{-4}\tau$ *(the same as in Fig. 9.6b), which assumed the bead had initial velocity* $v_{x0} = v_{y0} = \sqrt{D/\tau}$. *The* down-going *trajectory is what would have resulted if the initial velocity had instead been* $v_{x0} = v_{y0} = -\sqrt{D/\tau}$. *On the scale of this plot, the difference between the two trajectories is major.* (**b**) *But carrying out those two simulations using the much larger time increment* $\Delta t = 10^4\tau$, *as we did in Fig. 9.2b, yields trajectories that are indistinguishable from each other on the scale of this plot. We conclude that the* long-time *behavior of the bead's trajectory in the Langevin picture is insensitive to the bead's initial velocity.*

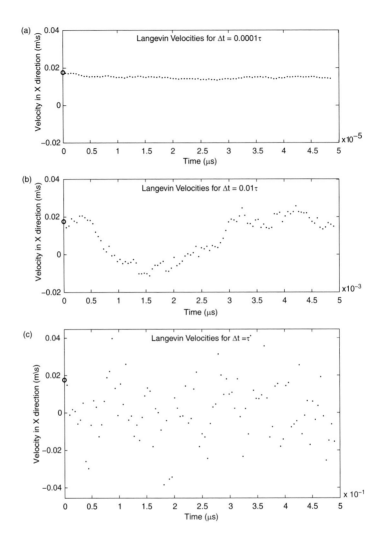

Fig. 9.9 *Plots of the x-component of the bead's velocity in the Langevin simulations in Figs 9.2b through 9.6b, but presented here in reverse order, from small Δt to large Δt. Notice that the scale of the time axis increases by two orders of magnitude on each successive plot, but the scale of the velocity axis is the same on all plots. The small circle on the velocity axis is the assumed value of the bead's initial x-velocity, the "thermal" value $\sqrt{k_\mathrm{B}T/m} = \sqrt{D/\tau}$ discussed at Eq. (9.26b). On the very short time-scale of **(a)**, the bead travels approximately ballistically. On the larger time-scale of **(b)**, the bead's velocity explores the one-standard deviation interval $\left[-\sqrt{D/\tau}, \sqrt{D/\tau}\right]$. On the still larger time-scales of **(c)** and **(d)**, the bead's velocity has become reasonably well "thermalized" in the two-standard deviation interval $\left[-2\sqrt{D/\tau}, 2\sqrt{D/\tau}\right]$. Finally, on the large time-scale **(e)**, a scale on which $\tau = 10^{-4}\Delta t$ seems "very small" and the velocity interval $\left[-2\sqrt{D/\tau}, 2\sqrt{D/\tau}\right]$ therefore seems "very large", the bead's velocity is starting to resemble Gaussian white noise (normal with mean 0 and variance ∞), as is tacitly assumed in Einstein's theory of diffusion (see Section 9.5).*

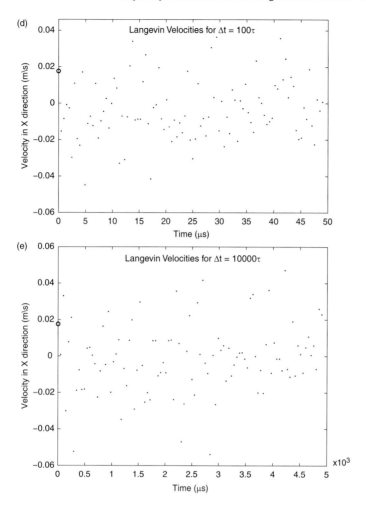

Fig. 9.9 *Continued*

indistinguishable from each other. This shows that the long-time behavior of the bead's trajectory in the Langevin picture is insensitive to the bead's initial velocity.

In proceeding from Fig. 9.2 to Fig. 9.6, the ratio of {the broken-line distance along the Einstein path} to {the broken-line distance along the Langevin path} steadily increases. Whereas those two distances are practically equal in Fig. 9.2, the distance along the Einstein path in Fig. 9.5a is about 16 times longer than the distance along the Langevin path in Fig. 9.5b. For the two trajectories in Fig. 9.6 the ratio of those distances has increased to 145. The physically unrealistic nature of the Einstein trajectories in these last two cases can be seen by dividing the length of each broken-line path by the total time simulated to get an estimate of the *minimum average speed* of the bead during that time ("minimum" because that broken line path is the shortest possible path connecting the observation points). Doing that, we find that a bead moving along the Einstein broken line in Fig. 9.5a must have had an average speed that equals the rms speed in an ideal gas of beads at temperature 4.5×10^4 K, a temperature that is over seven times hotter than the surface of the Sun. And the average speed for the Einstein trajectory in Fig. 9.6a corresponds to a temperature that is over 100 times hotter than that. Needless to say, such high temperatures combined with the high relative humidity that naturally exists in a water solution, would make for an extremely uncomfortable environment for the bead. This unrealistic behavior of the Einstein trajectory with decreasing Δt is of course a manifestation of the problem with the Einstein model of diffusion that was discussed in detail in Sections 4.2 and 4.4.

In a Langevin simulation, the instantaneous velocity of the bead is readily available since it gets simulated along with the bead's position by the Langevin updating formulas (9.22). Figure 9.9 shows the $V_x(t)$ versus t graphs corresponding to the Langevin trajectories in Figs 9.2b through 9.6b, but in *reverse order*, from small Δt to large Δt. The small circle at $t = 0$ in each plot indicates the initial value $v_{x0} = \sqrt{k_B T/m}$ assumed in Eq. (9.26b). For the $\Delta t = 10^{-4}\tau$ plot in Fig. 9.9a, the very small fluctuations in V_x about the initial value indicate approximate ballistic motion. Those small fluctuations together with like fluctuations in V_y, are responsible for the slight deviation of the trajectory in Fig. 9.7 away from the 45-degree line. With Δt increased to $10^{-2}\tau$ in Fig. 9.9b, V_x has time to explore more fully the one-standard deviation range $\left[-\sqrt{k_B T/m}, \sqrt{k_B T/m}\right]$ for a bead at room temperature. For the larger Δt values of τ in Fig. 9.9c and $10^2\tau$ in Fig. 9.9d, V_x has time to explore the two-standard deviation range $\left[-2\sqrt{k_B T/m}, 2\sqrt{k_B T/m}\right]$. On the largest time-scale in Fig. 9.9e where τ can be considered "infinitesimally small", $V_x(t)$ is beginning to resemble the Gaussian white-noise process $\Gamma(t)$, as we anticipated in our discussion following Eq. (9.18).

9.8 The relative motion of two solute molecules

Equation (3.19) shows that in Einstein's theory, a solute molecule with diffusion coefficient D_1 will *appear*, to an independently moving solute molecule with diffusion coefficient D_2, to be moving with diffusion coefficient $D_1 + D_2$. Since we know that Einstein's theory is a valid approximation to Langevin's theory for sufficiently large times, we may expect that this statement will be at least approximately true in Langevin's theory for times much greater than the larger of the two molecules' respective relaxation times, τ_1 and τ_2. But as we will now prove, in Langevin's theory the above statement about relative diffusional motion will be *exactly* true *only* if the two solute molecules have the *same relaxation time*.

To prove this, we first append in Eqs (9.20) molecule indices $i = 1, 2$ to V_x, X, D, and τ, and then we subtract the $i = 2$ equations from the $i = 1$ equations. For the case $\tau_1 = \tau_2 = \tau$, we find for the velocity $V_{12x}(t) \equiv V_{1x}(t) - V_{2x}(t)$ and position $X_{12}(t) \equiv X_1(t) - X_2(t)$ of solute molecule 1 relative to solute molecule 2,

$$V_{12x}(t+dt) = V_{12x}(t) - \frac{1}{\tau}V_{12x}(t)dt + \left(\frac{\sqrt{2D_1 dt}}{\tau}N_{1x}(t) - \frac{\sqrt{2D_2 dt}}{\tau}N_{2x}(t)\right), \quad (9.27a)$$

$$X_{12}(t+dt) = X_{12}(t) + V_{12x}(t)dt. \quad (9.27b)$$

Since $N_{1x}(t)$ and $N_{2x}(t)$ in Eq. (9.27a) are statistically independent normal random variables with means 0 and variances 1, it follows from the linear combination theorem (2.25) that the quantity in parentheses in Eq. (9.27a) can be written

$$\frac{\sqrt{2D_1 dt}}{\tau}N_1(0,1) - \frac{\sqrt{2D_2 dt}}{\tau}N_2(0,1) = N\left(0, \frac{2D_1 dt}{\tau^2} + \frac{2D_2 dt}{\tau^2}\right)$$

$$= N\left(0, \frac{2(D_1+D_2)dt}{\tau^2}\right)$$

$$= \frac{\sqrt{2(D_1+D_2)dt}}{\tau}N(0,1).$$

Equation (9.27a) can therefore be written

$$V_{12x}(t+dt) = V_{12x}(t) - \frac{1}{\tau}V_{12x}(t)dt + \frac{\sqrt{2(D_1+D_2)}}{\tau}N_{12x}(t)\sqrt{dt}, \quad (9.28)$$

where $N_{12x}(t)$ is a temporally uncorrelated unit normal random variable. Combining Eq. (9.28) with Eq. (9.27b), we obtain the bivariate Langevin equation (9.20) for a *single* solute molecule with diffusion coefficient $D_1 + D_2$ and relaxation time τ. This proves that two independently moving solute molecules that have the *same relaxation time* appear to each other to be moving with a diffusion coefficient that is the sum of their respective diffusion coefficients.

But if $\tau_1 \neq \tau_2$, the $D_1 + D_2$ rule for relative motion will not generally hold. The reason is that the $\tau_1 \neq \tau_2$ version of Eq. (9.27a) for $V_{12x}(t)$ cannot be brought into the form of the single-molecule velocity equation (9.20a) with modified forms for the constants D and τ. But since Langevin's theory approximately reduces to Einstein's theory for large times, we can expect that the $D_1 + D_2$ result will hold *approximately* for times much larger than $\max(\tau_1, \tau_2)$. To verify that expectation, and also to see in more detail how relative diffusional motion behaves when the relaxation times are not equal, let us consider the *equilibrium* mean-square x-displacement formula (9.5). Using the Einstein relation (9.7), and for simplicity taking $t_0 = 0$, we see that Eq. (9.5) implies for each of the two independently moving solute molecules,

$$\left\langle (X_i(t) - X_i(0))^2 \right\rangle_{eq} = 2D_i\left[t - \tau_i\left(1 - e^{-t/\tau_i}\right)\right] \quad (i = 1, 2). \quad (9.29)$$

Here the "eq" subscript means that the initial velocities of the two molecules have been averaged over their thermodynamically expected values for their common temperature. The square of the *relative* displacement is, by definition,

$$(X_{12}(t) - X_{12}(0))^2 \equiv [(X_1(t) - X_2(t)) - (X_1(0) - X_2(0))]^2$$
$$\equiv [(X_1(t) - X_1(0)) - (X_2(t) - X_2(0))]^2$$
$$\equiv (X_1(t) - X_1(0))^2 + (X_2(t) - X_2(0))^2$$
$$- 2(X_1(t) - X_1(0))(X_2(t) - X_2(0)).$$

We now take the *equilibrium average* of this equation. Since the motions of the two molecules are statistically independent and $\langle X_i(t) \rangle_{eq} = X_i(0)$, the equilibrium average of the last term in the above formula is zero; thus,

$$\left\langle (X_{12}(t) - X_{12}(0))^2 \right\rangle_{eq} = \left\langle (X_1(t) - X_1(0))^2 \right\rangle_{eq} + \left\langle (X_2(t) - X_2(0))^2 \right\rangle_{eq}.$$

Substituting on the right from Eqs (9.29) and then collecting terms, we conclude that

$$\left\langle (X_{12}(t) - X_{12}(0))^2 \right\rangle_{eq} = 2(D_1 + D_2)t - 2D_1\tau_1 \left(1 - e^{-t/\tau_1}\right) - 2D_2\tau_2 \left(1 - e^{-t/\tau_2}\right).$$

$$(9.30)$$

This formula for the *equilibrium* relative mean-square displacement is exact. In it, the last two terms on the right side evidently comprise the "Langevin correction" to the classical Einstein result (3.12) for the mean-square displacement.

For the special case $\tau_1 = \tau_2 = \tau$, Eq. (9.30) evidently reduces to

$$\left\langle (X_{12}(t) - X_{12}(0))^2 \right\rangle_{eq} = 2(D_1 + D_2)\left[t - \tau\left(1 - e^{-t/\tau}\right)\right] \quad (\tau_1 = \tau_2 = \tau). \quad (9.31)$$

This is evidently the single-molecule formula (9.29) with the diffusion coefficient D replaced by the sum of the diffusion coefficients of the two molecules—a result that we already inferred.

If $\tau_1 \neq \tau_2$, formula (9.30) cannot be brought into the form of the single-molecule formula (9.29). But let us see what Eq. (9.30) implies in the large and small time limits. In the large time limit $t \gg \max(\tau_1, \tau_2)$, the magnitude of the second term on the right side of Eq. (9.30) will evidently never be larger than $2D_1\tau_1$, and that is negligibly small compared to the $2D_1t$ portion of the first term. Similarly, the magnitude of the third term on the right side of Eq. (9.30) will never be larger than $2D_2\tau_2$, which is negligibly small compared to the $2D_2t$ portion of the first term. We conclude that in this limit, only the first term in Eq. (9.30) contributes:

$$\left\langle (X_{12}(t) - X_{12}(0))^2 \right\rangle_{eq} \approx 2(D_1 + D_2)t \quad (t \gg \max(\tau_1, \tau_2)). \quad (9.32)$$

Therefore, as expected, the *long-time* equilibrium relative mean-square displacement of two solute molecules behaves in Langevin's theory just like it does in Einstein's theory.

To see what Eq. (9.30) implies at the other extreme, namely $t \ll \min(\tau_1, \tau_2)$, we Taylor-expand both exponentials in Eq. (9.30) to second order in the small quantities

t/τ_i. Upon collecting terms, we find that the first-order terms in the two expansions are exactly cancelled by the first term on the right side of Eq. (9.30), leaving to lowest order in t,

$$\left\langle (X_{12}(t) - X_{12}(0))^2 \right\rangle_{\text{eq}} \approx \left(\frac{D_1}{\tau_1} + \frac{D_2}{\tau_2} \right) t^2.$$

But

$$\frac{D_1}{\tau_1} + \frac{D_2}{\tau_2} = \frac{k_{\text{B}}T/\gamma_1}{m_1/\gamma_1} + \frac{k_{\text{B}}T/\gamma_2}{m_2/\gamma_2} = k_{\text{B}}T \left(\frac{1}{m_1} + \frac{1}{m_2} \right) \equiv \frac{k_{\text{B}}T}{m_{12}},$$

where m_{12} is the reduced mass $m_1 m_2/(m_1 + m_2)$. Therefore,

$$\left\langle (X_{12}(t) - X_{12}(0))^2 \right\rangle_{\text{eq}} \approx \frac{k_{\text{B}}T}{m_{12}} t^2 = \overline{v_{12}^2}\, t^2 \quad (t \ll \min(\tau_1, \tau_2)). \tag{9.33}$$

The last step here invokes the fact, derived in Note 5 of Chapter 4, that the relative x-velocity of two molecules of masses m_1 and m_2 in thermal equilibrium is $\mathcal{N}(0, k_{\text{B}}T/m_{12})$, so $k_{\text{B}}T/m_{12}$ is the mean-square relative x-velocity that the two molecules would have in a dilute gas. Taking the square root of Eq. (9.33) then shows that, for very short times, the relative distance between two solute molecules increases linearly with time, just as we should expect for *ballistically* moving molecules.

Taken together, the results (9.32) and (9.33) give an interesting and convenient twist to the "$2(D_1 + D_2)\,t$" formula for the mean-square displacement in *relative* diffusional motion: On *long* time-scales where the motion is truly diffusional, the formula holds. On *short* time-scales it does *not* hold, but then we would not expect it to because we know that on short time-scales molecules move ballistically.

We should note that another approach to analyzing the relative variables $V_{12x}(t)$ and $X_{12}(t)$ would be to work with the exact bivariate-Langevin formulas for those two variables, namely formulas analogous to the ones for the single-molecule variables $V_x(t)$ and $X(t)$ in Eqs (8.29)–(8.31). It is straightforward to derive those formulas for $V_{12x}(t)$ and $X_{12}(t)$,[2] but since they are algebraically complicated and will not be needed elsewhere in this book, we will not exhibit them here.

Finally, it is reasonable to ask whether or not the Langevin deviations from the Einstein result for relative diffusional motion will have any impact on our derivation in Sections 3.7 and 4.8 of the formula for the stochastic reaction rate (4.33). The answer is no: That derivation invoked the Einstein result $D_{12} \equiv D_1 + D_2$ under conditions where the Einstein theory should be valid.

9.9   The velocity auto-covariance formula for D

The diffusion coefficient D made its debut in Chapter 1 as the proportionality constant in Fick's Law (1.2) between the net flux and the negative gradient of the molecular density. Concerns that this way of defining D is only "phenomenological", and constitutes a valid definition of D only if there is a non-zero density gradient, were dispelled by Einstein's result that the mean-square displacement of a *single* diffusing solute molecule in time Δt is $2D\Delta t$ regardless of the presence of any density gradients. That result, qualified by the caveat from Langevin's analysis that the result is accurate only when $\Delta t \gg \tau$, has since served as the operational definition of D for many experimental purposes. But Langevin's fuller description of the behavior of the *velocity* of a diffusing molecule admits yet another way of defining D, a way that makes no direct reference to the position of the molecule. This is the so-called *velocity auto-covariance formula*,

$$D = \int_0^\infty \langle V_x(t)\, V_x(t+t') \rangle_{t_0=-\infty}\, dt'. \tag{9.34}$$

Before deriving Eq. (9.34), let us be clear about what it says: We start observing the solute molecule at a time t which is sufficiently far removed from the preparation time in the infinite past ($t_0 = -\infty$) that its velocity $V_x(t)$ has effectively achieved the equilibrium Maxwell–Boltzmann distribution; thus,

$$V_x(t) = \mathcal{N}\left(0, \frac{k_{\mathrm{B}}T}{m}\right) = \mathcal{N}\left(0, \frac{D}{\tau}\right) \quad \text{(when } t_0 = -\infty\text{)}, \tag{9.35}$$

where the last equality invokes Eq. (9.19). We observe the molecule continuously after time t, and we plot the product $V_x(t)\, V_x(t+t')$ as a function of t' for all $t' > 0$. We then repeat this procedure many times, and from the many $V_x(t)\, V_x(t+t')$ versus t' curves thus obtained we compute and plot the *average* curve $\langle V_x(t)\, V_x(t+t') \rangle$ as a function of t'. Equation (9.34) asserts that the area bounded by that average curve and the positive t'-axis will be equal to the molecule's diffusion coefficient D. Notice that the integrand $\langle V_x(t)\, V_x(t+t') \rangle_{t_0=-\infty}$ in Eq. (9.34) is indeed the covariance $\mathrm{cov}\{V_x(t), V_x(t+t')\}_{t_0=-\infty}$, because $\langle V_x(t) \rangle_{t_0=-\infty} \langle V_x(t+t') \rangle_{t_0=-\infty}$ vanishes by Eq. (9.35).

To derive Eq. (9.34), we start with Eq. (8.35):

$$\frac{d}{dt_2} \langle V_x(t_1)V_x(t_2) \rangle = -\frac{1}{\tau} \langle V_x(t_1)V_x(t_2) \rangle \quad (t_0 \le t_1 \le t_2).$$

A relabeling of the variables $t_1 \to t$ and $t_2 \to t+t'$ transforms this equation into

$$\frac{d}{dt'} \langle V_x(t)V_x(t+t') \rangle = -\frac{1}{\tau} \langle V_x(t)V_x(t+t') \rangle \quad (t_0 \le t; 0 \le t').$$

The solution to this simple differential equation for the initial condition $\langle V_x(t)V_x(t+0) \rangle \equiv \langle V_x^2(t) \rangle$ is

$$\langle V_x(t)V_x(t+t') \rangle = \langle V_x^2(t) \rangle e^{-t'/\tau} \quad (t_0 \le t; 0 \le t').$$

Now we take $t_0 = -\infty$. Then Eq. (9.35) holds, so $\langle V_x^2(t) \rangle = D/\tau$, and the above equation becomes

$$\langle V_x(t)V_x(t+t') \rangle_{t_0=-\infty} = \frac{D}{\tau} e^{-t'/\tau} \quad (0 \le t'). \tag{9.36}$$

Integrating Eq. (9.36) over all $t' > 0$ then gives the velocity auto-covariance formula (9.34).

From an experimental standpoint, Eq. (9.36), whose integral gave us Eq. (9.34), would perhaps be more handy for determining the value of D than Eq. (9.34): Since a plot of $\ln \langle V_x(t)V_x(t+t') \rangle_{t_0=-\infty}$ versus t' will give us, according to Eq. (9.36), a straight line with slope $-1/\tau$ and y-intercept $\ln(D/\tau)$, then a measurement of the negative slope of that line will give us τ, and that value for τ together with a measurement of the y-intercept will give us D. Those two parameters τ and D, according to the exact updating formulas (9.22), completely determine the motion of a diffusing solute molecule in the Langevin picture if its initial position and velocity are specified.

9.10 The energetics of diffusion

Another benefit of the Langevin approach to diffusion is that it gives us quantitative insight into the energetics of the diffusion process. In particular, it shows that what gets "dissipated" in the fluctuation–dissipation theorem is energy.

To see that, we start by squaring the Langevin equation (9.20a) for $V_x(t)$, retaining on the right only terms up to first order in dt:

$$V_x^2(t+dt) = V_x^2(t) - \left(\frac{2}{\tau}\right) V_x^2(t)dt + 2\frac{\sqrt{2D}}{\tau}V_x(t)N_x(t)\sqrt{dt} + \frac{2D}{\tau^2}N_x^2(t)dt.$$

Averaging this equation, remembering that $V_x(t)$ and $N_x(t)$ are statistically independent, and also that $\langle N_x(t) \rangle = 0$ and $\langle N_x^2(t) \rangle = 1$, we get

$$\langle V_x^2(t+dt) \rangle = \langle V_x^2(t) \rangle - \left(\frac{2}{\tau}\right) \langle V_x^2(t) \rangle \, dt + \frac{2D}{\tau^2} dt.$$

Transposing the first term on the right, dividing through by dt, and taking the limit $dt \to 0^+$, we get

$$\frac{d}{dt} \langle V_x^2(t) \rangle = - \left(\frac{2}{\tau}\right) \langle V_x^2(t) \rangle + \frac{2D}{\tau^2}.$$

Finally, we multiply this last equation through by $m/2$, and then eliminate τ and D through the relations $\tau = m/\gamma$ and $D = k_B T/\gamma$. That gives

$$\frac{d}{dt} \langle \tfrac{1}{2}mV_x^2(t) \rangle = - \langle \gamma V_x^2(t) \rangle + \frac{\gamma k_B T}{m}. \tag{9.37}$$

Analogous equations follow for the y- and z-components of the velocity $\mathbf{V}(t)$ of the solute molecule. When those three equations are added together, the result is

$$\frac{d}{dt} \langle \tfrac{1}{2}m\mathbf{V}^2(t) \rangle = - \langle \gamma \mathbf{V}^2(t) \rangle + \frac{3\gamma k_B T}{m}. \tag{9.38}$$

Equation (9.38) evidently gives the time-rate-of-change of the *average kinetic energy* of the solute molecule. The equation tells us that that average energy has a *sink* and a *source*. The sink removes energy from the solute molecule at the average rate

$$\langle \gamma \mathbf{V}^2(t) \rangle = \langle (\gamma \mathbf{V}(t)) \cdot \mathbf{V}(t) \rangle.$$

This is just the force-times-velocity rate at which the solute molecule is doing work on the solvent molecules via the drag force component in Eq. (8.1). The energy thus dissipated by the solute molecule is transferred to the solvent molecules, and tends to raise the average speed of the solvent molecules and hence also their temperature. But that tendency, according to Eq. (9.38), is countered by the fact that the solvent molecules are simultaneously doing work on the solute molecule at an average rate of $(3\gamma k_{\mathrm{B}}T/m)$. Notice that both of these energy transfer rates are directly proportional to the drag coefficient.

When the solute molecule is *in equilibrium*, its average kinetic energy will be constant in time. In that case the derivative on the left side of Eq. (9.38) will be zero, so the two energy transfer rates on the right will then be equal in magnitude:

$$\langle \gamma \mathbf{V}^2(t) \rangle_{\mathrm{equil}} = \frac{3\gamma k_{\mathrm{B}}T}{m}.$$

Multiplying this equation through by $m/2\gamma$ gives

$$\langle \tfrac{1}{2}m\mathbf{V}^2(t) \rangle_{\mathrm{equil}} = \tfrac{3}{2}k_{\mathrm{B}}T. \tag{9.39}$$

This result agrees with the *equipartition of energy theorem* in statistical thermodynamics, which asserts that any system in "thermal equilibrium" will have $\frac{1}{2}k_{\mathrm{B}}T$ of average energy for each degree of freedom upon which the system's energy depends quadratically. In the present case, each molecule has three such degrees of freedom.

9.11   Are there "overdamped diffusing systems"?

Consider the Langevin equation (8.23):

$$m\frac{dV_x(t)}{dt} + \gamma V_x(t) = \sqrt{2\gamma k_{\mathrm{B}}T}\,\Gamma_x(t). \tag{9.40}$$

Suppose we somehow *know* that the values of the system parameters are such that the following strong inequality always holds:

$$\left| m\frac{dV_x(t)}{dt} \right| \ll |\gamma V_x(t)|. \tag{9.41}$$

Then the first term on the left side of Eq. (9.40) could be neglected relative to the second term, and solving for $V_x(t)$ would give

$$V_x(t) \approx \sqrt{\frac{2k_{\mathrm{B}}T}{\gamma}}\,\Gamma_x(t) = \sqrt{2D}\,\Gamma_x(t),$$

where the last step has invoked the Einstein relation (9.7). This last equation says that

$$\frac{dX(t)}{dt} \approx \sqrt{2D}\Gamma_x(t).$$
(9.42)

Since this equation is, according to Eq. (7.33b), the Langevin equation for a driftless Wiener process with diffusion constant $c = 2D$, then it would follow from the generic Wiener result (7.36) that

$$X(t) \approx \mathcal{N}\left(x_0, 2D(t - t_0)\right).$$
(9.43)

This of course is the classical Einstein diffusion formula.

The foregoing line of reasoning has led some to suppose that there exist solute–solvent systems whose parameter values cause condition (9.41) to *always* be satisfied, and whose time-evolution is therefore *always* described by Eq. (9.43). An inspection of condition (9.41) suggests that such systems would be ones that have a relatively large value of γ and a relatively small value of m, and which therefore might be called "high-friction/low-inertia" or "overdamped" systems. The latter appellation comes from an analogy with the classical *damped harmonic oscillator*—a mass m which is attached to a spring with a force constant k and moves in a viscous medium with drag coefficient γ. For that well-studied mechanical system it can be shown that, if the *ratio* of {the exponential decay time $\tau = m/\gamma$ of the oscillation amplitude} to {the undamped oscillation period $T = (2\pi)/\sqrt{k/m}$} is less than $(4\pi)^{-1}$, then the viscous damping will be so strong compared to the spring force that the oscillator, if released at rest from a point $x_0 > 0$, will asymptotically approach the origin without ever crossing the origin. The oscillator in that case is said to be "overdamped".

But there is a problem in making this analogy with diffusion: In the diffusion problem described by Eq. (9.40), there is nothing analogous to the undamped period T to which we can compare the relaxation time τ. To simply assert that γ is "very large" and/or m is "very small" inevitably begs the question, large or small *relative to what?* Notice that it makes *no* sense to say that "$\gamma \gg m$", because γ and m have different dimensions, so their magnitudes cannot be compared. (We can make the magnitude of γ either larger or smaller than the magnitude of m simply by changing our unit of mass.) On top of all that, the conclusion arrived at above, namely that Eq. (9.43) is a good approximation for *all* $t \geq t_0$, does *not* agree with our earlier analysis, which showed that no matter what the values of γ and m are, the solute molecule's motion *will* be accurately described by formula (9.43) whenever $t - t_0 \gg m/\gamma$, but *not* otherwise.

There is in fact a sense in which the above reasoning has a kernel of validity, but the argument must be drawn differently: Instead of starting with Eq. (8.23), we start with the ostensibly equivalent Eq. (8.24). Using relations among the system parameters which should by now be familiar, we can write Eq. (8.24) as

$$[V_x(t + dt) - V_x(t)]\tau + V_x(t)dt = \sqrt{2D}N_x(t)\sqrt{dt}.$$
(9.44)

Since $|V_x(t + dt) - V_x(t)|$ will generally be of the same order of magnitude as $|V_x(t)|$, then *if it is the case that*

$$dt \gg \tau, \tag{9.45}$$

a condition that we will discuss more fully below, then we would be entitled to drop the first term on the left side of Eq. (9.44) on the grounds that it is negligibly small compared to the second term. Doing that and then dividing through by dt would give us

$$V_x(t) \approx \sqrt{2D}\frac{N_x(t)}{\sqrt{dt}} \approx \sqrt{2D}\Gamma_x(t), \tag{9.46}$$

where the last step follows from the definition (7.20) of Gaussian white noise and the fact that dt is by definition close to zero. Since Eq. (9.46) is the same as Eq. (9.42), we thus obtain the same Einstein approximation (9.43) that we arrived at by the previous analysis. But the condition ensuring that result is now (9.45), not (9.41). The reason those two conditions are not equivalent is that the time derivative in (9.41) has already taken the limit $dt \to 0$, so it is not possible to satisfy condition (9.45) with any *fixed* value of $\tau \equiv m/\gamma$.

A proper understanding of condition (9.45) requires that we pay attention to the following subtle point: Whenever we describe a time-evolving system, we always envision a graph that plots the state of the system against a time axis *on which there is a time scale of our choosing*. And we always make the tacit assumption that any duration of time that is smaller than what appears to be a "point" on the scale of our time axis can be regarded as being "infinitesimally small". If our time axis spans several minutes, then dt might be as large as a millisecond, whereas if our time axis spans several millennia dt might be as large as a month. But there will always be some implied, if unstated, maximum allowed size for dt. In the present case, *if* that maximum value of dt satisfies condition (9.45), which we note properly compares quantities that have the same physical dimensions, *then* Eq. (9.43) will provide an acceptably accurate graph of how the position of the diffusing molecule changes with time. *On that time-scale*, the system will exhibit overdamped behavior. But it would be wrong to call the system *itself* overdamped, because on a much finer time-scale it will *not* exhibit overdamped behavior.

So, consistent with our earlier results for the small-time and large-time limits of the solution to Langevin's equations of motion for a solute molecule, we conclude that *there are no inherently overdamped systems that are always well described by Eq. (9.43)*. For *any* solute molecule, Eq. (9.43) *will* be accurate if $t - t_0 \gg m/\gamma$, but it will *not* be accurate otherwise.

Notes to Chapter 9

[1]The average total x-distance traveled by the solute molecule over a time $(t - t_0) \gg \tau$ will be the product of that time and the average speed of the molecule along the x-axis, which we can estimate as $\sqrt{\langle V_x^2(\infty) \rangle} = \sqrt{k_B T/m}$. Thus,

$$\sqrt{\frac{k_{\mathrm{B}}T}{m}} \cdot (t - t_0) = \sqrt{\frac{D\gamma}{m}}(t - t_0) = \sqrt{\frac{D}{\tau}}(t - t_0) = \sqrt{2D(t - t_0)}\sqrt{\frac{(t - t_0)}{2\tau}}.$$

The last expression shows that, when $(t - t_0) \gg \tau$, the solute molecule's average total x-distance traveled will be *much larger* than its rms x-displacement $\sqrt{2D(t - t_0)}$, as predicted by both Langevin's theory in this limit [see Eq. (9.6)] and also Einstein's theory [see Eq. (3.12)].

   [2]Appending molecule subscripts $i = 1, 2$ to Eqs (8.29) and (8.30), we may immediately conclude by the linear combination theorem for statistically independent normal random variables that the differences $V_{1x}(t) - V_{2x}(t)$ and $X_1(t) - X_2(t)$ must both be *normal* random variables. It remains only to determine their two means, their two variances, and their covariance. To do that, we first append subscripts $i = 1, 2$ to Eqs (8.31). Then by the result (2.21a), $\langle V_{12x}(t) \rangle$ will be the difference between the $i = 1$ and 2 versions of Eq. (8.31a), and $\langle X_{12}(t) \rangle$ will be the difference between the $i = 1$ and 2 versions of Eq. (8.31c). And by the result (2.21b), var $\{V_{12x}(t)\}$ will be the *sum* of the $i = 1$ and 2 versions of Eq. (8.31b), and var $\{X_{12}(t)\}$ will be the *sum* of the $i = 1$ and 2 versions of Eq. (8.31d). Finally, it is not hard to show from the definition of the covariance in Eq. (2.20) that if the pair of random variables (Y_1, Z_1) is statistically independent of the pair of random variables (Y_2, Z_2), then cov $\{(Y_1 - Y_2), (Z_1 - Z_2)\}$ will be the *sum* of cov $\{Y_1, Z_1\}$ and cov $\{Y_2, Z_2\}$; therefore, cov $\{V_{12x}(t), X_{12}(t)\}$ will be the *sum* of the $i = 1$ and 2 versions of Eq. (8.31e). With these five moments in hand, the bivariate normal random variables $V_{12x}(t)$ and $X_{12}(t)$ are fully determined. Indeed, we can even construct from formulas (2.29) explicit exact updating formulas for numerical simulation, just as we did to get the single-molecule updating formulas (9.22).

10
Diffusion in an external force field

In this chapter we will consider what happens when diffusing solute molecules are subjected to an externally applied force $F_e(x)$. We will assume that this force is conservative, in that it can be derived from a potential energy function:

$$F_e(x) = -\frac{dU_e(x)}{dx}.\tag{10.1}$$

We will begin by reviewing the traditional derivation of the *Smoluchowski equation*, which is a generalization of the classical diffusion equation to accommodate such an external force. We will illustrate the Smoluchowski equation by using it to describe a rudimentary mechanism for gradient-sensing bacterial chemotaxis. After taking note of some weaknesses in the traditional derivation of the Smoluchowski equation, we will describe a more satisfactory approach to the external force problem which is made possible by Langevin's theory of diffusion, and which leads to the *Kramers equation*. We will give a careful critique of what are commonly called the "low friction" and "high friction" limits of the Kramers equation, and we will show that in the latter limit we can obtain the Smoluchowski equation as an approximation, although with some subtle caveats. We will conclude by treating in detail the simple case in which $F_e(x)$ is a non-zero constant.

10.1 The Smoluchowski equation—a Fickian derivation

The traditional way of analyzing the problem of a diffusing molecule in an external force field uses the following Fickian line of reasoning to derive a generalized form of the standard diffusion equation (1.4): The external force $F_e(x)$ acting on a solute molecule at x is assumed to impart to the molecule an average velocity $\bar{v}_e(x)$, which is such that the resulting average drag force $-\gamma\bar{v}_e(x)$ on the molecule, where γ is the previously defined drag coefficient, exactly balances the impressed force $F_e(x)$; thus, $\bar{v}_e(x) = F_e(x)/\gamma$. Since solute molecules are present at x at time t with density $\rho(x,t)$, this force-induced motion of those molecules produces an *average flux* in the positive x-direction of

$$J_e(x,t) = \bar{v}_e(x)\rho(x,t) = \frac{1}{\gamma}F_e(x)\rho(x,t).\tag{10.2}$$

The *total* average flux at x is then assumed to be the sum of this externally induced average flux plus the omnipresent Fickian flux (1.2), which arises from collisions of the solute molecules with the solvent molecules:

$$J_{\text{tot}}(x,t) = \frac{1}{\gamma} F_{\text{e}}(x)\rho(x,t) + \left(-D\frac{\partial\rho(x,t)}{\partial x} \right). \tag{10.3}$$

Upon substituting Eq. (10.3) into the continuity equation (1.1), namely,

$$\frac{\partial\rho(x,t)}{\partial t} = -\frac{\partial J_{\text{tot}}(x,t)}{dx}, \tag{10.4}$$

we obtain

$$\frac{\partial\rho(x,t)}{\partial t} = -\frac{1}{\gamma}\frac{\partial}{\partial x}\left[F_{\text{e}}(x)\rho(x,t)\right] + D\frac{\partial^2\rho(x,t)}{\partial x^2}. \tag{10.5}$$

Equation (10.5), which is evidently a generalization of the standard diffusion equation (1.4), is called the *Smoluchowski equation*.[1] That name is also applied to the equation obtained by replacing $\rho(x,t)$ in Eq. (10.5) with $P_X(x,t|x_0,t_0)$, in the manner of Einstein's extension (3.9) of the classical diffusion equation to describe the *stochastic* motion of a *single* solute molecule:

$$\frac{\partial P_X(x,t|x_0,t_0)}{\partial t} = -\frac{1}{\gamma}\frac{\partial}{\partial x}\left[F_{\text{e}}(x)P_X(x,t|x_0,t_0)\right] + D\frac{\partial^2 P_X(x,t|x_0,t_0)}{\partial x^2}. \tag{10.6}$$

When we compare the form of this equation to the generic forward Fokker–Planck equation (7.24), we see that Eq. (10.6) is technically a forward Fokker–Planck equation. As such, it defines a continuous Markov process $X(t)$ with drift function $A(x,t) = F_{\text{e}}(x)/\gamma$ and diffusion function $D(x,t) = 2D$. Therefore, according to the generic Langevin equation (7.13), the position $X(t)$ of the diffusing solute molecule in this Smoluchowski scenario satisfies the Langevin equation

$$X(t+dt) = X(t) + \frac{F_{\text{e}}(X(t))}{\gamma}dt + \sqrt{2D}N(t)\sqrt{dt}, \tag{10.7}$$

where $N(t)$ is as usual a temporally uncorrelated normal random variable with mean 0 and variance 1. Notice that if $F_{\text{e}} \equiv 0$, Eq. (10.7) properly reduces to the $\Delta t = dt$ version of the Einstein updating formula (4.11a) for diffusion in the *absence* of an external force.

If the functional form of $F_{\text{e}}(x)$ is such that the continuous Markov process defined by Eqs (10.6) and (10.7) is "stable" in the sense of Eq. (7.25), then it follows from Eq. (7.26) that the solution to Eq. (10.6) will approach, as $t - t_0 \to \infty$, the time-independent "equilibrium" solution

$$P_X^{\text{eq}}(x) = \frac{K}{2D}\exp\left(\int^x \frac{2[F_{\text{e}}(x')/\gamma]}{2D}dx'\right),$$

where K is a normalization constant. Evaluating the integral here is easy using the definition of the potential function in Eq. (10.1), and gives

$$P_X^{\text{eq}}(x) = K'\exp\left(-\frac{U_{\text{e}}(x)}{\gamma D}\right), \tag{10.8a}$$

where K' is again a normalization constant. Using the Einstein relation (9.7), we can also write this result as

$$P_X^{\text{eq}}(x) = K' \exp\left(-\frac{U_{\text{e}}(x)}{k_{\text{B}}T}\right). \tag{10.8b}$$

Equation (10.8b) has the form of the *Boltzmann distribution* of statistical thermodynamics. Qualitatively, it implies that at equilibrium, the solute molecule is most likely to be found near points where the potential energy function is a minimum. Its simplest and perhaps most common application is for a solution of solute molecules of mass m in a uniform gravitational field: With "up" in the $+x$ direction, $F_{\text{e}}(x) = -mg$ and $U_{\text{e}}(x) = mgx$, so $P_X^{\text{eq}}(x) \propto \exp\left(-mgx/k_{\text{B}}T\right)$. The implied exponential fall-off of solute concentration with height in a beaker containing a solution in equilibrium provided an early way of experimentally estimating the value of Boltzmann's constant k_{B}, and hence also Avogadro's number (which is equal to the ideal gas constant R divided by k_{B}).

Interestingly, Einstein's original derivation of his famous relation (9.7) essentially argued that the equilibrium solution (10.8a) of the Smoluchowski equation should, for purely thermodynamic reasons, agree with the classical Boltzmann equilibrium distribution formula (10.8b). Such agreement would evidently require γD to be equal to $k_{\text{B}}T$, so Einstein concluded that relation (9.7) had to be true. By contrast, in Langevin's approach to diffusion, we made the connection to the system's temperature T via the equilibrium *velocity* distribution, requiring the equilibrium prediction (8.20) of the Langevin equation for the solute molecule's velocity to conform to the requirement (8.21) of statistical thermodynamics.

Despite the physical plausibility of the $t \to \infty$ solution (10.8b) of the Smoluchowski equation, we will see later that the Smoluchowski equation has problems for finite t. That should not be surprising for two reasons: First, we know from Chapter 9 that the Einstein $F_{\text{e}} \equiv 0$ version of Eq. (10.7), basically Eq. (4.11a) with $\Delta t = dt$, is *physically* accurate only if $dt \gg \tau$ (see for instance Figs 9.4 to 9.6), and it seems unlikely that that condition will be softened by the presence of an external force. Second, whereas Einstein's derivation of the $F_{\text{e}} \equiv 0$ version of Eq. (10.5) studiously avoided any mention of the diffusing molecule's *velocity*, which as we saw in Chapter 4 is ill-defined in Einstein's approach, the above derivation of Eq. (10.5) needs the velocity to estimate the average flux (10.2) induced by the external force. We will return to these matters in Section 10.3, after we see how Langevin's approach to diffusion deals with the external force field. But first we will look at an illustrative application of the Smoluchowski equation.

10.2 An application: a rudimentary type of gradient-sensing chemotaxis

Suppose a "one-dimensional bacterium" of length δ diffuses along the x-axis with diffusion coefficient D and drag coefficient γ. Suppose further that on each end of this bacterium is a *receptor* for a certain *chemo-attractant* (CA) molecule, and that whenever a CA molecule comes into contact with a receptor on the bacterium, a response is triggered in the bacterium that causes its center to move *toward* the contacted receptor. (For *chemo-repellant* molecules, the bacterium would instead move *away* from the contacted receptor, and the analysis would be the same except for a change in the sign of a modeling parameter which will be introduced shortly; however, for simplicity we will confine our discussion here to the chemo-attractant case only.) Suppose finally that CA molecules are present in the solution containing our bacterium with an average density of $\eta(x)$, and this density does not change appreciably during the time that we observe the motion of the bacterium. We would like to develop a mathematical model for the bacterium's "chemotaxic" motion. But doing that will require that we be more explicit about how the bacterium responds when a CA molecule contacts one of its receptors.

Perhaps the simplest modeling of the bacterium's response to the surrounding CA molecules would be to assume that, through some mechanism that we do not try to specify, an *average pulling force* gets applied to the bacterium at each receptor. This force acts at the site of the receptor, is directed away from the bacterium's center, and has a magnitude that is equal to some constant α times the concentration of CA molecules at the receptor. Therefore, if the center of the bacterium is at x, so that its left and right receptors are at $x - \frac{1}{2}\delta$ and $x + \frac{1}{2}\delta$ respectively (see Fig. 10-1), the total average chemo-attractive force on a bacterium is

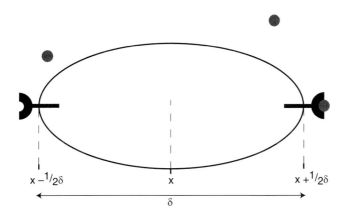

Fig. 10.1 *An idealized model of a one-dimensional bacterium, which has at each end a receptor for a chemo-attractant molecule. The bacterium is assumed to undergo ordinary diffusion. But if a chemo-attractant molecule (indicated here by a gray sphere) happens to attach to one of the receptors, the model assumes that the bacterium will promptly respond by moving toward the newly occupied receptor. Regardless of the detailed mechanism of the induced movement, the bacterium will, over the long run, be more likely to move in the direction of increasing concentration of chemo-attractant molecules, since the receptor that is located at the higher concentration end will be more likely to encounter a chemo-attractant molecule. The resulting motion is a modified form of diffusion that is described by the Keller–Segel equation (10.12).*

$$F_{\text{CA}}(x) = +\alpha\eta(x + \tfrac{1}{2}\delta) - \alpha\eta(x - \tfrac{1}{2}\delta)$$
$$= \alpha\left[\eta(x) + \eta'(x)\left(\tfrac{1}{2}\delta\right)\right] - \alpha\left[\eta(x) + \eta'(x)\left(-\tfrac{1}{2}\delta\right)\right]$$
$$F_{\text{CA}}(x) = \alpha\delta\eta'(x). \tag{10.9}$$

In the second line, we Taylor-expanded $\eta(x \pm \tfrac{1}{2}\delta)$ to first order in δ (the prime denotes differentiation with respect to x), and in so doing made the assumption that the bacterium is small enough that the CA density η varies linearly over the bacterium's length δ; in practice this is almost always so, because bacteria are so small. Now letting $P_{\text{b}}(x,t)$ denote the PDF of the x-coordinate of the bacterium's center at time t, we can straightaway substitute $F_{\text{CA}}(x)$ for the external force $F_{\text{e}}(x)$ in the Smoluchowski equation (10.6) to get

$$\frac{\partial P_{\text{b}}(x,t)}{\partial t} = -\frac{1}{\gamma}\frac{\partial}{\partial x}\left[(\alpha\delta\eta'(x))\,P_{\text{b}}(x,t)\right] + D\frac{\partial^2 P_{\text{b}}(x,t)}{\partial x^2}. \tag{10.10}$$

Letting

$$\chi = \frac{\alpha\delta}{\gamma} = \frac{\alpha\delta D}{k_{\text{B}}T}, \tag{10.11}$$

where the last equality invokes the Einstein relation (9.7), Eq. (10.10) can be written

$$\frac{\partial P_{\text{b}}(x,t)}{\partial t} = D\frac{\partial^2 P_{\text{b}}(x,t)}{\partial x^2} - \chi\frac{\partial}{\partial x}\left[\eta'(x)P_{\text{b}}(x,t)\right]. \tag{10.12}$$

Equation (10.12) is known as the *Keller–Segel equation*. It shows how the standard diffusion equation for the bacterium is modified by the presence of chemo-attractant molecules under the stated modeling assumptions. But we should note that many bacteria (e.g., *E. coli*) that appear to "sense the CA gradient" in order to perform chemotaxis are *not* believed to do so by the simple mechanism just described. The parameter χ in the Keller–Segel equation is called the *coefficient of chemotaxis*; it is positive for a chemo-attractant, and negative for a chemo-repellant.

An arguably more sophisticated model of gradient-sensing chemotaxis can be devised by replacing the "average force" assumption in the preceding model with the following two assumptions: (i) If a CA molecule collides with a receptor on the bacterium, the bacterium immediately responds by *moving a distance Δ* in the direction from its center toward the contacted receptor. (ii) The *probability* that a CA molecule will collide with a receptor in the next infinitesimally small time dt is directly proportional to dt times {the average density of CA molecules at the receptor}.

Letting β denote the proportionality constant in assumption (ii), then if the bacterium's center is at x, so that its left and right receptors are at $x - \tfrac{1}{2}\delta$ and $x + \tfrac{1}{2}\delta$ respectively, the probability $p_{\pm}(x)dt$ that the bacterium will jump to $x \pm \Delta$ in the next dt is

$$p_{\pm}(x)dt = \beta\eta\big(x \pm \tfrac{1}{2}\delta\big)\,dt. \tag{10.13}$$

From this it follows from formula (2.37) that the *average distance* the bacterium will move in the next dt is

$$
\begin{aligned}
\overline{\Delta x_{dt}} &= p_-(x)dt \cdot (-\Delta) + p_+(x)dt \cdot (+\Delta) \\
&= \Delta \left[p_+(x)dt - p_-(x)dt \right] \\
&= \Delta \left[\beta\eta(x + \tfrac{1}{2}\delta)dt - \beta\eta(x - \tfrac{1}{2}\delta)dt \right] \\
&= \Delta\beta dt \left[\{\eta(x) + \eta'(x) \cdot (\tfrac{1}{2}\delta)\} - \{\eta(x) + \eta'(x) \cdot (-\tfrac{1}{2}\delta)\} \right] \\
\overline{\Delta x_{dt}} &= \beta\Delta\delta\eta'(x)dt.
\end{aligned}
\tag{10.14}
$$

Again we have assumed that the CA density η varies approximately linearly over a distance equal to the bacterium's length δ. It follows that the *average velocity* of the bacterium at x due to the presence of the CA molecules is

$$
\bar{v}_{\mathrm{b}}(x) \equiv \frac{\overline{\Delta x_{dt}}}{dt} = \beta\Delta\delta\eta'(x).
\tag{10.15}
$$

Now we argue along the lines of the derivation of the Smoluchowski equation in Section 10.1: The average velocity $\bar{v}_{\mathrm{b}}(x)$ gives rise to a *flux* in the probability density of the bacterium's x-coordinate of

$$
J_{\mathrm{CA}}(x,t) \equiv \bar{v}_{\mathrm{b}}(x)P_{\mathrm{b}}(x,t) = \beta\Delta\delta\eta'(x)P_{\mathrm{b}}(x,t).
\tag{10.16}
$$

The *total* probability flux is then assumed to be the sum of this CA-induced flux and the omnipresent flux (3.20), which arises from collisions of the bacterium with solvent molecules:

$$
J_{\mathrm{tot}}(x,t) = \beta\Delta\delta\eta'(x)P_{\mathrm{b}}(x,t) + \left(-D\frac{\partial P_{\mathrm{b}}(x,t)}{\partial x} \right).
\tag{10.17}
$$

Upon substituting Eq. (10.17) into the continuity equation (3.21) for the probability density of the bacterium, namely

$$
\frac{\partial P_{\mathrm{b}}(x,t)}{\partial t} = -\frac{\partial J_{\mathrm{tot}}(x,t)}{dx},
\tag{10.18}
$$

we obtain once again the Keller–Segel equation (10.12), but now with the coefficient of chemotaxis given by

$$
\chi = \beta\Delta\delta.
\tag{10.19}
$$

Both formulas (10.19) and (10.11) have χ proportional to the cell length δ. The different proportionality constants, namely $\beta\Delta$ in (10.19) and α/γ in (10.11), reflect the difference in the assumed physical mechanisms that give rise to chemotaxis in those two models. Of course, whether either of these two models, or any other model, of gradient-sensing chemotaxis applies to any real bacterium would have to be determined by a careful inspection of the chemotaxic mechanism of the bacterium in question.

10.3   The Langevin equation for a solute molecule in an external force field

The status of Langevin's dynamical equation (8.23) as Newton's Second Law for a single solute molecule implies that we can take the external force $F_e(x)$ into account in Langevin's theory simply by making it an additional force term on the right side of that equation:

$$m\frac{dV_x(t)}{dt} = -\gamma V_x(t) + \sqrt{2\gamma k_B T}\,\Gamma_x(t) + F_e(X(t)). \tag{10.20}$$

Eliminating the parameters γ and T in favor of τ and D by way of the definition $\tau = m/\gamma$ and the Einstein relation $D = k_B T/\gamma$, we see that the x-motion of the solute molecule is now governed by the pair of equations

$$\frac{dV_x(t)}{dt} = -\frac{1}{\tau}V_x(t) + \frac{1}{m}F_e(X(t)) + \frac{\sqrt{2D}}{\tau}\Gamma_x(t), \tag{10.21a}$$

$$\frac{dX(t)}{dt} = V_x(t). \tag{10.21b}$$

These equations evidently have the canonical "white-noise" form of the bivariate Langevin equation (7.29), with $M = 2$, $X_1 = V_x$, and $X_2 = X$. Recalling from Eq. (7.20) the connection $\Gamma_x(t) = N_x(t)/\sqrt{dt}$ between Gaussian white noise and the temporally uncorrelated normal random variable with mean 0 and variance 1, we see that the corresponding "standard form" Langevin equation is [cf. Eq. (7.28)]

$$V_x(t + dt) = V_x(t) - \frac{1}{\tau}V_x(t)dt + \frac{1}{m}F_e(X(t))\,dt + \frac{\sqrt{2D}}{\tau}N_x(t)\sqrt{dt}, \tag{10.22a}$$

$$X(t + dt) = X(t) + V_x(t)dt. \tag{10.22b}$$

In Langevin's approach to diffusion, Eqs (10.22) completely define the functions $V_x(t)$ and $X(t)$ for all $t > t_0$, given the initial conditions $V_x(t_0) = v_{x0}$ and $X(t_0) = x_0$.

The rest of this chapter will be devoted to developing the consequences of Eqs (10.22). We may begin by pointing out two obvious implications: First, if the function F_e depends explicitly on x, then the right side of the updating formula (10.22a) will depend on $X(t)$. Since $X(t)$ contains information about the past of V_x that is not contained in V_x's present value $V_x(t)$, then V_x, like X, will not by itself be Markov. But the *pair* (V_x, X) is still a bivariate Markov process, because the two values $V_x(t)$ and $X(t)$ *together* are sufficient to determine both $V_x(t + dt)$ and $X(t + dt)$.

Second, Eqs (10.22) immediately give us an extension of the *approximate* $F_e \equiv 0$ updating formulas (9.21) for numerical simulation: By simply replacing dt in Eqs (10.22) with a finite but "suitably small" Δt, we obtain

$$v_x(t + \Delta t) \approx v_x(t) - \frac{1}{\tau}v_x(t)\Delta t + \frac{1}{m}F_e(x(t))\,\Delta t + \frac{\sqrt{2D}}{\tau}n_x\sqrt{\Delta t}, \tag{10.23a}$$

$$x(t + \Delta t) \approx x(t) + v_x(t)\Delta t, \tag{10.23b}$$

where n_x denotes a sample value of $\mathcal{N}(0, 1)$. Since the more restrictive $F_e \equiv 0$ formulas (9.21) require $\Delta t \ll \tau$ for accuracy, that condition will surely apply here too. But we now have the additional constraint that Δt should be small enough that the factor $F_e(x(t))$ in Eq. (10.23a) does not change appreciably over any time interval of size Δt. In Section 10.8 we will derive *exact* updating formulas, valid for any $\Delta t > 0$, for the special case in which F_e is a *non-zero constant*.

10.4 The Kramers equation

A comparison of Eqs (10.22) with the canonical multivariate Langevin equations (7.28) shows that, for our bivariate continuous Markov process $[V_x(t), X(t)]$, the functions $A_i(x_1, x_2) \equiv A_i(v_x, x)$ and $b_{ij}(x_1, x_2) \equiv b_{ij}(v_x, x)$ in Eq. (7.28) are

$$\begin{cases} A_1(v_x, x) = -\left(\dfrac{1}{\tau}v_x - \dfrac{1}{m}F_e(x)\right), & A_2(v_x, x) = v_x, \\ \\ b_{11}(v_x, x) = \dfrac{\sqrt{2D}}{\tau}, & b_{12}(v_x, x) = b_{21}(v_x, x) = b_{22}(v_x, x) = 0. \end{cases}$$

It follows that the functions D_i and C_{ij} in Eqs (7.30) are

$$D_1(v_x, x) = \frac{2D}{\tau^2}, \quad D_2(v_x, x) = 0, \quad C_{12}(v_x, x) = 0.$$

When these functions are substituted into the generic multivariate Fokker–Planck equation (7.31), we find that the bivariate Fokker–Planck equation for the process (V_x, X) defined by Eqs (10.22) is

$$\frac{\partial P_{V_x,X}}{\partial t} = \frac{\partial}{\partial v_x}\left[\left(\frac{v_x}{\tau} - \frac{F_e(x)}{m}\right)P_{V_x,X}\right] - v_x\frac{\partial P_{V_x,X}}{\partial x} + \frac{D}{\tau^2}\frac{\partial^2 P_{V_x,X}}{\partial v_x^2}. \tag{10.24}$$

Here, $P_{V_x,X} \equiv P_{V_x,X}(v_x, x; t|v_{x0}, x_0; t_0)$ is the joint PDF of $V_x(t)$ and $X(t)$, given that $V_x(t_0) = v_{x0}$ and $X(t_0) = x_0$. Equation (10.24) is called the *Kramers equation*.[2]

Like the bivariate Langevin equation (10.22) to which it is mathematically equivalent, the Kramers equation (10.24) completely describes the time evolution of the bivariate continuous Markov process (V_x, X). In the literature, the Kramers equation appears in several different forms. For example, if we use the Einstein formula and the definition of τ to eliminate the parameters D and τ in Eq. (10.24) in favor of the parameters γ and T, then a modest rearrangement of terms yields

$$\frac{\partial P_{V_x,X}}{\partial t} + v_x\frac{\partial P_{V_x,X}}{\partial x} + \frac{F_e}{m}\frac{\partial P_{V_x,X}}{\partial v_x} = \frac{\gamma}{m}\frac{\partial}{\partial v_x}\left(v_x P_{V_x,X} + \frac{k_BT}{m}\frac{\partial P_{V_x,X}}{\partial v_x}\right). \tag{10.25}$$

This form of the Kramers equation has some intriguing features that we will discuss in Section 10.6.

If $F_e \equiv 0$, the Kramers equation would of course describe the situation treated in Chapters 8 and 9. The results we obtained in those two chapters imply that the solution $P_{V_x,X}$ to the $F_e \equiv 0$ Kramers equation is the *bivariate normal* PDF in Eq. (2.28) in which the means, variances, and covariance are given by Eqs (8.31).

As might be expected, increasingly complicated functional forms for $F_e(x)$ will make the Kramers equation increasingly more difficult to solve. It is sometimes easier to solve the Langevin equation (10.22) than the Kramers equation (10.24), which is what we did in the $F_e \equiv 0$ case in Chapter 8. In Section 10.8 we will again follow that route to obtain the solution for the case in which $F_e(x)$ is a non-zero constant.

But there is one important general result of the Kramers equation that can be obtained rather easily: By straightforwardly evaluating the partial derivatives in Eq. (10.24), we can show that a time-*independent* solution of that equation is

$$P^{eq}_{V_x, X}(v_x, x) = K \exp \left(-\frac{\tau v_x^2}{2D} - \frac{\tau U_e(x)}{mD} \right) \equiv K \exp \left(-\frac{\frac{1}{2}mv_x^2 + U_e(x)}{k_B T} \right), \quad (10.26)$$

where K is a normalizing constant. It turns out that for most problems of physical interest—specifically, for problems in which the normalization constant K is finite—all time-*dependent* solutions of Eq. (10.24) will *approach* this time-independent form as $t \to \infty$. In such cases, $P^{eq}_{V_x, X}$ can be said to describe the *equilibrium* distribution of $V_x(\infty)$ and $X(\infty)$. The nature of that equilibrium distribution can be seen more clearly by writing Eq. (10.26) in the form

$$P^{eq}_{V_x, X}(v_x, x) = K \exp \left(-\frac{(v_x - 0)^2}{2 (k_B T/m)} \right) \cdot \exp \left(-\frac{U_e(x)}{k_B T} \right).$$

The fact that this joint PDF is a product of a function of v_x only times a function of x only implies that $V_x(\infty)$ and $X(\infty)$ are *statistically independent*. The form of the v_x-factor implies that $V_x(\infty)$ is a *normal* random variable with mean 0 and variance $k_B T/m$:

$$V_x(\infty) = \mathcal{N} \left(0, \frac{k_B T}{m} \right). \quad (10.27)$$

And the form of the x-factor implies that the PDF of the equilibrium distribution for the position $X(\infty)$ of the diffusing molecule is

$$P^{eq}_X(x) = K' \exp \left(-\frac{U_e(x)}{k_B T} \right), \quad (10.28)$$

where the normalizing constant K' is different from the one in Eq. (10.26). This agrees with the result (10.8b) that we inferred from the Smoluchowski equation. What all this means is that, as $t \to \infty$, $V_x(t)$ and $X(t)$ become statistically independent of each other, with $V_x(\infty)$ conforming to the Maxwell–Boltzmann velocity distribution, and $X(\infty)$ conforming to the standard Boltzmann distribution of statistical thermodynamics. It follows that the time-*dependent* solution of the Kramers equation (10.24) will merely tell us *how* $V_x(t)$ and $X(t)$ approach their equilibrium distributions from their respective initial sure values v_{x0} and x_0.

It is easy to confirm the validity of the results (10.27) and (10.28) for the case $F_e(x) \equiv 0$ by comparing with the results of Chapter 8: Equation (10.27) is exactly the $t \to \infty$ prediction of Eqs (8.29), (8.31a), and (8.31b). And since $F_e(x) \equiv 0$ implies that

$U_e(x)$ is independent of x, then Eq. (10.28) implies that $P_X^{eq}(x)$ will be independent of x, and since x ranges from $-\infty$ to $+\infty$, that flat x-distribution will have an infinite variance—exactly as predicted by Eq. (8.31d) for $t \to \infty$. Finally, the result that $V_x(\infty)$ and $X(\infty)$ are statistically independent agrees with the implication of Eq. (8.33) that corr $\{V_x(\infty), X(\infty)\} = 0$.

10.5 Energetics revisited

Formula (9.38) for the average energy of a solute molecule when $F_e(x) \equiv 0$ has a straightforward generalization for a non-zero force field. To see how that comes about, we begin by squaring Eq. (10.22a), retaining only terms up to first order in dt:

$$V_x^2(t + dt) = V_x^2(t) - \frac{2}{\tau} V_x^2(t) dt + \frac{2}{m} V_x(t) F_e(X(t)) dt$$

$$+ 2\left(\frac{\sqrt{2D}}{\tau}\right) V_x(t) N_x(t) \sqrt{dt} + \left(\frac{\sqrt{2D}}{\tau}\right)^2 N_x^2(t) dt.$$

Taking the average of this equation, remembering that $V_x(t)$ and $N_x(t)$ are statistically independent and also that $\langle N_x(t)\rangle = 0$ and $\langle N_x^2(t)\rangle = 1$, we get

$$\langle V_x^2(t + dt)\rangle = \langle V_x^2(t)\rangle - \frac{2}{\tau}\langle V_x^2(t)\rangle dt + \frac{2}{m}\langle V_x(t) F_e(X(t))\rangle dt + \frac{2D}{\tau^2} dt.$$

Next we replace the function F_e with the negative derivative of the potential function U_e in accordance with Eq. (10.1), and we eliminate the parameters D and τ in favor of the parameters γ and T using the Einstein formula and the definition of τ:

$$\langle V_x^2(t + dt)\rangle = \langle V_x^2(t)\rangle - \frac{2}{m}\gamma\langle V_x^2(t)\rangle dt - \frac{2}{m}\langle V_x(t) U_e'(X(t))\rangle dt + \frac{2\gamma k_B T}{m^2} dt.$$

Transposing the first term on the right, multiplying through by $m/(2dt)$, and then taking the limit $dt \to 0^+$, we get

$$\frac{d}{dt}\langle \tfrac{1}{2}m V_x^2(t)\rangle = -\langle \gamma V_x^2(t)\rangle - \langle V_x(t) U_e'(X(t))\rangle + \frac{\gamma k_B T}{m}. \qquad (10.29)$$

This is the same as Eq. (9.37) except for a new term involving U_e. To clarify the effect of that term, we use Eq. (10.22b) to write

$$U_e(X(t + dt)) = U_e(X(t) + V_x(t)dt)$$

$$= U_e(X(t)) + U_e'(X(t)) \cdot V_x(t)dt.$$

Averaging this gives

$$\langle U_e(X(t + dt))\rangle = \langle U_e(X(t))\rangle + \langle U_e'(X(t)) V_x(t)\rangle dt.$$

Transposing the first term on the right, dividing through by dt, and then taking the limit $dt \to 0^+$, we get

$$\frac{d}{dt} \langle U_e \left(X(t) \right) \rangle = \langle U_e' \left(X(t) \right) V_x(t) \rangle. \tag{10.30}$$

Substituting this into Eq. (10.29) and bringing all time derivatives to the left side gives

$$\frac{d}{dt} \langle \tfrac{1}{2} m V_x^2(t) + U_e(X(t)) \rangle = - \langle \gamma V_x^2(t) \rangle + \frac{\gamma k_B T}{m}. \tag{10.31}$$

Finally, we add to this equation the two equations for the y- and z-components, which do not involve U_e, and so conclude that

$$\frac{d}{dt} \langle \tfrac{1}{2} m \mathbf{V}^2(t) + U_e(X(t)) \rangle = - \langle \gamma \mathbf{V}^2(t) \rangle + \frac{3 \gamma k_B T}{m}. \tag{10.32}$$

This generalization of Eq. (9.38) simply replaces the solute molecule's average *kinetic* energy with its average *total* energy, thereby taking account of the external potential energy field. The interpretation of the two terms on the right are the same as described in connection with Eq. (9.38): The negative "sink" term is the average rate at which energy is being transferred from the solute molecule to the solvent molecules, and the positive "source" term is the average rate at which the solvent molecules are transferring energy to the solute molecule. When the system is in equilibrium the time derivative on the left side of Eq. (10.32) is zero, and the consequent equality in magnitude of the sink and source terms again yields Eq. (9.39). Thus, the equipartition of energy theorem of statistical mechanics is unaffected by the external force field.

10.6 Some interesting aspects of the uninteresting limit $\gamma \to 0$

As we saw in Chapter 8 [cf. Eqs (8.23) and (8.25)], the parameter γ controls the strength of the interactions between a solute molecule and its surrounding solvent molecules: Since the dissipative drag force on the solute molecule is proportional to γ and the randomly fluctuating force is proportional to $\sqrt{\gamma}$, the limit $\gamma \to 0$ is equivalent to removing the solvent molecules from the solution. While this is not a particularly interesting limit of diffusion theory, it is instructive to see how the equations we have developed behave in that limit. As regards the average energy of the solute molecule, it follows immediately from Eq. (10.32) that when $\gamma = 0$ that average energy is constant in time.

The $\gamma \to 0$ limit of the bivariate Langevin equation (10.21) becomes obvious if Eq. (10.21a) is written in the equivalent form (10.20):

$$m \frac{dV_x(t)}{dt} = F_e \left(X(t) \right), \quad \frac{dX(t)}{dt} = V_x(t) \quad (\gamma = 0). \tag{10.33}$$

These coupled ordinary differential equations define *deterministic* functions $V_x(t)$ and $X(t)$, and those functions evolve in time in exactly the way we expect in Newtonian mechanics when the solvent molecules are subject to only the force F_e.

The $\gamma \to 0$ limit of the Kramers equation is equally obvious from its form (10.25):

$$\frac{\partial P_{V_x,X}}{\partial t} + v_x \frac{\partial P_{V_x,X}}{\partial x} + \frac{F_e}{m} \frac{\partial P_{V_x,X}}{\partial v_x} = 0 \quad (\gamma = 0). \tag{10.34}$$

Not obvious, however, is how this equation for the joint PDF of the two random variables $V_x(t)$ and $X(t)$ is consistent with the deterministic result (10.33). To show that it is, let us temporarily simplify our notation by dropping the subscripts on the velocity variable and the joint PDF. What we will prove is this: The $\gamma = 0$ Kramers equation,

$$\frac{\partial P(v,x;t)}{\partial t} + v \frac{\partial P(v,x;t)}{\partial x} + \frac{F_e(x)}{m} \frac{\partial P(v,x;t)}{\partial v} = 0, \tag{10.35}$$

has the time-dependent solution

$$P(v,x;t) = \delta(v - v_t)\delta(x - x_t), \tag{10.36}$$

where v_t and x_t are the deterministic functions of t that satisfy the $\gamma = 0$ Langevin equation (10.33),

$$\frac{dv_t}{dt} = \frac{F_e(x_t)}{m}, \quad \frac{dx_t}{dt} = v_t. \tag{10.37}$$

The form of the joint PDF (10.36) tells us, for reasons explained in Sections 2.3 and 2.4, that V and X are the respective *sure* variables v_t and x_t, and Eqs (10.37) tell us that those sure variables satisfy the $\gamma = 0$ bivariate Langevin equation (10.33). Therefore, this result implies consistency between the Langevin equation and the Kramers equation in the limit $\gamma = 0$.

The proof of the above result is not hard, but it requires acknowledging the possibly surprising fact that the Dirac delta function $\delta(x)$ actually has a *derivative* $\delta'(x)$ that is no less legitimate than $\delta(x)$ itself.[3] That being so, we can evaluate each of the three terms in Eq. (10.35) for the function $P(v,x;t)$ in Eq. (10.36) as follows:

$$\frac{\partial P(v,x;t)}{\partial t} = \delta(v - v_t) \frac{\partial}{\partial t} \delta(x - x_t) + \delta(x - x_t) \frac{\partial}{\partial t} \delta(v - v_t)$$

$$= \delta(v - v_t)\delta'(x - x_t)\left(-\frac{dx_t}{dt}\right) + \delta(x - x_t)\delta'(v - v_t)\left(-\frac{dv_t}{dt}\right);$$

$$v \frac{\partial P(v,x;t)}{\partial x} = v\delta(v - v_t)\delta'(x - x_t) \equiv v_t\delta(v - v_t)\delta'(x - x_t);$$

$$\frac{F_e(x)}{m} \frac{\partial P(v,x;t)}{\partial v} = \frac{F_e(x)}{m}\delta(x - x_t)\delta'(v - v_t) \equiv \frac{F_e(x_t)}{m}\delta(x - x_t)\delta'(v - v_t).$$

Adding these three equations gives

$$\frac{\partial P(v, x; t)}{\partial t} + v \frac{\partial P(v, x; t)}{\partial x} + \frac{F_e(x)}{m} \frac{\partial P(v, x; t)}{\partial v}$$

$$= \left(-\frac{dx_t}{dt} + v_t \right) \delta(v - v_t) \delta'(x - x_t)$$

$$+ \left(-\frac{dv_t}{dt} + \frac{F_e(x_t)}{m} \right) \delta(x - x_t) \delta'(v - v_t).$$

The right side of this equation vanishes identically by virtue of Eqs (10.37), so we conclude that the function (10.36) indeed satisfies the $\gamma = 0$ Kramers equation (10.35).

Another way to have answered the question of why the $\gamma = 0$ Kramers equation is equivalent to the $\gamma = 0$ Langevin equation would be to simply invoke the mathematical equivalence for *all* $\gamma \geq 0$ of the Fokker–Planck equation (10.24) [or (10.25)] and the Langevin equation (10.22). For, as noted in Chapter 7, the derivation of the canonical Fokker–Planck equation (7.24) from the canonical Langevin equation (7.13) invokes only the laws of probability theory, and makes no approximations or extra assumptions. That same mathematically rigorous equivalence exists between the canonical *multivariate* Langevin equation (7.28) and the canonical multivariate Fokker–Planck equation (7.31), of which our present problem is a specific two-variable example. In earlier times this equivalence was not widely appreciated, and attempts were made to justify the Kramers equation instead on the basis of the Liouville equation of classical mechanics. Not only is that unnecessary, it is actually misguided.[4]

Again we emphasize that there is no *practical* advantage in describing the motion of the solute molecule for $\gamma = 0$ as a delta-function peak that moves about in the velocity–position plane in the deterministic manner prescribed by Newton's Second Law. Nevertheless, it is reassuring to see that the general Kramers equation (10.25) properly gives that description when $\gamma = 0$.

10.7   The large-γ limit: the Smoluchowski equation revisited

Of more practical interest than the $\gamma = 0$ limit examined in the preceding section is the other extreme, which we will call the "large-γ" limit. But here we must be careful: As discussed in connection with the force-free case in Section 9.11, although we know what $\gamma = 0$ means, the assertion that "γ is large" is meaningless unless we can say large compared to what. Also, we must remember that comparing values of dynamical quantities makes sense only if those quantities have the same dimensional units.

As a first attempt to examine the "large-γ" limit, we start by multiplying Eq. (10.21a) through by m and then transposing a term to get

$$m \frac{dV_x(t)}{dt} + \gamma V_x(t) = F_e \left(X(t) \right) + \gamma \sqrt{2D} \Gamma_x(t). \tag{10.38}$$

Now, *if* we could say that γ is "so large" and/or m is "so small" that

$$\left| m\frac{dV_x(t)}{dt} \right| \ll |\gamma V_x(t)|, \tag{10.39}$$

then the first term on the left side of Eq. (10.38) could be omitted, and that equation would approximate to

$$V_x(t) \doteq \frac{1}{\gamma}F_e\left(X(t)\right) + \sqrt{2D}\Gamma_x(t).$$

Since $V_x(t) = dX(t)/dt$, this would imply that

$$\frac{dX(t)}{dt} \doteq \frac{1}{\gamma}F_e\left(X(t)\right) + \sqrt{2D}\Gamma_x(t). \tag{10.40}$$

Equation (10.40) is, according to Eq. (7.21), a Langevin equation. It defines a continuous Markov process with drift function $A(x,t) = F_e(x)/\gamma$ and diffusion function $D(x,t) = 2D$. Therefore, according to the generic Fokker–Planck equation (7.24), $X(t)$ will be (approximately) described by the Fokker–Planck equation

$$\frac{\partial P_X(x,t|x_0,t_0)}{\partial t} \doteq -\frac{1}{\gamma}\frac{\partial}{\partial x}\left[F_e(x)P_X(x,t|x_0,t_0)\right] + D\frac{\partial^2 P_X(x,t|x_0,t_0)}{\partial x^2}. \tag{10.41}$$

This is none other than the Smoluchowski equation (10.6).

But there is a problem with this "derivation" of the Smoluchowski equation: The assumption that γ can be "so large" and/or m "so small" that condition (10.39) will be satisfied is not tenable. That's because $V_x(t)$, being a continuous Markov process, has an undefined (i.e., infinite) derivative; therefore, for any finite γ and non-zero m the inequality in (10.39) will inevitably point the other way. This is the same difficulty that we discussed in Section 9.11. But as we showed there, we can fix the above derivation by arguing differently.

As discussed in Section 7.5, the white-noise form Langevin equation (7.21) is really just a mnemonic for the standard-form Langevin equation (7.13). So, instead of starting the above analysis by using the white-noise form Langevin equation (10.21a) to write down Eq. (10.38), let us start by using the equivalent standard-form Langevin equation (10.22a),

$$V_x(t+dt) = V_x(t) - \frac{1}{\tau}V_x(t)dt + \frac{1}{m}F_e\left(X(t)\right)dt + \frac{\sqrt{2D}}{\tau}N_x(t)\sqrt{dt}.$$

Here, $N_x(t)$ is a temporally uncorrelated normal random variable with mean 0 and variance 1. Multiplying this equation through by $\tau \equiv m/\gamma$ and rearranging gives

$$[V_x(t+dt) - V_x(t)]\,\tau + V_x(t)dt = \frac{1}{\gamma}F_e\left(X(t)\right)dt + \sqrt{2D}N_x(t)\sqrt{dt}. \tag{10.42}$$

Now, *if* we are working on a time-scale for which an "infinitesimally small" time can nevertheless be much larger than τ, i.e., if our time-scale is such that

$$dt \gg \tau, \tag{10.43}$$

then the first term on the left side of Eq. (10.42) can be neglected relative to the second term, and after dividing through by dt we will be left with

$$V_x(t) \approx \frac{1}{\gamma} F_e\left(X(t)\right) + \sqrt{2D} \frac{N_x(t)}{\sqrt{dt}} \approx \frac{1}{\gamma} F_e\left(X(t)\right) + \sqrt{2D}\,\Gamma_x(t). \tag{10.44}$$

Here, the last step follows from Eq. (7.20) and the fact that dt is by definition "small". Since $V_x(t) \equiv dX(t)/dt$, Eq. (10.44) is the same as the white-noise Langevin equation (10.40), the result of the previous (flawed) analysis. And that Langevin equation implies the Fokker–Planck equation (10.41), which is the Smoluchowski equation.

This modified argument constitutes a "proper" derivation of the Smoluchowski equation (10.41) from the Kramers equation (10.24). Unlike the derivation based on the specious condition (10.39), or for that matter the heuristic derivation based on Fick's Law given in the first section of this chapter, this derivation makes it clear that the Smoluchowski equation will be valid *only* on time-scales for which condition (10.43) is satisfied—i.e., only when the parameter τ is much smaller than what we would call an "infinitesimally small time". It would thus appear that the "large-γ" limit in the title of this section would be more properly described as a *large time-scale* limit; because, what we are really assuming here is that the system is being observed on a time-scale where the relaxation time τ can be regarded as being smaller than infinitesimally small.

The physical validity of the Smoluchowski equation *on large time-scales* is corroborated by the fact that the infinite-time "equilibrium" solution of that equation is the same Boltzmann formula (10.28) that is implied by the Kramers equation. But condition (10.43) has an important implication for doing *numerical simulations* of the process defined by the Smoluchowski equation: Since the Smoluchowski equation (10.41), when viewed as a Fokker–Planck equation, is equivalent to the white-noise Langevin equation (10.40), then the Smoluchowski equation is also equivalent to the corresponding standard-form Langevin equation,

$$X(t+dt) = X(t) + \frac{1}{\gamma} F_e\left(X(t)\right) dt + \sqrt{2D}\,N_x(t)\sqrt{dt}. \tag{10.45}$$

Indeed, Eq. (10.45) is just Eq. (10.42) under condition (10.43), and with $V_x(t)dt$ replaced by $X(t+dt) - X(t)$ in accordance with Eq. (10.22b). Now, a simple but approximate way to simulate the process defined by Eq. (10.45) would be to just replace the infinitesimal dt by a finite but "suitably small" Δt. That would give the approximate updating formula

$$x(t+\Delta t) \approx x(t) + \frac{1}{\gamma} F_e\left(x(t)\right) \Delta t + \sqrt{2D\Delta t}\, n_x, \tag{10.46}$$

where n_x is a sample value of $\mathcal{N}(0, 1)$. In the case $F_e \equiv 0$, this formula is the Einstein updating formula (4.11a). And in that case, we know that the formula is *mathematically exact* for all Δt, but *physically accurate only if $\Delta t \gg \tau$*. For the more general case of an x-dependent function F_e, the condition $\Delta t \gg \tau$ for physical accuracy remains in effect via condition (10.43). But unlike the Einstein updating formula

(4.11a), Eq. (10.46) is *not* in general mathematically exact, because for Eq. (10.46), we must take Δt *small enough* that F_e does not change appreciably over the time interval $[t, t + \Delta t]$.

These considerations imply the following rather curious situation: In order for the approximate updating formula (10.46) to produce a simulated trajectory that is accurate *from the perspective of the Smoluchowski equation* (10.41), it will be necessary to take Δt small enough that F_e stays approximately constant over that time. *That might require taking Δt smaller than τ.* But the resulting trajectory, although accurate from the perspective of the Smoluchowski equation, will be *physically* accurate only if it is *plotted* on a time-scale that is large enough that τ can be considered infinitesimally small. The same situation arose in connection with the $F_e = 0$ version of the Smoluchowski equation, namely Eq. (3.9): In that case the updating formula (4.11a) is accurate from the perspective of Eq. (3.9) for arbitrarily small Δt, but the resulting trajectory will be *physically* accurate only if it is *plotted* on a time-scale that is large enough that τ can be considered infinitesimally small.

For numerical simulations, it would probably always be better to use the coupled pair of updating formulas (10.23) suggested by the bivariate Langevin equation (10.22) instead of the single updating formula (10.46) suggested by the Smoluchowski equation through its equivalent Eq. (10.45). Even though formulas (10.23) are slightly more time-consuming, they should be more accurate. We thus conclude that, although the Smoluchowski equation is a legitimate approximation to the Kramers equation, its limitations should be carefully weighed before using it in any practical context.

10.8 A constant external force field in the Langevin picture

We will conclude this chapter by examining in detail the case in which $F_e(x) \equiv f_e$, a constant. An example would be a solute molecule in a uniform gravitational potential $U_e(x) = mgx$, in which case $f_e = -mg$. Putting $F_e(x) \equiv f_e$ in the joint Langevin equations (10.22) gives

$$V_x(t + dt) = V_x(t) - \frac{1}{\tau}V_x(t)dt + \frac{f_e}{m}dt + \frac{\sqrt{2D}}{\tau}N_x(t)\sqrt{dt}, \qquad (10.47\text{a})$$

$$X(t + dt) = X(t) + V_x(t)dt. \qquad (10.47\text{b})$$

We want to solve these coupled equations for the unrestricted motion of a solute molecule on the x-axis, given the initial conditions $V_x(t_0) = v_{x0}$ and $X(t_0) = x_0$.

It is easy to establish from Eqs (10.47) that both $V_x(t)$ and $X(t)$ will be *normal* random variables for all $t \geq t_0$: Since $N_x(t) = \mathcal{N}(0, 1)$, and since any sure variable a can be written $\mathcal{N}(a, 0)$, then for $t = t_0$ the right sides of both Eqs (10.47) will be linear combinations of normal random variables. Therefore, by Eq. (2.30), $V_x(t_0 + dt)$ and $X(t_0 + dt)$ will both be normal random variables. The same reasoning then establishes that $V_x(t_0 + 2dt)$ and $X(t_0 + 2dt)$ will both be normal random variables, and normality for all $t \geq t_0$ follows by induction. However, the two normal random variables $V_x(t)$ and $X(t)$ will *not* be statistically independent of each other, owing to the dependence of $X(t + dt)$ on $V_x(t)$ in Eq. (10.47b).

As discussed in Eqs (2.28) and (2.29), two statistically dependent normal random variables will be completely specified by their two means, their two variances, and their covariance. We will now derive from Eqs (10.47) explicit formulas for those five moments. In so generalizing the $f_e = 0$ formulas (8.31), we will discover that *only the formulas for the two means get changed*; the variance and covariance formulas remain the same as in the $f_e = 0$ case. For the reader in a hurry, the new formulas for the two means are exhibited in Eqs (10.53a) and (10.53c).

To compute $\langle V_x(t) \rangle$, we start by averaging Eq. (10.47a), using the fact that $\langle N_x(t) \rangle = 0$. After transposing the first term on the right, dividing through by dt, and then taking the limit $dt \to 0$, we obtain the ordinary differential equation

$$\frac{d \langle V_x(t) \rangle}{dt} = -\frac{1}{\tau} \langle V_x(t) \rangle + \frac{f_e}{m}. \tag{10.48}$$

The solution to this equation for the initial condition $\langle V_x(t_0) \rangle = v_{x0}$ is the function displayed in Eq. (10.53a), as can easily be verified by computing its derivative.

To compute $\mathrm{var}\{V_x(t)\} \equiv \langle V_x^2(t) \rangle - \langle V_x(t) \rangle^2$, we start by taking the derivative of both sides of this equation to get

$$\frac{d \, \mathrm{var}\{V_x(t)\}}{dt} \equiv \frac{d \langle V_x^2(t) \rangle}{dt} - 2 \langle V_x(t) \rangle \frac{d \langle V_x(t) \rangle}{dt}$$

$$= \frac{d \langle V_x^2(t) \rangle}{dt} - 2 \langle V_x(t) \rangle \left(-\frac{1}{\tau} \langle V_x(t) \rangle + \frac{f_e}{m} \right),$$

$$\frac{d \, \mathrm{var}\{V_x(t)\}}{dt} = \frac{d \langle V_x^2(t) \rangle}{dt} + \frac{2}{\tau} \langle V_x(t) \rangle^2 - \frac{2 f_e}{m} \langle V_x(t) \rangle,$$

where the second line has invoked Eq. (10.48). To evaluate the derivative term in the last equation, we square Eq. (10.47a), retaining only terms up to first order in dt:

$$V_x^2(t+dt) = V_x^2(t) - \frac{2}{\tau} V_x^2(t) dt + \frac{2 f_e}{m} V_x(t) dt$$

$$+ \frac{2\sqrt{2D}}{\tau} V_x(t) N_x(t) \sqrt{dt} + \frac{2D}{\tau^2} N_x^2(t) dt.$$

Averaging this equation, using $\langle V_x(t) N_x(t) \rangle = 0$ and $\langle N_x^2(t) \rangle = 1$, then transposing the first term on the right, dividing through by dt, and taking the limit $dt \to 0$, we get

$$\frac{d \langle V_x^2(t) \rangle}{dt} = -\frac{2}{\tau} \langle V_x^2(t) \rangle + \frac{2 f_e}{m} \langle V_x(t) \rangle + \frac{2D}{\tau^2}.$$

Substituting this expression into the above equation for the derivative of the variance, we obtain after algebraic simplification the ordinary differential equation

$$\frac{d \, \mathrm{var}\{V_x(t)\}}{dt} = -\frac{2}{\tau} \mathrm{var}\{V_x(t)\} + \frac{2D}{\tau^2}. \tag{10.49}$$

The solution to this equation for the initial condition var $\{V_x(t_0)\} = 0$ is the function displayed in Eq. (10.53b), as can easily be verified by computing its derivative.

To compute $\langle X(t) \rangle$, we start by taking the average of Eq. (10.47b). Transposing the first term on the right, dividing through by dt, and then taking the limit $dt \to 0$, we obtain

$$\frac{d \langle X(t) \rangle}{dt} = \langle V_x(t) \rangle. \tag{10.50}$$

Substituting on the right the solution to Eq. (10.48), which is displayed in Eq. (10.53a), we obtain an explicit differential equation for $\langle X(t) \rangle$. Its solution for the initial condition $\langle X(t_0) \rangle = x_0$ is the function displayed in Eq. (10.53c), as can be verified by computing its derivative.

To compute cov $\{V_x(t), X(t)\} \equiv \langle V_x(t)X(t) \rangle - \langle V_x(t) \rangle \langle X(t) \rangle$, we start by taking the derivative of both sides of this equation to get

$$\frac{d}{dt} \text{cov} \{V_x(t), X(t)\} \equiv \frac{d}{dt} \langle V_x(t)X(t) \rangle - \frac{d \langle V_x(t) \rangle}{dt} \langle X(t) \rangle - \langle V_x(t) \rangle \frac{d \langle X(t) \rangle}{dt}$$

$$= \frac{d}{dt} \langle V_x(t)X(t) \rangle - \left(-\frac{1}{\tau} \langle V_x(t) \rangle + \frac{f_e}{m} \right) \langle X(t) \rangle$$

$$- \langle V_x(t) \rangle \langle V_x(t) \rangle,$$

where the last step has invoked Eqs (10.48) and (10.50). Collecting terms, this is

$$\frac{d}{dt} \text{cov} \{V_x(t), X(t)\} = \frac{d}{dt} \langle V_x(t)X(t) \rangle + \frac{1}{\tau} \langle V_x(t) \rangle \langle X(t) \rangle - \frac{f_e}{m} \langle X(t) \rangle - \langle V_x(t) \rangle^2.$$

To evaluate the derivative term on the right, we start by taking the product of Eqs (10.47a) and (10.47b). That gives, to first order in dt,

$$V_x(t + dt)X(t + dt) = V_x(t)X(t) - \frac{1}{\tau} V_x(t)X(t)dt + \frac{f_e}{m} X(t)dt$$

$$+ \frac{\sqrt{2D}}{\tau} N_x(t)X(t)\sqrt{dt} + V_x^2(t)dt.$$

Now we average this equation, remembering that $N_x(t)$ has mean 0 and is statistically independent of $X(t)$. Then we transpose the first term on the right, divide through by dt, and take the limit $dt \to 0$; that gives

$$\frac{d}{dt} \langle V_x(t)X(t) \rangle = -\frac{1}{\tau} \langle V_x(t)X(t) \rangle + \frac{f_e}{m} \langle X(t) \rangle + \langle V_x^2(t) \rangle.$$

Substituting this expression into the above equation for the derivative of the covariance, we find after collecting terms that

$$\frac{d}{dt} \text{cov} \{V_x(t), X(t)\} = -\frac{1}{\tau} \text{cov} \{V_x(t), X(t)\} + \text{var} \{V_x(t)\}. \tag{10.51}$$

Into this equation we substitute, for the last term on the right, the explicit solution to Eq. (10.49), which is displayed in Eq. (10.53b). That gives an explicit differential equation for $\text{cov}\{V_x(t), X(t)\}$ whose solution, for the initial condition $\text{cov}\{V_x(t_0), X(t_0)\} = 0$, is the function displayed in Eq. (10.53e), as can be verified by computing its derivative.

Finally, to compute $\text{var}\{X(t)\} \equiv \langle X^2(t) \rangle - \langle X(t) \rangle^2$, we start by taking the derivative of both sides of this equation to get

$$\frac{d \, \text{var}\{X(t)\}}{dt} \equiv \frac{d\langle X^2(t) \rangle}{dt} - 2 \langle X(t) \rangle \frac{d\langle X(t) \rangle}{dt}$$

$$= \frac{d\langle X^2(t) \rangle}{dt} - 2 \langle X(t) \rangle \langle V_x(t) \rangle,$$

where the last step invokes Eq. (10.50). To evaluate the derivative term on the right, we take the square of Eq. (10.47b), retaining as usual only terms up to first order in dt:

$$X^2(t + dt) = X^2(t) + 2V_x(t)X(t)dt.$$

Averaging this equation, transposing the first term on the right, dividing through by dt, and taking the limit $dt \to 0$, we get

$$\frac{d\langle X^2(t) \rangle}{dt} = 2 \langle V_x(t)X(t) \rangle.$$

Substituting this expression into the above equation for the derivative of the variance, we obtain

$$\frac{d}{dt} \text{var}\{X(t)\} = 2\text{cov}\{V_x(t), X(t)\}. \tag{10.52}$$

Substituting on the right side of this equation the solution to Eq. (10.51), which is displayed in Eq. (10.53e), we obtain an explicit differential equation for $\text{var}\{X(t)\}$. Its solution for the initial condition $\text{var}\{X(t_0)\} = 0$ is the function displayed in Eq. (10.53d), as can be verified by computing its derivative.

The results of the foregoing calculations are collected below:

$$\langle V_x(t) \rangle = v_{x0}e^{-(t-t_0)/\tau} + \frac{f_e}{\gamma}\left(1 - e^{-(t-t_0)/\tau}\right), \tag{10.53a}$$

$$\text{var}\{V_x(t)\} = \frac{D}{\tau}\left(1 - e^{-2(t-t_0)/\tau}\right), \tag{10.53b}$$

$$\langle X(t) \rangle = x_0 + \tau\left(v_{x0} - \frac{f_e}{\gamma}\right)\left(1 - e^{-(t-t_0)/\tau}\right) + \frac{f_e}{\gamma}(t - t_0), \tag{10.53c}$$

$$\text{var}\{X(t)\} = 2D\left[(t - t_0) - 2\tau\left(1 - e^{-(t-t_0)/\tau}\right) + \tfrac{1}{2}\tau\left(1 - e^{-2(t-t_0)/\tau}\right)\right], \tag{10.53d}$$

$$\text{cov}\{V_x(t), X(t)\} = D\left(1 - e^{-(t-t_0)/\tau}\right)^2. \tag{10.53e}$$

Comparing these results with the $f_e = 0$ results (8.31), and taking account of the Einstein relation $D\gamma = k_B T$ and the definition $\tau \equiv m/\gamma$, we see that the formulas for the two variances and the covariance are exactly the same. The formulas for the two means, (10.53a) and (10.53c), differ from their respective $f_e = 0$ counterparts (8.31a) and (8.31c); as expected, those differences disappear when $f_e = 0$.

To summarize, we have proved that $V_x(t)$ and $X(t)$ for a solute molecule in a constant external force field $F_e(x) = f_e$ are *normal* random variables,

$$V_x(t) = \mathcal{N}\left(\langle V_x(t)\rangle, \mathrm{var}\{V_x(t)\}\right), \tag{10.54a}$$

$$X(t) = \mathcal{N}\left(\langle X(t)\rangle, \mathrm{var}\{X(t)\}\right), \tag{10.54b}$$

whose respective means and variances are as given in Eqs (10.53). But these two normal random variables are not statistically independent, because their covariance (10.53e) is not identically zero.

Equations (10.53a) and (10.53b) imply that $V_x(t)$ relaxes exponentially, in a time of order τ, from its initial sure value v_{x0} to the steady-state random variable

$$V_x(\infty) = \mathcal{N}\left(\frac{f_e}{\gamma}, \frac{k_B T}{m}\right). \tag{10.55}$$

Notice that $V_x(\infty)$ has the same "thermal variance" $k_B T/m$ as in the force-free case; however, it has a non-zero mean, $\langle V_x(\infty)\rangle = f_e/\gamma$, which is just the speed that gives a balance between the average drag force $\gamma \langle V_x(\infty)\rangle$ and the external force f_e. Equations (10.53c) and (10.53d) imply that the corresponding large-t behavior of $X(t)$ is

$$X(t \to \infty) \approx \mathcal{N}\left(x_0 + v_{x0}\tau + \frac{f_e}{\gamma}(t - t_0), \, 2D(t - t_0)\right). \tag{10.56}$$

This shows that the asymptotic variance in the position increases in the same linear way with time as in the $f_e = 0$ case. But now the asymptotic mean increases with a constant "velocity" f_e/γ.

Since corr $\{V_x(t), X(t)\}$ is by Eq. (2.23) the ratio of cov $\{V_x(t), X(t)\}$ to $\sqrt{\mathrm{var}\{V_x(t)\} \cdot \mathrm{var}\{X(t)\}}$, neither of which depends on f_e, then it is the same as in the $f_e = 0$ case; see Eqs (8.32) and (8.33). Also, it is easy to show by using the argument in Section 8.6 for the Langevin equation (10.45) that the two-time auto-correlations corr $\{V_x(t_1), V_x(t_2)\}$ and corr $\{X(t_1), X(t_2)\}$ for $t_0 \leq t_1 \leq t_2$ are likewise independent of f_e; see Eqs (8.37) and (8.42). Finally, we note that exact numerical simulations of a solute molecule in a constant external force field can be carried out by using the bivariate normal simulation formulas (7.49) with the means, variances, and covariance as given in Eqs (10.53).

This concludes our three-chapter presentation of Langevin's theory of diffusion. In our next and final chapter, we will return to Einstein's theory of diffusion, which from the perspective of the last three chapters is what Langevin's theory simplifies to when we view diffusion on a time-scale that is large compared to τ. In that final chapter, we will adopt a perspective on the motion of a solute molecule that is intriguingly different from the perspective we have taken thus far.

Notes to Chapter 10

[1]Marian Smoluchowski (1872–1917) was a Polish physicist who, in his too-brief lifetime (he died at age 45), accomplished a great deal. Equations (10.5) and (10.6), which bear his name, were first published by him in a paper with the (translated) title "On Brownian molecular motion under the influence of external forces and their relationship to the generalized diffusion equation", *Annalen der Physik* **353**:1103–1112 (1916). Another equation in physics that bears his name is the Smoluchowski coagulation equation, which describes the time-evolving distribution of particle sizes in a coagulating aerosol or colloid. Born in Austria, Smoluchowski was raised and educated in Vienna, where his father was a high-ranking government official. His well-educated and artistic mother saw to it that he received the best education Vienna had to offer; largely owing to her, he became a capable pianist and painter. He eventually developed strong passions for physics, experimental as well as theoretical, and serious mountain climbing. His undergraduate and graduate studies were taken at the University of Vienna, and his Ph.D. was awarded there in 1895 with honors that won him a two-year scholarship for post-doctoral work abroad. He went first to Paris, then to Glasgow (where he worked with Lord Kelvin), and finally to Berlin. Upon returning to the University of Vienna, he advanced through a succession of increasingly prestigious positions that took him to universities in Lwow (1899–1913) and Krakow (1913–1917). But just as he was about to assume the presidency of the University of Krakow (now the Jagiellonian University) in 1917, he died in a dysentery epidemic. An interesting account of Smoluchowski's life and some of his scientific accomplishments can be found in the article "Marian Smoluchowski and the theories of probabilities in physics" by Stanislaw Ulam in the *American Journal of Physics* **25**:475–481 (1956).

In 1906, Smoluchowski published a paper on Brownian motion that reported results he had obtained independently of Einstein using different reasoning: "On the kinetic theory of Brownian molecular motion and of suspensions", *Annalen der Physik* **326**:756–780 (1906). Although this work by Smoluchowski was published after Einstein's 1905 papers, and its result for the mean-square displacement contained an erroneous overall factor of (64/27), it was regarded by most people at the time, including Einstein, as a solid and welcomed contribution. Smoluchowski had delayed publishing his work until he could perform some experimental tests of it—something which Einstein did not bother to do. That prompted Abraham Pais to aver, in his book *Subtle Is the Lord: The Science and the Life of Albert Einstein* (Oxford University Press, 1982), that "if Smoluchowski had been only an outstanding theoretical physicist, and not a fine experimentalist as well, he would probably have been the first to publish a quantitative theory of Brownian motion". Pais's book also describes several independent but parallel researches of Smoluchowski and Einstein, in particular Smoluchowski's explanation in 1908 of the phenomenon of critical opalescence in a second-order phase transition. Pais notes that six surviving letters between Einstein and Smoluchowski "all show cordiality and great mutual respect between the two men". Indeed, Marian Smoluchowski seems to have been not only respected and admired, but genuinely liked by everyone who knew him.

[2]Hendrik A. Kramers (1894–1952) was a Dutch physicist who made important contributions to both statistical mechanics and quantum mechanics. Equation (10.24), or more accurately Eq. (10.25) expressed in slightly different variables, was derived in his paper "Brownian motion in a field of force and the diffusion model of chemical reactions", *Physica* **7**:284–304 (1940). The far-reaching influence of that paper can be seen in the article "Reaction-rate theory: fifty years after Kramers", *Reviews of Modern Physics* **62**:251–341 (1990), by P. Hänggi, P. Talkner, and M. Borkovec. An interesting personal account of the life and work of Kramers can be found in the article "Kramers's contributions to statistical mechanics", *Physics Today*, Sep. 1988, pp. 26–33, by Max Dresden.

[3]The Dirac delta function $\delta(x)$ can be defined by the statement that, for any function $f(x)$,

$$\int_{-\infty}^{\infty} f(x)\delta(x)dx = f(0).$$

The derivative $\delta'(x)$ is similarly defined by the statement that, for any *differentiable* function $f(x)$,

$$\int_{-\infty}^{\infty} f(x)\delta'(x)dx = -f'(0).$$

The rationale for this definition can be seen by evaluating the integral on the left by parts:

$$\int_{-\infty}^{\infty} f(x)\delta'(x)dx = \int_{x=-\infty}^{\infty} f(x)d[\delta(x)]$$

$$= f(x)\delta(x)|_{x=-\infty}^{\infty} - \int_{x=-\infty}^{\infty} \delta(x)d[f(x)]$$

$$= 0 - \int_{-\infty}^{\infty} \delta(x)f'(x)dx$$

$$= -f'(0),$$

where the last step uses the definition of $\delta(x)$.

[4]To understand why attempts were made to justify the Kramers equation on the basis of the Liouville equation, and why in the end those attempts were unsuccessful, we need to recall a few basic facts about the Liouville equation.

For a mechanical system consisting of a single particle of mass m moving along the x-axis in a conservative external force field $F_e(x)$, the Liouville equation describes the motion of an ensemble of "phase points" $(x_t, p_t \equiv mv_t)$ of an infinitely large number of such systems in the two-dimensional position–momentum "phase plane". These phase points move about independently of each other, each according to Newton's equations (10.37); they differ from one another only in their initial location in the phase plane. If $\rho(x, p; t)$ denotes the density of phase points at the point (x, p) at time t, then it can be proved that the Newtonian motion of the phase points is such that ρ obeys the *Liouville equation*:

$$\frac{\partial \rho(x,p;t)}{\partial t} + \frac{\partial \rho(x,p;t)}{\partial x}\frac{dx}{dt} + \frac{\partial \rho(x,p;t)}{\partial p}\frac{dp}{dt} = 0.$$

It can further be shown that this equation implies that the phase points move about on the phase plane in the manner of an incompressible fluid. Since for our system we have $dx/dt = v$ and $dp/dt = F_e(x)$, then the Liouville equation can also be written

$$\frac{\partial \rho(x,p;t)}{\partial t} + v\frac{\partial \rho(x,p;t)}{\partial x} + F_e(x)\frac{\partial \rho(x,p;t)}{\partial p} = 0.$$

Recalling that $p \equiv mv$, we see that this equation for ρ is identical to the $\gamma = 0$ Kramers equation (10.34) for $P_{V_x,X}$. It is therefore tempting to infer a physically meaningful connection between those two equations, and to speculate on how the right side of the *general* ($\gamma > 0$) Kramers equation (10.25) might arise as a consequence of changes in the incompressible flow of the phase points that are induced by interactions with the solvent molecules.

But closer inspection reveals that the connection between the Kramers equation and the Liouville equation is superficial: The Liouville equation describes *determinis-tically* moving phase points, and the randomness in its density function ρ arises solely from the randomness in the *initial seeding* of those phase points in the position–momentum plane. But in the case of the Kramers equation, the initial seeding of the phase points is *deterministic*: every system starts at the same point (x_0, mv_{x0}) in the phase plane. But thereafter, every system phase point moves *stochastically*, as each solute molecule is buffeted about by the surrounding solvent molecules whose motions we make no attempt to track. Given this profound difference in the underlying dynamics that drive ρ in the Liouville equation and $P_{V_x,X}$ in the Kramers equation, there is no rational basis for supposing that generalizations in either should be applicable to the other.

11
The first-passage time approach

Most of our work in earlier chapters has focused on the problem of computing where a diffusing solute molecule will be at a specified future time. In this concluding chapter, we will take a look at the subtly different problem of computing *when* a diffusing solute molecule will *first* reach a specified location. This is known as the first-passage time problem. Within the framework of the Einstein model of diffusion (the stochastically reinterpreted classical diffusion equation), the first-passage time problem has been solved analytically in only a few relatively simple cases. In the context of the Langevin model the problem is even more challenging, and apparently no exact results are known at present. In this chapter we will consider the first-passage time problem only within the framework of the Einstein model, and we will examine only its two simplest forms. Our aim will be to illuminate the concepts and issues involved, and to demonstrate some solution strategies. In the course of doing that, we will uncover some results that will deepen our understanding of the hypothesis (5.3) that underlies the discrete-stochastic model of diffusion.

11.1 The basic first-passage time problem

There are two forms of the *basic* first-passage time problem. In their simplest versions, both concern a solute molecule undergoing Einstein diffusion on the x-axis with diffusion coefficient D, and both assume that at time $t = 0$ the molecule is at some specified point x_0 inside some specified interval $[a, b]$. The first form of the problem asks what can be said about the random variable

$$T^{(a,r)}(x_0; a, b) \equiv \text{the time at which the solute molecule, if released at time}$$
$$0 \text{ from } x_0 \in [a, b] \text{ with } x = b \text{ a } \textit{reflecting} \text{ boundary, will} \quad (11.1)$$
$$\textit{first} \text{ reach } x = a.$$

The superscript (a,r) means that in this case the lower boundary $x = a$ plays the role of an *absorbing* boundary, since the molecule's arrival there terminates its trajectory so far as the definition (11.1) is concerned, while the upper boundary $x = b$ is a *reflecting* boundary. But note that the $x = a$ boundary might not actually be absorbing, since the fate of the solute molecule *after* it arrives at $x = a$ does not matter to the first-passage time defined above. This problem can be thought of as a one-dimensional version of the problem in which a solute molecule lies between a stationary object at $x = a$ and a reflecting wall at $x = b$, and we want to know when the solute molecule will first make contact with the object at $x = a$.

The second form of the first-passage problem asks what can be said about the random variable

$$T^{(\text{a,a})}(x_0; a, b) \equiv \text{the time at which the solute molecule, if released at time}$$
$$\text{0 from } x_0 \in [a, b], \text{ will } \textit{first} \text{ reach } \textit{either } x = a \text{ or } x = b. \quad (11.2)$$

Here the superscript (a,a) means that the upper and lower boundaries of the interval $[a, b]$ *both* act like *absorbing* boundaries so far as the definition (11.2) is concerned, although in actuality each might be a reflecting boundary or even no boundary at all. This problem is a one-dimensional version of the situation in which a solute molecule lies inside a cell whose walls are at $x = a$ and $x = b$, and we want to know when the solute molecule will first come into contact with *either* wall.

Ideally, we would like to find for each of these random variables $T(x_0; a, b)$ its probability density function (PDF), since that would tell us everything knowable about the random variable. In particular, if the integral of the PDF, namely the cumulative distribution function (CDF), could be found and its inverse computed, we could use the standard inversion method of Section 2.8 to numerically generate exact sample values of $T(x_0; a, b)$. That could be very useful for simulation purposes. At the very least though, we would like to compute the mean and variance of $T(x_0; a, b)$.

11.2   Limitations of the usual simulation approaches

Surprisingly, the basic Einstein simulation formula (4.11a) is of limited help in quantifying the random variables $T(x_0; a, b)$ in Eqs (11.1) and (11.2). It might be thought that we could use that formula to simulate the motion of the solute molecule until the absorption event occurs, and then simply note the time at which that happens. To see why that is not as easy as it sounds, consider doing this for the "a,r" problem in Eq. (11.1), where a is an absorbing boundary and b is a reflecting boundary. In a typical time step Δt that starts from some point $x_t \in (a, b)$, the updating formula (4.11a) will take the solute molecule to some new point $x_{t+\Delta t} = x_t + \sqrt{2D\Delta t} n_x$, where n_x is a sample of $\mathcal{N}(0, 1)$. In the special case $b \to \infty$, we could use results we derived in Section 4.1 to argue as follows: If $x_{t+\Delta t} \leq a$, then the molecule can be considered to have been absorbed at a in the time interval $[t, t + \Delta t)$. Alternatively, if $x_{t+\Delta t} > a$, then the molecule can be considered to be at $x_{t+\Delta t}$ at time $t + \Delta t$ with probability [cf. Eq. (4.9) with $x_\text{a} \to a$, $x_t^\text{tent} \to x_{t+\Delta t}$, and $x_0 \to x_t$]

$$1 - \exp\left(-\frac{(x_{t+\Delta t} - a)(x_t - a)}{D\,\Delta t}\right);$$

otherwise, the molecule will have been absorbed at a in the time interval $[t, t + \Delta t)$. Notice that this result for $b \to \infty$, although exact, gives us the time of absorption, or first passage to $x = a$, only to within an uncertainty of Δt. But if b is *finite*, these results will *not* be valid. To appreciate that fact, consider the extreme case in which the reflecting boundary b is only slightly larger than x_t. With the molecule thus denied the freedom to escape to the infinite region above x_t, its probability of being absorbed at a in time $[t, t + \Delta t)$ will surely be much larger than in the $b \to \infty$ case.

The basic difficulty here is that, although the simple Einstein updating formula $x_{t+\Delta t} = x_t + \sqrt{2D\Delta t}\,n_x$ will be correct *if* the molecule did not encounter either boundary point a or b in the time interval $[t, t + \Delta t)$, we have no generally exact simple procedure for deciding whether or not that is so when more than one boundary point is present. It seems likely that errors incurred in using the single-boundary point strategies anyway could be mitigated by taking Δt "small" when the molecule gets "close" to a boundary point—a strategy that would also mitigate the overriding problem that this method for determining the first-passage time can never give a result whose uncertainty is less than Δt. But the strategy of taking Δt "small" to ensure that a reflection or absorption does *not* happen in time $[t, t + \Delta t)$ seems rather counterproductive given that the goal here is to carry out the simulation until some such event *does* happen. Also, we cannot take Δt *arbitrarily* small, since that would violate the condition $\Delta t \gg \tau$, which we showed in Section 9.2 is required in order for the Einstein updating formula to be *physically* valid.

Surprisingly, some of these shortcomings of the Einstein simulation formula can be avoided by using the discrete-stochastic simulation approach, as embodied in the single-molecule simulation algorithm of Section 6.4. In that approach, we would begin by covering the x-axis interval $[a, b]$ with a suitably large number M of contiguous cells of equal length l in such a way that x_0 is at the center of one of the cells, and any *absorbing* boundary is at the center of another cell, and any *reflecting* boundary is at a boundary between two cells. Then, with $\kappa_l = D/l^2$, we would start the simulation at time $t_0 = 0$ with the solute molecule somewhere inside the cell containing x_0; note that we cannot be more specific about the molecule's location in that cell, because in the discrete-stochastic model solute molecules within a given cell are assumed to be distributed in a randomly uniform way. We would then run the simulation until the molecule *either* (i) first enters the cell containing a for the "a,r" problem (11.1), *or* (ii) first enters either the cell containing a or the cell containing b for the "a,a" problem (11.2). Since in this simulation procedure the time steps are chosen automatically by the simulation algorithm [see Eq. (6.15)], there will be no ambiguity as to the time at which the first-passage event occurs. A single run would suffice to generate a sample value of $T(x_0; a, b)$, and an ensemble of such runs would provide data from which a frequency histogram could be constructed that would numerically estimate the PDF of $T(x_0; a, b)$. But of course, this procedure entails errors arising from the finiteness of the cell length l. Subject to the requirement $l \gg \sqrt{D\tau}$ discussed in Section 9.4, the procedure should become more and more accurate as l is taken smaller and smaller, and hence as M is taken larger and larger. But unfortunately, the computation time to the absorption event will be a rapidly increasing function of M. Thus, although the discrete-stochastic simulation procedure might be useful for testing purposes—and we shall use it in that way later in this chapter—it is in the end a rather cumbersome way of quantifying first-passage times.

11.3   A direct analytical approach

A direct analytical approach to quantifying the first-passage time $T(x_0; a, b)$ in the context of Einstein's model of diffusion would focus on solving Einstein's Fokker–Planck equation (3.9) for $t_0 = 0$,

$$\frac{\partial P_X(x,t|x_0,0)}{\partial t} = D\frac{\partial^2 P_X(x,t|x_0,0)}{\partial x^2}, \tag{11.3}$$

subject to the initial condition

$$P_X(x,0|x_0,0) = \delta(x-x_0) \quad (x_0 \in [a,b]), \tag{11.4}$$

and to boundary conditions that are appropriate for the respective definitions (11.1) and (11.2). It is in imposing those boundary conditions that we part company with the simulation formula (4.11a), which was obtained by solving Eq. (11.3) assuming unrestricted access to the entire x-axis. In the case of $T^{(a,r)}(x_0; a, b)$ the boundary conditions should reflect the fact that a is an absorbing boundary and b is a reflecting boundary, and in the case of $T^{(a,a)}(x_0; a, b)$ the boundary conditions should reflect the fact that a and b are both absorbing boundaries. The boundary conditions for Einstein's diffusion equation (11.3) at perfectly reflecting and absorbing boundaries are given in Eqs (3.26) and (3.27), respectively. Thus, for $T^{(a,r)}(x_0; a, b)$ we have the boundary conditions

$$P_X^{(a,r)}(a,t|x_0,0) = 0 \quad (\forall t \geq 0), \tag{11.5a}$$

$$\left.\frac{\partial P_X^{(a,r)}(x,t|x_0,0)}{\partial x}\right|_{x=b} = 0 \quad (\forall t \geq 0). \tag{11.5b}$$

And for $T^{(a,a)}(x_0; a, b)$ we have the boundary conditions

$$P_X^{(a,a)}(a,t|x_0,0) = 0 \quad (\forall t \geq 0), \tag{11.6a}$$

$$P_X^{(a,a)}(b,t|x_0,0) = 0 \quad (\forall t \geq 0). \tag{11.6b}$$

But even if we could solve the partial differential equation (11.3) subject to the initial condition (11.4) and the boundary conditions (11.5) or (11.6), that would *not* immediately give us the information about the first-passage times (11.1) and (11.2) that we are after, because the PDF of $X(t)$ is not the PDF of the first-passage time $T(x_0; a, b)$. To address that concern, we consider for each of the solutions $P_X^{(a,r)}(x,t|x_0,0)$ and $P_X^{(a,a)}(x,t|x_0,0)$ to Eq. (11.3), the corresponding function

$$G(x_0,t;a,b) \equiv \int_a^b P_X(x,t|x_0,0)\,dx. \tag{11.7}$$

By the addition law of probability, $G(x_0,t;a,b)$ is the probability that the solute molecule will be *somewhere* inside the interval $[a,b]$ at time t, given that it was at point x_0 at time 0. But to say that the solute molecule is somewhere inside the interval $[a,b]$ at time t is to say that the molecule has *not yet been absorbed by time t*, and hence that the first-passage time $T(x_0; a, b)$ is *greater* than t. Thus, the probability that the molecule is *not* inside $[a,b]$ at time t must be the probability that the first-passage time $T(x_0; a, b)$ is *less* than t:

$$1 - G(x_0,t;a,b) = \text{Prob}\,\{T(x_0;a,b) \leq t\}. \tag{11.8}$$

This last equation implies, by the definition of the cumulative distribution function (CDF) in Eq. (2.34), that $1 - G(x_0, t; a, b)$, considered as a function of t, is the CDF of the first passage time $T(x_0; a, b)$.

As discussed in Section 2.8, the most direct way to generate a sample value of any random variable Y is to set its CDF $F_Y(y)$ equal to a sample value u of the unit-interval uniform random variable $\mathcal{U}(0, 1)$, and then solve for y. Doing that for the random variable $T(x_0; a, b)$, whose CDF is $1 - G(x_0, t; a, b)$, we conclude that the value of t found by solving the equation

$$G(x_0, t; a, b) = 1 - u \tag{11.9}$$

can be regarded as *an exact sample value* of the first-passage time $T(x_0; a, b)$. Thus we have here, at least in principle, a way of obtaining samples of the first-passage times (11.1) and (11.2) that avoids all the difficulties connected with the simulation approaches described in Section 11.2.

Since the derivative of any random variable's CDF is its PDF, then the PDF of $T(x_0; a, b)$ is the derivative of $1 - G(x_0, t; a, b)$ with respect to t:

$$P_T(t; x_0, a, b) = -\frac{\partial G(x_0, t; a, b)}{\partial t}. \tag{11.10}$$

In general, the nth moment $T_n(x_0; a, b) \equiv \int_0^\infty t^n P_T(t; x_0, a, b)\, dt$ of $T(x_0; a, b)$ may or may not exist for $n \geq 1$. If it does exist, Eq. (11.10) tells us that it can be computed as

$$T_n(x_0; a, b) = \int_0^\infty t^n \left(-\frac{\partial G(x_0, t; a, b)}{\partial t} \right) dt \quad (n = 0, 1, \ldots). \tag{11.11}$$

If the first two moments $T_1(x_0; a, b)$ and $T_2(x_0; a, b)$ exist, then their computation from Eq. (11.11) will give us the mean and variance of $T(x_0; a, b)$:

$$\langle T(x_0; a, b) \rangle = T_1(x_0; a, b), \tag{11.12a}$$

$$\text{var}\left\{ T(x_0; a, b) \right\} = T_2(x_0; a, b) - T_1^2(x_0; a, b). \tag{11.12b}$$

Thus we see that this direct analytical approach allows us to compute, from the probability function $G(x_0, t; a, b)$ defined in Eq. (11.7), every existing moment of the first-passage time $T(x_0; a, b)$. In the following sections, we will see that computing with $G(x_0, t; a, b)$ turns out to be simpler than one might expect.

11.4   A little help from the backward Fokker–Planck equation

The foregoing analysis makes it clear that to quantify the first-passage time $T(x_0; a, b)$ analytically, we need to compute the function $G(x_0, t; a, b)$. That would appear to be a two-step task: First we have to solve Eq. (11.3) for $P_X(x, t|x_0, 0)$, subject to the initial condition (11.4) and the boundary conditions (11.5) or (11.6). Then we have to integrate that function over x to get $G(x_0, t; a, b)$ in Eq. (11.7). We will show now that this two-step task can be simplified to a one-step task by using, instead of the forward Fokker–Planck equation (11.3), the corresponding *backward* Fokker–Planck equation (see Section 7.13). Since in our problem $A(x_0) \equiv 0$, $D(x_0) \equiv 2D$, and $t_0 = 0$, then by Eq. (7.56) the backward Fokker–Planck equation corresponding to the forward Fokker–Planck equation (11.3) is

$$\frac{\partial}{\partial t} P_X(x, t | x_0, 0) = D \frac{\partial^2 P_X(x, t | x_0, 0)}{\partial x_0^2} \quad (t \geq 0). \tag{11.13}$$

This equation differs from Eq. (11.3) in that the second derivative on the right side is taken with respect to x_0 instead of x. That crucial difference allows us to straightforwardly integrate Eq. (11.13) over x from a to b to obtain, by virtue of the definition (11.7),

$$\frac{\partial}{\partial t} G(x_0, t; a, b) = D \frac{\partial^2 G(x_0, t; a, b)}{\partial x_0^2} \quad (t \geq 0). \tag{11.14}$$

Equation (11.14) is evidently a partial differential equation for $G(x_0, t; a, b)$ of the same "diffusion type" as the partial differential equation (11.3) for $P_X(x, t | x_0, 0)$. Note that the function $G(x_0, t; a, b)$ in Eq. (11.14) is *not* a PDF; it is a *probability*. Nevertheless, if we can solve Eq. (11.14) subject to appropriate initial and boundary conditions, we can obtain the function $G(x_0, t; a, b)$ in a single step. But what are those appropriate initial and boundary conditions?

The initial condition for Eq. (11.14) can easily be inferred by substituting Eq. (11.4) into the definition (11.7), and then carrying out the integration over x using Eq. (2.7). That gives

$$G(x_0, t = 0; a, b) = 1. \tag{11.15}$$

The physical import of this relation is clear: At time $t = 0$, the diffusing molecule is with probability 1 located somewhere inside the interval $[a, b]$.

Deducing the boundary conditions for Eq. (11.14) requires a little more effort. Suppose that at time $t = 0$ we place the solute molecule at the absorbing boundary a; i.e., suppose we take $x_0 = a$. Then the solute molecule will be immediately removed from the x-axis. Therefore, the probability of finding the molecule inside the interval $[a, b]$ at any later time will be zero:

$$G(x_0 = a, t; a, b) = 0 \quad (\forall t \geq 0; \text{ if } a \text{ is absorbing}).$$

This gives us the lower boundary condition for the first-passage time $T^{(a,r)}(x_0; a, b)$, and also the lower and upper boundary conditions for $T^{(a,a)}(x_0; a, b)$. To infer the reflecting boundary condition at b for $T^{(a,r)}(x_0; a, b)$, we will compare the values of $1 - G(x_0, t; a, b)$, the probability that the solute molecule initially at x_0 *will* have been absorbed by time t, for the two x_0-values $x_0 = b$ and $x_0 = b - dx_0$, where dx_0 is a positive infinitesimal: For $x_0 = b$, $1 - G(x_0, t; a, b)$ gives the probability that the molecule will travel the net distance $(b - a)$ from b to a before time t. For $x_0 = b - dx_0$, the molecule will have *two* path choices to get to a before time t: either an *initially* left-going route of net distance $(b - a - dx_0)$; or an *initially* right-going route of net distance $(b - a + dx_0)$ which involves an immediate reflection off the wall at *b*. Since the net probability flow in the immediate vicinity of a perfectly reflecting wall is zero [see Eqs (3.25) and (3.26)], then these two routes from $b - dx_0$ to a, one initially left-going and the other initially right-going, will be *equally likely* for the solute molecule. Therefore, the average net distance that needs to be covered by the molecule at $x_0 = b - dx_0$ to reach point a will be the *equally weighted average* of $(b - a - dx_0)$ and $(b - a + dx_0)$. But that average is just $(b - a)$, which is the same as if the molecule had started at $x_0 = b$. We conclude that the two absorption probabilities $1 - G(x_0 = b, t; a, b)$ and $1 - G(x_0 = b - dx_0, t; a, b)$ should be *equal*, at least to first order in dx_0. The consequent equality of $G(x_0 = b, t; a, b)$ and $G(x_0 = b - dx_0, t; a, b)$ to first order in dx_0, together with the definition of the derivative, implies that

$$\left. \frac{\partial G(x_0, t; a, b)}{\partial x_0} \right|_{x_0=b} = 0 \quad (\forall t \geq 0; \text{ if } b \text{ is reflecting}).$$

On the basis of these observations, we conclude that for the first-passage time $T^{(\mathrm{a,r})}(x_0; a, b)$, the function $G^{(\mathrm{a,r})}(x_0, t; a, b)$ will be the solution to Eq. (11.14) that satisfies the initial condition (11.15) and the boundary conditions

$$G^{(\mathrm{a,r})}(x_0 = a, t; a, b) = 0, \tag{11.16a}$$

$$\left. \frac{\partial G^{(\mathrm{a,r})}(x_0, t; a, b)}{\partial x_0} \right|_{x_0=b} = 0. \tag{11.16b}$$

And for the first-passage time $T^{(\mathrm{a,a})}(x_0; a, b)$, the function $G^{(\mathrm{a,a})}(x_0, t; a, b)$ will be the solution to Eq. (11.14) that satisfies the initial condition (11.15) and the boundary conditions

$$G^{(\mathrm{a,a})}(x_0 = a, t; a, b) = 0, \tag{11.17a}$$

$$G^{(\mathrm{a,a})}(x_0 = b, t; a, b) = 0. \tag{11.17b}$$

Using the backward Fokker–Planck equation thus allows us to compute $G(x_0, t; a, b)$ in a single step. In the next section, we will see how $G(x_0, t; a, b)$ allows us to develop equations for efficiently computing the moments of the first-passage time.

11.5   Formulas for the moments of the first-passage time

Another useful consequence of the explicit partial differential equation (11.14) for $G(x_0, t; a, b)$ is a closed set of ordinary differential equations for the *moments* $T_n(x_0; a, b)$ ($n = 0, 1, \ldots$) of the first-passage time $T(x_0; a, b)$. Of particular interest is the *first* moment $T_1(x_0; a, b)$, the *mean* first-passage time. As noted earlier, $T_n(x_0; a, b)$ for a particular value of n may or may not exist, but when it does it can be computed from Eq. (11.11). The form of Eq. (11.11) shows that the existence of $T_n(x_0; a, b)$ requires that the integral there be convergent, and that in turn requires the integrand in Eq. (11.11) to go to zero as $t \to \infty$ faster than t^{-1}:

$$t^n \left(-\frac{\partial G(x_0, t; a, b)}{\partial t} \right) \xrightarrow[t \to \infty]{} 0 \text{ faster than } t^{-1}.$$

This requires $\partial G / \partial t \to 0$ as $t \to \infty$ faster than t^{-n-1}, and hence that

$$G(x_0, t; a, b) \xrightarrow[t \to \infty]{} 0 \text{ faster than } t^{-n}. \tag{11.18}$$

This then is the condition for the *existence* of $T_n(x_0; a, b)$. Notice it implies that if $T_n(x_0; a, b)$ exists, then so do all the lower order moments, $T_{n-1}(x_0; a, b)$, $T_{n-2}(x_0; a, b)$, ..., $T_0(x_0; a, b)$; because if $G(x_0, t; a, b) \to 0$ as $t \to \infty$ faster than t^{-n}, then $G(x_0, t; a, b) \to 0$ faster than $t^{-(n-1)}$.

For $n = 0$, Eq. (11.18) is simply the requirement that

$$G(x_0, t = \infty; a, b) = 0, \tag{11.19}$$

or in words, the requirement that the solute molecule will for certain eventually be removed from the interval (a, b). Since for $n = 0$ Eq. (11.11) integrates to

$$T_0(x_0; a, b) = - [G(x_0, \infty; a, b) - G(x_0, 0; a, b)],$$

then substituting from Eqs (11.19) and (11.15) on the right side gives the result $T_0(x_0; a, b) = 1$. That merely tells us that $T(x_0; a, b)$ is an "ordinary" random variable, in that the area under its PDF is 1.

For the more interesting case $n \geq 1$, we start by assuming that $T_n(x_0; a, b)$ exists, so that condition (11.18) holds and Eq. (11.11) is meaningful. Integrating Eq. (11.11) once by parts then gives

$$T_n(x_0; a, b) = -t^n G(x_0, t; a, b)|_{t=0}^{t=\infty} + \int_0^\infty G(x_0, t; a, b) n t^{n-1} dt \quad (n = 1, 2, \ldots).$$

The integrated term on the right vanishes at the lower limit because of Eq. (11.15), and also at the upper limit because of Eq. (11.18). Therefore,

$$\int_0^\infty t^{n-1} G(x_0, t; a, b) dt = \frac{1}{n} T_n(x_0; a, b) \quad (n = 1, 2, \ldots). \tag{11.20}$$

We can now use Eqs (11.20) and (11.11) to convert the *partial* differential equation (11.14) for $G(x_0, t; a, b)$ into a set of *coupled ordinary* differential equations for the moments $T_n(x_0; a, b)$ of $T(x_0; a, b)$. To that end, we first multiply Eq. (11.14) through by t^{n-1}, and then integrate the result over all t:

$$\int_0^\infty \left(t^{n-1} \frac{\partial}{\partial t} G(x_0, t; a, b) \right) dt = D \frac{\partial^2}{\partial x_0^2} \int_0^\infty t^{n-1} G(x_0, t; a, b) dt \quad (n = 1, 2, \ldots).$$

Into this equation, we substitute on the *left* side from Eq. (11.11), and on the *right* side from Eq. (11.20):

$$-T_{n-1}(x_0; a, b) = D \frac{\partial^2}{\partial x_0^2} \left(\frac{1}{n} T_n(x_0; a, b) \right) \quad (n = 1, 2, \ldots).$$

Since the existence of $T_n(x_0; a, b)$ for any $n \geq 1$ implies that $T_0(x_0; a, b) = 1$, a simple rearrangement of this last equation gives us the set of coupled equations

$$\frac{d^2 T_1(x_0; a, b)}{dx_0^2} = -\frac{1}{D}, \tag{11.21a}$$

$$\frac{d^2 T_n(x_0; a, b)}{dx_0^2} = -\frac{n}{D} T_{n-1}(x_0; a, b) \quad (n = 2, 3, \ldots). \tag{11.21b}$$

Equations (11.21) can, at least in principle, be solved recursively for all the existing moments of the first-passage time $T(x_0; a, b)$. However, each of these second-order ordinary differential equations will require *two boundary conditions*. Those can be inferred from the boundary conditions on $G(x_0, t; a, b)$ in Eqs (11.16) and (11.17) for the specific first-passage times $T^{(a,r)}(x_0; a, b)$ and $T^{(a,a)}(x_0; a, b)$. Thus, to get the boundary conditions on the moments of $T^{(a,r)}(x_0; a, b)$, we first put $x_0 = a$ in Eq. (11.20) and then invoke the boundary condition (11.16a) to get

$$T_n^{(a,r)}(x_0 = a; a, b) = 0 \quad (n \geq 1). \tag{11.22a}$$

Next, we differentiate Eq. (11.20) with respect to x_0, then set $x_0 = b$, and invoke the boundary condition (11.16b) to get

$$\left. \frac{dT_n^{(a,r)}(x_0; a, b)}{dx_0} \right|_{x_0 = b} = 0 \quad (n \geq 1). \tag{11.22b}$$

In the same way, we find from the boundary conditions (11.17) on $G^{(a,a)}(x_0, t; a, b)$ the following boundary conditions on the moments of $T^{(a,a)}(x_0; a, b)$:

$$T_n^{(a,a)}(x_0 = a; a, b) = 0 \quad (n \geq 1), \tag{11.23a}$$

$$T_n^{(a,a)}(x_0 = b; a, b) = 0 \quad (n \geq 1). \tag{11.23b}$$

Solving Eqs (11.21) subject to the aforementioned boundary conditions will give us all the moments of the first-passage time that exist. In the next section, we will solve the first two moment equations to obtain explicit expressions for the mean and variance of the two specific first-passage times $T^{(a,r)}(x_0; a, b)$ and $T^{(a,a)}(x_0; a, b)$.

11.6 Explicit solutions for the mean and variance

The mean of the first-passage time (11.1) will be the solution $T_1^{(a,r)}(x_0; a, b)$ of Eq. (11.21a) that satisfies the two boundary conditions (11.22) for $n = 1$. It is clear that the solution of the differential equation (11.21a) will be a quadratic in x_0, since that is the only function whose second derivative is a non-zero constant. Therefore, we can take $T_1^{(a,r)}(x_0; a, b)$ to be the general quadratic form $\alpha x_0^2 + \beta x_0 + \gamma$, and then choose the three constants α, β, and γ so that, with a prime denoting differentiation with respect to x_0, the following three conditions are satisfied: $T_1^{(a,r)'}(a; a, b) = 0$, as required by boundary condition (11.22a); $T_1^{(a,r)'}(b; a, b) = 0$, as required by boundary condition (11.22b); and $T_1^{(a,r)''}(x_0; a, b) = -D^{-1}$, as required by the differential equation (11.21a). Solving these three equations simultaneously for α, β, and γ yields

$$T_1^{(a,r)}(x_0; a, b) = \frac{(x_0 - a)}{2D}[2(b - a) - (x_0 - a)], \tag{11.24}$$

as can easily be verified. This is the *average time* for the solute molecule to *first* reach $x = a$ from $x_0 \in [a, b]$, given that $x = b$ is a reflecting wall.

Likewise, the mean of the first-passage time (11.2) will be the solution $T_1^{(a,a)}(x_0; a, b)$ of Eq. (11.21a) that satisfies the two boundary conditions (11.23) for $n = 1$. Once again, any solution to Eq. (11.21a) will be a quadratic in x_0; thus, we can take $T_1^{(a,a)}(x_0; a, b) = \alpha x_0^2 + \beta x_0 + \gamma$, and then choose the three constants α, β, and γ so that the following three conditions are satisfied: $T_1^{(a,a)}(a; a, b) = 0$, as required by Eq. (11.23a); $T_1^{(a,a)}(b; a, b) = 0$, as required by Eq. (11.23b); and $T_1^{(a,a)''}(x_0; a, b) = -D^{-1}$, as required by Eq. (11.21a). Solving these three equations simultaneously for α, β, and γ yields

$$T_1^{(a,a)}(x_0; a, b) = \frac{1}{2D}(x_0 - a)(b - x_0). \tag{11.25}$$

This is the *average time* for the solute molecule to *first* reach *either* $x = a$ or $x = b$ from $x_0 \in [a, b]$.

The *variance* of $T(x_0; a, b)$, which is given by Eq. (11.12b), could now be computed by first solving the differential equation (11.21b) for $T_2(x_0; a, b)$, using in that equation the appropriate solution for $T_1(x_0; a, b)$, and then subtracting $T_1^2(x_0; a, b)$. But it turns out to be easier to derive and solve a differential equation for the variance itself. To obtain that equation, we differentiate Eq. (11.12b) twice with respect to x_0 to get, suppressing for now the dependence on a and b:

$$\frac{d^2 \text{var}\{T(x_0)\}}{dx_0^2} = \frac{d}{dx_0}\left(\frac{dT_2(x_0)}{dx_0} - 2T_1(x_0)\frac{dT_1(x_0)}{dx_0}\right)$$

$$= \frac{d^2 T_2(x_0)}{dx_0^2} - 2\left(\frac{dT_1(x_0)}{dx_0}\right)^2 - 2T_1(x_0)\frac{d^2 T_1(x_0)}{dx_0^2}$$

$$= -\frac{2}{D}T_1(x_0) - 2\left(\frac{dT_1(x_0)}{dx_0}\right)^2 - 2T_1(x_0)\left(-\frac{1}{D}\right).$$

In the last step, we have substituted into the first and third terms from formulas (11.21b) and (11.21a) respectively. The first and third terms evidently cancel each other, so we are left with

$$\frac{d^2 \text{var}\{T(x_0; a, b)\}}{dx_0^2} = -2\left(\frac{dT_1(x_0; a, b)}{dx_0}\right)^2. \tag{11.26}$$

By substituting the appropriate explicit solution for $T_1(x_0; a, b)$ into the right side of Eq. (11.26), we will get an explicit ordinary differential equation for $\text{var}\{T(x_0; a, b)\}$. But solving that equation will again require imposing appropriate boundary conditions on $\text{var}\{T(x_0; a, b)\}$. From the definition of the variance in Eq. (11.12b), it is easy to show that the boundary conditions on the first two moments of $T^{(a,r)}(x_0; a, b)$ in Eq. (11.22) imply the following boundary conditions on the variance of $T^{(a,r)}(x_0; a, b)$:

$$\text{var}\{T^{(a,r)}(x_0 = a; a, b)\} = 0, \tag{11.27a}$$

$$\frac{d \, \text{var}\{T^{(a,r)}(x_0; a, b)\}}{dx_0} \bigg|_{x_0=b} = 0. \tag{11.27b}$$

Similarly, the boundary conditions on the first two moments of $T^{(a,a)}(x_0; a, b)$ in Eq. (11.23) imply the following boundary conditions on the variance of $T^{(a,a)}(x_0; a, b)$:

$$\text{var}\{T^{(a,a)}(x_0 = a; a, b)\} = 0, \tag{11.28a}$$

$$\text{var}\{T^{(a,a)}(x_0 = b; a, b)\} = 0. \tag{11.28b}$$

So, to get $\text{var}\{T^{(a,r)}(x_0; a, b)\}$, we substitute the expression (11.24) for $T_1^{(a,r)}(x_0; a, b)$ into the right side of Eq. (11.26) to get

$$\frac{d^2 \text{var}\{T^{(a,r)}(x_0; a, b)\}}{dx_0^2} = -\frac{2}{D^2}(b - x_0)^2. \tag{11.29}$$

As can easily be verified, the solution to this equation that satisfies the two boundary conditions (11.27) is

$$\text{var}\{T^{(a,r)}(x_0; a, b)\} = \frac{1}{6D^2} \left[(b - a)^4 - (b - x_0)^4\right]. \tag{11.30}$$

And to obtain $\text{var}\{T^{(a,a)}(x_0; a, b)\}$, we substitute the expression (11.25) for $T_1^{(a,a)}(x_0; a, b)$ into the right side of Eq. (11.26) to get

$$\frac{d^2 \text{var}\{T^{(a,a)}(x_0; a, b)\}}{dx_0^2} = -\frac{1}{2D^2}(b + a - 2x_0)^2. \tag{11.31}$$

As can easily be verified, the solution to this equation that satisfies the two boundary conditions (11.28) is

$$\text{var}\{T^{(a,a)}(x_0; a, b)\} = \frac{1}{96D^2} \left[(b - a)^4 - (b + a - 2x_0)^4\right]. \tag{11.32}$$

We have now seen how to compute the mean and variance of two typical first-passage times $T^{(a,r)}(x_0; a, b)$ and $T^{(a,a)}(x_0; a, b)$ by making use of the backward Fokker–Planck equation. In the next section, we will see how this approach can help us answer a few lingering questions about the discrete-stochastic approach to diffusion.

11.7 Implications for the discrete-stochastic approach

In the discrete-stochastic approach to diffusion presented in Chapters 5 and 6, the domain of the diffusing solute molecule on the x-axis is divided into M contiguous cells of equal length l, and it is hypothesized that $\kappa_l dt$, with $\kappa_l = D/l^2$, gives the probability that a randomly chosen molecule in any cell will jump to a particular adjacent cell in the next infinitesimal time interval dt [see Eqs (5.3) and (5.8)]. In earlier chapters we have examined from several perspectives the question of how accurately this "kappa hypothesis" replicates the Einstein model of diffusion. One concern in that regard which we mentioned in Section 5.5 but did not pursue in detail is this: Immediately after a solute molecule has jumped from cell i to cell $i+1$ (see Fig. 11.1), wouldn't the molecule more likely be *closer* to cell i than to cell $i+2$, especially if l is "large", and therefore more likely to jump next back to cell i instead of to the other adjacent cell $i+2$? If that were so, then the discrete-stochastic hypothesis would be inaccurate in assigning equal probabilities to the two possible next jumps $i+1 \rightarrow i$ and $i+1 \rightarrow i+2$. And certainly, neither the master equations nor the stochastic simulation algorithms derived for the discrete-stochastic model in Chapter 6 make any allowance for such a "past-remembering" effect. So is this an inaccuracy that we need to guard against, perhaps by taking l even smaller?

A first-passage time analysis can help us answer this question. Consider first the situation in which a solute molecule of interest is stipulated to be "inside cell M" in Fig. 11.1a, a cell that is bounded on its right by a reflecting boundary. What can we say about when that molecule will next jump to cell $M - 1$? The kappa hypothesis asserts that $\kappa_l dt$ is the probability that the molecule will make that jump in the next

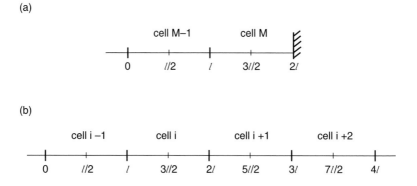

Fig. 11.1 *Axis scalings used in defining first-passage times between adjacent cells of length l.* (**a**) *For a first passage from the center of end cell M to the center of cell $M-1$, the initial point is $x_0 = \frac{3}{2}l$, and the $[a, b]$ interval is $[\frac{1}{2}l, 2l]$ with a absorbing and b reflecting. The mean first-passage time according to Einstein's theory is given in Eq. (11.33).* (**b**) *For a first passage from the center of interior cell i to the center of either cell $i-1$ or cell $i+1$, the initial point is $x_0 = \frac{3}{2}l$, and the $[a, b]$ interval is $[\frac{1}{2}l, \frac{5}{2}l]$ with a and b both absorbing. The mean first-passage time according to Einstein's theory is given in Eq. (11.34).*

dt, and that in turn implies that the *average waiting time* for that jump is $1/\kappa_l$. But one could instead look at the average time for the molecule, starting at the *center* of cell M, to *first* reach the *center* of cell $M-1$. From the geometry of Fig. 11.1a, we see that the latter is the mean first-passage time $T_1^{(a,r)}(x_0; a, b)$ in Eq. (11.24) with $x_0 = 3l/2$, $a = l/2$, and $b = 2l$:

$$T_1^{(a,r)}(\tfrac{3}{2}l; \tfrac{1}{2}l, 2l) = \frac{\left(\tfrac{3}{2}l - \tfrac{1}{2}l\right)}{2D} \left[2\left(2l - \tfrac{1}{2}l\right) - \left(\tfrac{3}{2}l - \tfrac{1}{2}l\right)\right] = \frac{l^2}{D} = \frac{1}{\kappa_l}. \tag{11.33}$$

Thus, the {*mean first-passage time* of a solute molecule from the *center* of cell M to the *center* of cell $M-1$} is numerically equal to the discrete-stochastic model's {*average waiting time* for a solute molecule in cell M to jump to cell $M-1$}. So the average time for the jump of a solute molecule from cell M to cell $M-1$ in the discrete-stochastic model is equal to the average time for a first passage of the molecule from *well inside* cell M (its center) to *well inside* cell $M-1$ (its center). And if the jump of the solute molecule is to a point that is *well inside* cell $M-1$, then its *next* jump should be equally likely to be to the left or the right, contrary to the concern expressed above.

Let's see if this conclusion holds if the solute molecule is initially inside an *interior* cell i, as shown in Fig. 11.1b. According to the kappa hypothesis, the probability that a solute molecule in cell i will jump next to *either* cell $i-1$ *or* cell $i+1$ in the next dt is $2\kappa_l dt$. That in turn implies that the average time before the molecule jumps out of cell i to either of those two adjacent cells is $(2\kappa_l)^{-1}$. From the geometry of Fig. 11.1b, we see that the *mean first-passage time* for a solute molecule to diffuse from the *center* of cell i to *either* the *center* of cell $i-1$ *or* the *center* of cell $i+1$ is given by $T_1^{(a,a)}(x_0; a, b)$ in Eq. (11.25) with $x_0 = 3l/2$, $a = l/2$, and $b = 5l/2$:

$$T_1^{(a,a)}(\tfrac{3}{2}l; \tfrac{1}{2}l, \tfrac{5}{2}l) = \frac{1}{2D}\left(\tfrac{3}{2}l - \tfrac{1}{2}l\right)\left(\tfrac{5}{2}l - \tfrac{3}{2}l\right) = \frac{l^2}{2D} = \frac{1}{2\kappa_l}. \tag{11.34}$$

Once again, {the center-to-center mean first-passage time} is numerically equal to {the average waiting time in the discrete-stochastic model}. So once again it seems reasonable to interpret a "jump out of cell *i*" in the discrete-stochastic approach to mean a jump from *well inside* cell i to *well inside* either of the two adjacent cells. And that implies that there should be no bias in favor of cell i on the following jump.

But these encouraging results become a bit less encouraging when we look a little deeper into the matter. For the situation in Fig. 11.1a, the discrete-stochastic hypothesis implies not only that $1/\kappa_l$ is the mean of the waiting time for a solute molecule to jump out of cell M, but also that that waiting time is *exponentially distributed* [see the discussion leading to Eq. (6.15)]. That in turn implies that the standard deviation of the waiting time is equal to its mean. But when we use formula (11.30) to compute the standard deviation of the first-passage time from the center of cell M to the center of cell $M-1$, we get

$$\text{sdev}\left\{T^{(a,r)}(\tfrac{3}{2}l; \tfrac{1}{2}l, 2l)\right\} = \sqrt{\frac{1}{6D^2}\left[\left(2l - \tfrac{1}{2}l\right)^4 - \left(2l - \tfrac{3}{2}l\right)^4\right]} = \sqrt{\frac{5}{6}}\frac{1}{\kappa_l}. \tag{11.35}$$

This is only 91% of the mean in Eq. (11.33). Similarly, when we use formula (11.32) to compute the standard deviation of the first-passage time from the center of any interior cell i to the center of either of its adjacent cells $i-1$ and $i+1$, we get

$$
\mathrm{sdev}\left\{T^{(\mathrm{a,a})}\left(\tfrac{3}{2}l;\tfrac{1}{2}l,\tfrac{5}{2}l\right)\right\} = \sqrt{\frac{1}{96D^2}\left[\left(\tfrac{5}{2}l-\tfrac{1}{2}l\right)^4 - \left(\tfrac{5}{2}l+\tfrac{1}{2}l-2\tfrac{3}{2}l\right)^4\right]} = \sqrt{\frac{2}{3}}\frac{1}{2\kappa_l},
$$

(11.36)

which is only 82% of the mean in Eq. (11.34). It follows that *these two first-passage times cannot be exponentially distributed.*

To get a clearer picture of what is going on here, we need to examine the probability density functions of these two first-passage times. One way to compute those PDFs was presented in Section 11.4: First solve the partial differential equation (11.14) for $G(x_0, t; a, b)$, subject to the initial condition (11.15) and the boundary conditions (11.16) or (11.17); then compute the t-derivative in Eq. (11.10). Not seeing a way to carry out those computations analytically, we have settled for a *numerical* computation, using the arbitrarily chosen parameter values $D = 9$ and $l = 3$. The results for the $M \to M - 1$ case are shown in Fig. 11.2a. Here the solid curve is the theory's prediction for the PDF of the first-passage time from the center of cell M to the center of cell $M - 1$. For comparison, we show as the dashed curve the exponential PDF with the same mean. Results for a first passage from the center of an interior cell i to the center of either adjacent cell are shown in this same format in Fig. 11.2b.

The plots in Fig. 11.2 make it clear that the PDFs of the two center-to-center first-passage times whose means are given in Eqs (11.33) and (11.34) are *not exponential*, as are the hypothesized waiting times in the discrete-stochastic model of diffusion. Does this imply an inconsistency in the kappa hypothesis which underlies the discrete-stochastic model, and which predicts exponentially distributed waiting times?

Not necessarily. What these results tell is that {the first-passage time from the center of the present cell to the center of an adjacent cell} is *not the same* as {the waiting time in the discrete-stochastic approach for a jump from the present cell to an adjacent cell}. But this is not terribly surprising in view of the fact that there is ambiguity in how one would go about measuring the "waiting time" in the discrete-stochastic approach. For example, the waiting time for a jump from cell M to cell $M - 1$ does not specify *from where* in cell M, and *to where* in cell $M - 1$, the molecule is to jump. By contrast, the first-passage time from the center of cell M to the center of cell $M - 1$ is quite unambiguous in that regard. Nevertheless, we saw in Sections 5.3, 5.4, and 6.2 that there are several other compelling arguments which indicate that the kappa hypothesis *is* consistent with Einstein's diffusion equation, at least provided the cell size is sufficiently small.

An intriguing way to explore this issue would be to use the single-molecule *stochastic simulation algorithm* of Section 6.4, which *assumes* the kappa hypothesis, to estimate the first-passage time PDFs in Fig. 11.2. We briefly discussed in Section

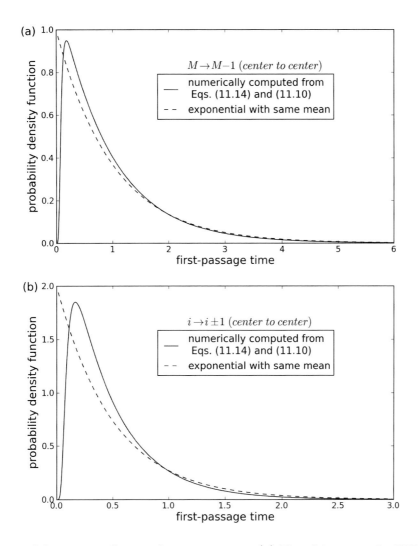

Fig. 11.2 *Cell-center to cell-center first-passage times.* **(a)** *The solid curve is the PDF of the first-passage time from the center of cell M in Fig. 11.1a to the center of cell $M - 1$ for $D = 9$ and $l = 3$. The dashed curve is the exponential PDF with the same mean. The true PDF was computed by first numerically solving Eq. (11.14) for $G(x_0, t; a, b)$ subject to initial condition (11.15) and boundary conditions (11.16), taking $x_0 = \frac{3}{2}l$, $a = \frac{1}{2}l$, and $b = 2l$, and then numerically evaluating the derivative in Eq. (11.10). The numerical solution to Eq. (11.14) was obtained using MATLAB's built-in solver pdepe, which essentially converts a partial differential equation to a coupled set of ordinary differential equations using a second-order spatial discretization based on a fixed set of user specified nodes.* **(b)** *The solid curve is the PDF of the first-passage time from the center of cell i in Fig. 11.1b to the center of either cell $i - 1$ or cell $i + 1$ for $D = 9$ and $l = 3$. The dashed curve is the exponential PDF with the same mean. The true PDF was computed by first using pdepe to numerically solve Eq. (11.14) for $G(x_0, t; a, b)$ subject to initial condition (11.15) and boundary conditions (11.17), taking $x_0 = \frac{3}{2}l$, $a = \frac{1}{2}l$, and $b = \frac{5}{2}l$, and then numerically evaluating the derivative in Eq. (11.10). The message of the plots in this figure is that, even though the center-to-center first-passage times have the same means as the discrete-stochastic approach's "waiting times" for a jump to an adjacent cell, the first-passage times do not share with the waiting times the property of being exponentially distributed.*

(a)

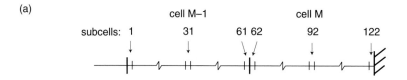

(b)

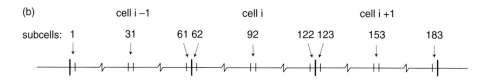

Fig. 11.3 *Dividing the cells in Fig. 11.1 into subcells that will be used by the discrete-stochastic single-molecule simulation procedure to estimate the first-passage time PDFs in Fig. 11.2. Each cell is divided into 61 subcells of length l/61. (**a**) For estimating the true PDF in Fig. 11.2a, each simulation run will be started with the molecule in subcell 92 (the center of cell M), and will then proceed, with no up-going transitions allowed from subcell 122, until the molecule first enters subcell 31 (the center of cell M − 1). (**b**) For estimating the true PDF in Fig. 11.2b, each simulation run will be started with the molecule in subcell 92 (the center of cell i), and will be terminated when the molecule first enters either subcell 31 (the center of cell i − 1) or subcell 153 (the center of cell i + 1).*

11.2 how that could be done. To estimate the PDF of the first-passage time from the center of the end cell M to the center of cell $M − 1$, we first overlay those two cells with a mesh of many smaller *subcells* as shown in Fig. 11.3a. Using a large number of subcells ensures that the output of our simulations will have a large number of data points; furthermore, choosing an odd number of subcells ensures that there will be a subcell at the exact center of the cell. With these factors in mind, we will cover each cell with exactly 61 subcells, in such a way that the left edge of subcell 1 coincides with the left edge of cell $M − 1$, subcell 31 is at the center of cell $M − 1$, subcell 92 is at the center of cell M, and the right reflecting boundary of cell M is also the right reflecting boundary of subcell 122. Since there are 61 subcells per cell and the length of each cell is l, then the length of each subcell is $l' = l/61$. Assuming as in Fig. 11.2 that $D = 9$ and $l = 3$, we thus have for the *cells*, $\kappa_l = D/l^2 = 1$. But for a simulation run over the *subcells*, we must use the kappa value

$$\kappa_{l'} = D/l'^2 = \kappa_l(61)^2 = (61)^2. \tag{11.37}$$

Using this in the single-molecule simulation algorithm of Section 6.4, we made 10^6 simulation runs. Each run started at time 0 with the solute molecule in subcell 92 (the center of cell M), and when the molecule *first* reached subcell 31 (the center of cell $M − 1$) the run was terminated and the simulated time t for that run was recorded. As a check, we verified that the mean and standard deviation of the 10^6 t-values obtained were statistically consistent with the mean and standard deviation predicted by Eqs (11.33) and (11.35) for $D = 9$ and $l = 3$. Figure 11.4a shows, superimposed on the two curves in Fig. 11.2a, a normalized histogram of the 10^6 t-values obtained in these simulation runs. (Normalization was accomplished by dividing the population of each time bin in the histogram by the product of the bin width and the total number of runs.) Evidently, the PDF as estimated by these discrete-stochastic simulations agrees extremely well with the PDF predicted by first-passage time theory. So even though the first-passage time's non-exponential character makes it *different* from the "waiting time" in the discrete-stochastic model of diffusion, there is no indication that the discrete-stochastic model produces results that are inconsistent with the Einstein diffusion equation. On the contrary, the excellent agreement in Fig. 11.4a between our simulation results, which are based on the kappa hypothesis, and our first-passage time theory results, which are not based on the kappa hypothesis, indicates that the kappa hypothesis is consistent with the Einstein diffusion equation.

A simulation estimate of the PDF of the first-passage time from the center of an *interior* cell i to either the center of cell $i - 1$ or the center of cell $i + 1$ produced similar results. As indicated in Fig. 11.3b, we divided each of those three cells into 61 subcells of length $l' = l/61$. We started each simulation run in subcell 92 (the center of cell i), and we terminated the run when *either* subcell 31 (the center of cell $i - 1$) or subcell 153 (the center of cell $i + 1$) was first reached. For an ensemble of 10^6 such runs, we verified that the mean and standard deviation of the run times thus obtained were statistically consistent with the theoretical predictions of Eqs (11.34) and (11.36) for $D = 9$ and $l = 3$. A normalized histogram of those run times is shown in Fig. 11.4b, superimposed on the two curves of Fig. 11.2b. Once again, the agreement between the discrete-stochastic simulation results and first-passage time theory is excellent.

11.8   An "averaged" first-passage time

Is there a first-passage time other than the "cell-center to cell-center" one we considered in the last section which would be a better approximation to the waiting time of the discrete-stochastic approach? The fact that the discrete-stochastic approach tacitly assumes that the solute molecules are distributed *uniformly* over the cells (see the discussion in Section 5.1) suggests that some kind of "cell-uniform to cell-uniform" first-passage time might be a better approximation. Referring to Fig. 11.1a, perhaps a better approximation to the waiting time in cell M before a jump to cell $M - 1$ would be the first-passage time $T^{(a,r)}(x_0; a, 2l)$ where x_0 is uniformly random inside cell M, i.e., $x_0 = \mathcal{U}(l, 2l)$, and a is uniformly random inside cell $M - 1$, i.e., $a = \mathcal{U}(0, l)$. Similarly, referring to Fig. 11.1b, perhaps a better approximation to the waiting time in an interior cell i before a jump to either cell $i - 1$ or cell $i + 1$ would be the first-passage time $T^{(a,a)}(x_0; a, b)$ with $x_0 = \mathcal{U}(l, 2l)$, $a = \mathcal{U}(0, l)$, and $b = \mathcal{U}(2l, 3l)$.

As a preliminary test of these conjectures, let us compute the *means* of these alternative first-passage times. Referring again to Fig. 11.1a, we know that $T_1^{(a,r)}(x_0; a, 2l)$ from Eq. (11.24) gives the mean first-passage time from any point x_0 in cell M to any point a in cell $M - 1$. Therefore, the mean first-passage time from a *uniformly*

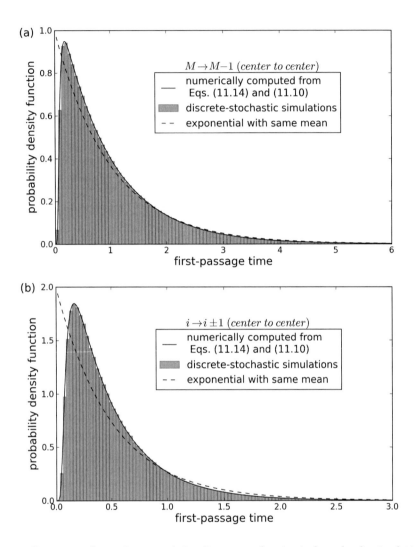

Fig. 11.4 *Comparing the predictions of the discrete-stochastic single-molecule simulation procedure to the predictions of first-passage time theory. (**a**) The solid curve is the PDF that was plotted in Fig. 11.2a for the first-passage time from the center of cell M to the center of cell $M - 1$. The gray area is a normalized frequency histogram of the first-passage times obtained in 10^6 simulation runs of the kind described in the caption of Fig. 11.3a. Even though the simulation algorithm assumes exponentially distributed waiting times for cell-to-cell jumps, the simulation results are in excellent agreement with the non-exponential PDF predicted by first-passage time theory. (**b**) The solid curve is the PDF that was plotted in Fig. 11.2b for the first-passage time from the center of interior cell i to the center of either cell $i - 1$ or cell $i + 1$. The gray area is a normalized frequency histogram of the first-passage times obtained in 10^6 simulation runs of the kind described in the caption of Fig. 11.3b. Once again, the discrete-stochastic simulation results are in excellent agreement with the non-exponential results of first-passage time theory. The message here is that the cell-center to cell-center first-passage time is not the same dynamical variable as the discrete-stochastic model's waiting time for the next jump, but that does not imply a problem with the discrete-stochastic model.*

random point in cell M to a *uniformly random* point in cell $M - 1$ will be the *average* of $T_1^{(a,r)}(x_0; a, 2l)$ over $x_0 = \mathcal{U}(l, 2l)$ and $a = \mathcal{U}(0, l)$:

$$
\begin{aligned}
T_1(M \to M - 1) &= \int_{a=0}^{l} \frac{da}{l} \int_{x_0=l}^{2l} \frac{dx_0}{l} T_1^{(a,r)}(x_0; a, 2l) \\
&= \frac{1}{l^2} \int_{a=0}^{l} da \int_{x_0=l}^{2l} dx_0 \frac{(x_0 - a)}{2D} \left[2(2l - a) - (x_0 - a) \right].
\end{aligned}
$$

The integrations here are straightforwardly performed, and upon setting $D = \kappa_l l^2$ we find

$$
T_1(M \to M - 1) = \frac{1}{\kappa_l}. \tag{11.38}
$$

Thus we see that this "uniform-to-uniform" mean first-passage time is exactly equal to the "center-to-center" mean first-passage time considered earlier, and hence also exactly equal to the mean waiting time in the discrete-stochastic approach. Referring next to Fig. 11.1b, we know that $T_1^{(a,a)}(x_0; a, b)$ in Eq. (11.25) gives the mean first-passage time from any point x_0 in cell i to *either* any point a in cell $i - 1$ or any point b in cell $i + 1$; therefore, the mean first-passage time from a *uniformly random* point inside cell i to *either* a *uniformly random* point inside cell $i - 1$ or a *uniformly random* point inside cell $i + 1$ should be given by the average of $T_1^{(a,a)}(x_0; a, b)$ over $x_0 = \mathcal{U}(l, 2l)$, $a = \mathcal{U}(0, l)$, and $b = \mathcal{U}(2l, 3l)$:

$$
\begin{aligned}
T_1(i \to i \pm 1) &= \int_{a=0}^{l} \frac{da}{l} \int_{x_0=l}^{2l} \frac{dx_0}{l} \int_{b=2l}^{3l} \frac{db}{l} T_1^{(a,a)}(x_0; a, b) \\
&= \frac{1}{l^3} \int_{a=0}^{l} da \int_{x_0=l}^{2l} dx_0 \int_{b=2l}^{3l} db \frac{1}{2D} (x_0 - a)(b - x_0).
\end{aligned}
$$

The integrations again are straightforwardly performed, and upon setting $D = \kappa_l l^2$ we find

$$
T_1(i \to i \pm 1) = \left(\frac{11}{12} \right) \frac{1}{2\kappa_l}. \tag{11.39}
$$

This is slightly smaller than $1/(2\kappa_l)$, the mean waiting time in cell i that is hypothesized by the discrete-stochastic approach. That is not exactly the result we were looking for. But what really matters is how the corresponding PDF looks.

It would be very challenging to compute analytically from first-passage time theory the above "averaged" PDFs. But it requires only a few minor changes in the simulation programs we used to get the histograms in Fig. 11.4 to estimate those averaged PDFs: For the $M \to M - 1$ simulations over the subcells in Fig. 11.3a, instead of starting each run in subcell $n_0 = 92$ and terminating the run when subcell $n_1 = 31$ is first reached, we will draw two sample values u_0 and u_1 of $\mathcal{U}(0, 1)$, and then start each run in subcell $n_0 = [62 + 61 \cdot u_0]$, where $[x]$ denotes the greatest integer in x [see Eq. (2.41)], and we will terminate the run when subcell $n_1 = [1 + 61 \cdot u_1]$ is first reached. Likewise for the $i \to i \pm 1$ simulations over the subcells in Fig. 11.3b, we will start each run in subcell $n_0 = [62 + 61 \cdot u_0]$ and terminate the run when either subcell $n_1 = [1 + 61 \cdot u_1]$ or subcell $n_2 = [123 + 61 \cdot u_2]$ is first reached, where u_2 is yet another sample of $\mathcal{U}(0, 1)$. As a check on this simulation procedure, we can verify that the means of the run times are statistically consistent with the theoretical predictions in Eqs (11.38) and (11.39).

Normalized frequency histograms from this new set of 10^6 simulation runs are shown in Fig. 11.5, along with the same exponential comparison curves that appear in Figs 11.2 and 11.4. Although the "uniform-to-uniform" first-passage time PDFs given by the histograms are still not exponential, they are evidently *much closer* to exponential than the "center-to-center" first-passage-time PDFs in Fig. 11.4. In particular, the strikingly non-exponential decrease in the PDFs to 0 as $t \to 0$ seems now to be gone.

11.9   Some conclusions

What conclusions can we draw from all this? First, all of the first-passage time results obtained here suggest that the magnitude of the "average waiting time" in the discrete-stochastic approach for a solute molecule to jump to an adjacent cell typifies the time it takes the molecule to get *well inside* the adjacent cell, not the time it takes the molecule to just reach that cell. Therefore, the concern that the solute molecule's next jump might more likely be to its previous cell than to the other adjacent cell is without foundation.

Second, the discrete-stochastic approach's waiting time does *not* appear to be identical to *any* first-passage time. But the consequent implication that we cannot rigorously derive hypothesis (5.3) from first-passage time theory should *not* be viewed as a shortcoming of the discrete-stochastic approach. Indeed, to the several compelling arguments presented in Sections 5.3, 5.4 and 6.2 showing that hypothesis (5.3) is consistent with Einstein's diffusion equation when the cell size is sufficiently small, we may now add the excellent agreement found in Fig. 11.4 between the first-passage time theory predictions and the discrete-stochastic simulation predictions.

Third, these investigations have illustrated once again the benefits that can come from exploiting the synergism between theory and simulation: In Fig. 11.4, results obtained from rigorous first-passage time theory (the solid curves) were used to validate simulation results obtained using the discrete-stochastic approach (the histograms). That validation then gave us confidence in the correctness of the discrete-stochastic simulation estimates of the "averaged" first-passage time PDFs (the histograms) in Fig. 11.5, results which would have been very difficult to obtain using the formal analytical machinery of first-passage time theory.

Finally, we should keep in mind that all of the analysis in this chapter has been predicated on the *Einstein* model of diffusion, which we know breaks down for time intervals that are not large compared to the relaxation time $\tau = m/\gamma$. This naturally leads us to wonder if a first-passage time analysis can be carried out for the more accurate Langevin model of diffusion. That should be possible, but it would be

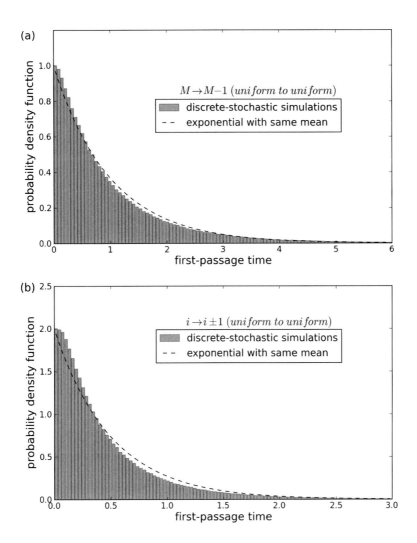

Fig. 11.5 *Discrete-stochastic simulation predictions for the PDFs of cell-uniform to cell-uniform first-passage times. Reassured by the accurate performance of the discrete-stochastic single-molecule simulation algorithm in Fig. 11.4, we use here a slightly modified version of that simulation procedure to estimate the PDFs of some first-passage times that would be very difficult to compute directly from first-passage time theory. (**a**) Each of the 10^6 simulation runs used in making this normalized frequency histogram of first-passage times was started with the diffusing molecule in a subcell in Fig. 11.3a whose index was chosen randomly and uniformly from the integers in [62, 122]. The run was terminated when the molecule first reached a subcell whose index was chosen randomly and uniformly from the integers in [1, 61]. The run time is thus the first-passage time from a point randomly uniform inside cell M to a point randomly uniform inside cell M − 1. The PDF of this first-passage time still differs from the exponential form shown by the dashed curve, but it is evidently closer to exponential than the "cell-center to cell-center" first-passage time PDF in Fig. 11.4a. (**b**) An analogous estimate of the PDF of the first-passage time from a point randomly uniform inside cell i in Fig. 11.3b to a point randomly uniform inside either cell i − 1 or cell i + 1. Here each simulation run was started with the diffusing molecule in a subcell whose index was chosen randomly and uniformly from the integers in [62, 122], and the run was terminated when the molecule first reached either a subcell whose index was chosen randomly and uniformly from the integers in [1, 61] or a subcell whose index was chosen randomly and uniformly from the integers in [123, 183].*

considerably more difficult than for the Einstein model. The relevant forward Fokker–Planck equation for the Langevin model would be the Kramers equation (10.24) with the external force F_e set to zero. The corresponding backward Fokker–Planck equation could be straightforwardly inferred from Eq. (7.58). But since those equations describe the simultaneous evolution of the *two* processes $V_x(t)$ and $X(t)$, both of those processes would need to be taken into account when formulating the appropriate first-passage criterion. We will not pursue that more challenging problem here.

Index